Harper College Library
3 2158

MW01346320

A Historian Looks Back

The Calculus as Algebra and Selected Writings

Part I was formerly published as
The Calculus as Algebra: J.-L. Lagrange, 1736–1813
by Garland Publishing, New York, 1990.

Library of Congress Catalog Card Number 2010927261

ISBN 978-0-88385-572-0

Printed in the United States of America

Current Printing (last digit):
10 9 8 7 6 5 4 3 2 1

A Historian Looks Back

The Calculus as Algebra and Selected Writings

By

Judith V. Grabiner

Pitzer College

Published and Distributed by

The Mathematical Association of America

SPECTRUM SERIES

The Spectrum Series of the Mathematical Association of America was so named to reflect its purpose: to publish a broad range of books including biographies, accessible expositions of old or new mathematical ideas, reprints and revisions of excellent out-of-print books, popular works, and other monographs of high interest that will appeal to a broad range of readers, including students and teachers of mathematics, mathematical amateurs, and researchers.

777 Mathematical Conversation Starters, by John de Pillis

99 Points of Intersection: Examples—Pictures—Proofs, by Hans Walser. Translated from the original German by Peter Hilton and Jean Pedersen.

Aha Gotcha and Aha Insight, by Martin Gardner

All the Math That's Fit to Print, by Keith Devlin

The Calculus as Algebra and Selected Writings, by Judith V. Grabiner

Calculus Gems: Brief Lives and Memorable Mathematics, by George F. Simmons

Carl Friedrich Gauss: Titan of Science, by G. Waldo Dunnington, with additional material by Jeremy Gray and Fritz-Egbert Dohse

The Changing Space of Geometry, edited by Chris Pritchard

Circles: A Mathematical View, by Dan Pedoe

Complex Numbers and Geometry, by Liang-shin Hahn

Cryptology, by Albrecht Beutelspacher

The Early Mathematics of Leonhard Euler, by C. Edward Sandifer

The Edge of the Universe: Celebrating 10 Years of Math Horizons, edited by Deanna Haunsperger and Stephen Kennedy

Euler and Modern Science, edited by N. N. Bogolyubov, G. K. Mikhailov, and A. P. Yushkevich. Translated from Russian by Robert Burns.

Euler at 300: An Appreciation, edited by Robert E. Bradley, Lawrence A. D'Antonio, and C. Edward Sandifer

Five Hundred Mathematical Challenges, Edward J. Barbeau, Murray S. Klamkin, and William O. J. Moser

The Genius of Euler: Reflections on his Life and Work, edited by William Dunham

The Golden Section, by Hans Walser. Translated from the original German by Peter Hilton, with the assistance of Jean Pedersen.

The Harmony of the World: 75 Years of Mathematics Magazine, edited by Gerald L. Alexanderson with the assistance of Peter Ross

A Historian Looks Back: The Calculus as Algebra and Selected Writings, by Judith Grabiner

History of Mathematics: Highways and Byways, Amy Dahan-Dalmedico and Jeanne Peiffer. Translated by Sanford Segal.

How Euler Did It, by C. Edward Sandifer

Is Mathematics Inevitable? A Miscellany, edited by Underwood Dudley

I Want to Be a Mathematician, by Paul R. Halmos

Journey into Geometries, by Marta Sved

JULIA: a life in mathematics, by Constance Reid

The Lighter Side of Mathematics: Proceedings of the Eugène Strens Memorial Conference on Recreational Mathematics & Its History, edited by Richard K. Guy and Robert E. Woodrow

Lure of the Integers, by Joe Roberts

Magic Numbers of the Professor, by Owen O'Shea and Underwood Dudley

Magic Tricks, Card Shuffling, and Dynamic Computer Memories: The Mathematics of the Perfect Shuffle, by S. Brent Morris

Martin Gardner's Mathematical Games: The entire collection of his Scientific American columns

The Math Chat Book, by Frank Morgan

Mathematical Adventures for Students and Amateurs, edited by David Hayes and Tatiana Shubin. With the assistance of Gerald L. Alexanderson and Peter Ross.

Mathematical Apocrypha, by Steven G. Krantz

Mathematical Apocrypha Redux, by Steven G. Krantz

Mathematical Carnival, by Martin Gardner

Mathematical Circles Vol I: In Mathematical Circles Quadrants I, II, III, IV, by Howard W. Eves

Mathematical Circles Vol II: Mathematical Circles Revisited and Mathematical Circles Squared, by Howard W. Eves

Mathematical Circles Vol III: Mathematical Circles Adieu and Return to Mathematical Circles, by Howard W. Eves

Mathematical Circus, by Martin Gardner

Mathematical Cranks, by Underwood Dudley

Mathematical Evolutions, edited by Abe Shenitzer and John Stillwell

Mathematical Fallacies, Flaws, and Flimflam, by Edward J. Barbeau

Mathematical Magic Show, by Martin Gardner

Mathematical Reminiscences, by Howard Eves

Mathematical Treks: From Surreal Numbers to Magic Circles, by Ivars Peterson

Mathematics in Historical Context, by Jeff Suzuki

Mathematics: Queen and Servant of Science, by E.T. Bell

Memorabilia Mathematica, by Robert Edouard Moritz

Musings of the Masters: An Anthology of Mathematical Reflections, edited by Raymond G. Ayoub

New Mathematical Diversions, by Martin Gardner

Non-Euclidean Geometry, by H. S. M. Coxeter

Numerical Methods That Work, by Forman Acton

Numerology or What Pythagoras Wrought, by Underwood Dudley

Out of the Mouths of Mathematicians, by Rosemary Schmalz

Penrose Tiles to Trapdoor Ciphers... and the Return of Dr. Matrix, by Martin Gardner

Polyominoes, by George Martin

Power Play, by Edward J. Barbeau

Proof and Other Dilemmas: Mathematics and Philosophy, edited by Bonnie Gold and Roger Simons

The Random Walks of George Pólya, by Gerald L. Alexanderson

Remarkable Mathematicians, from Euler to von Neumann, Ioan James

The Search for E.T. Bell, also known as John Taine, by Constance Reid

Shaping Space, edited by Marjorie Senechal and George Fleck

Sherlock Holmes in Babylon and Other Tales of Mathematical History, edited by Marlow Anderson, Victor Katz, and Robin Wilson

Student Research Projects in Calculus, by Marcus Cohen, Arthur Knoebel, Edward D. Gaughan, Douglas S. Kurtz, and David Pengelley

Symmetry, by Hans Walser. Translated from the original German by Peter Hilton, with the assistance of Jean Pedersen.

The Trisectors, by Underwood Dudley

Twenty Years Before the Blackboard, by Michael Stueben with Diane Sandford

Who Gave You the Epsilon? and Other Tales of Mathematical History, edited by Marlow Anderson, Victor Katz, and Robin Wilson

The Words of Mathematics, by Steven Schwartzman

MAA Service Center
P.O. Box 91112
Washington, DC 20090-1112
800-331-1622 FAX: 301-206-9789

Contents

---

* Recipient of the Lester R. Ford Award

† Recipient of the Carl B. Allendoerfer Award

Introduction

In 1869, Darwin's champion Thomas Henry Huxley praised scientific experiment and observation over dogmatism. But his criticisms extended also to the textbook view of mathematics. Huxley said, "The mathematician starts with a few simple propositions, the proof of which is so obvious that they are called self-evident, and the rest of his work consists of subtle deductions from them.... Mathematics... knows nothing of observation, nothing of experiment." In reply, the great English algebraist J. J. Sylvester spoke about what he knew from his own work: Mathematics "unceasingly call[s] forth the faculties of observation and comparison... it has frequent recourse to experimental trials and verification... it affords a boundless scope for the exercise of the highest efforts of imagination and invention." [**3**, 204]

As a historian of mathematics, I'm with Sylvester. I have long been interested in what mathematicians actually do, and how mathematics actually has developed. I have nothing against textbooks and logically structured subjects. It is just that they represent the finished product, not the creativity that produced it. The past, as L. P. Hartley said in opening his novel *The Go-Between*, "is a foreign country: they do things differently there." [**2**, 1] Historians provide guidebooks to that past, although mathematicians tend to be more interested in knowing how that foreign past became transformed into the mathematics we know and teach today.

Mathematics is incredibly rich and mathematicians have been unpredictably ingenious. Therefore, the history of mathematics is not rationally reconstructible. It must be the subject of empirical investigation. I address the results of my own investigations not principally to other historians, but to mathematicians and teachers of mathematics. The present volume brings together much of what I have tried to say to this audience.

In *The Calculus as Algebra: J.-L. Lagrange, 1736–1813*, I show what Lagrange's mathematical practice was like, in order to understand the genesis of the rigorous analysis of Cauchy, Bolzano, and Weierstrass. For Lagrange, the calculus was not about rates of change or ratios of differentials, or even about limits as then understood. Lagrange thought that the calculus should be reduced to "the algebraic analysis of finite quantities." This sounds as though he was about to introduce deltas and epsilons. But instead he believed that there was an algebra of infinite series, and that every function had a power-series expansion except perhaps at finitely many isolated points. Lagrange *defined* the derivative as the coefficient of the linear term in the function's power-series expansion. Why he thought this was justified tells us both about his philosophy of mathematics and about the way many mathematicians practiced their subject in the eighteenth century. Euler, for example, did marvelous things by what we would now call the carefree formal manipulation of infinite series, infinite products, and infinite continued fractions. But Lagrange found something else in infinite series as well. He imported what we now call delta-epsilon techniques from the 18th-century study of approximations into some of his proofs about the concepts of the calculus. He was the first to attempt to prove, let alone to use inequalities in so doing, statements like "a function with a positive derivative on an interval is increasing there." He justified many results of calculus using inequalities, including the mean-value theorems for derivatives and integrals, and the Lagrange remainder of the Taylor series. This, and much more, helped build Cauchy's work in the 1820s on the foundations of analysis.

The story is instructive in many ways, especially for teachers. Students often have difficulties like those encountered by the mathematicians who invented key ideas. For example, "You say 'can be made as close as you like' is the same as 'equals the limit,' but how can a bunch of *inequalities* somehow turn into two things being *equal*?" is an excellent question. Bishop Berkeley's criticisms of arguments about the limit of the ratio of two quantities when the quantities vanish—and the inadequacy of eighteenth-century attempts to answer him in his own terms—recur among very good students. Knowledge of the history of the relevant mathematical concepts can help the teacher understand what is troubling the student. And the history also lets those of us who learned a subject with ease appreciate which concepts are inherently hard; it's no accident that it took a hundred and eighty years from the understanding of the derivative by Newton and Leibniz to the proofs of Weierstrass.

Students can see something else as well. Mathematicians solve problems, ask questions, explore fuzzy ideas, make false starts or go in the wrong direction. They can be right for the wrong reasons, like Lagrange. And the "right" definitions often come only at the end. Thus following difficult and crooked paths to problem solution makes the student part of the mathematical community. Historians, mathematicians, and students can all gain from understanding Lagrange's grappling with the calculus as algebra.

Both historians and mathematicians can contribute to the history of mathematics, as I argue in the earliest of the articles reprinted here, "The Mathematician, the Historian, and the History of Mathematics," but they may do this in different ways. Historians in general, as I said in that paper, ask two types of questions: What was the past like? And, how did the present come to be? Mathematicians more often tend to be interested in the second question, historians in the first. But these two approaches complement one another, and I have tried to use both in my own work.

For instance, contrast the older and the modern understandings of the derivative. "The Changing Concept of Change: The Derivative from Fermat to Weierstrass" shows how a key mathematical idea can develop quite differently than in the way we now understand it. Properties of varying quantities reaching their maximum or minimum values were discovered and used by people like Fermat and Descartes, but these men did not see their calculations as examples of calculating rates of change. When Newton and Leibniz instead explained how such problems were special cases of fluxions or of differential quotients, they opened the door for research into previously inconceivable topics like differential equations. Understanding and proving the properties of derivatives in terms of a sufficiently powerful and precise concept of limit came even later, and teasing out the distinction between pointwise and uniform convergence to a limit came later still. Beginning the study of rigorous calculus with the delta-epsilon definitions of limit and derivative turns this history on its head. Recapturing the history helps us motivate our students' rediscovery of the concepts, and also makes us marvel at the brilliance of our predecessors, who created these concepts out of such chaos.

In "Who Gave You the Epsilon?" I do another thing historians do: making something familiar in the present look strange and unexpected, and only then showing the path by which it came to be. I imagine a student asking how anybody could ever have imagined explaining "the car is going 50 miles per hour" in delta-epsilon terms. I try to answer this by explaining what the calculus was like before Cauchy, and then how different it looked afterwards. Contrast Euler's successful discoveries by adding, multiplying, and transforming infinite series, even divergent ones, with Cauchy's "A divergent series has no sum" and Abel's lamenting "Can you imagine anything more horrible" than to claim that $1^n - 2^n + 3^n - 4^n + \cdots = 0$? This change in the agreed-on rules of the game is an example of what historians have come to call a paradigm shift. Such a change requires an explanation, which the article presents. The role of approximations in bringing about the change is signaled by the notation "epsilon," which, I argue, comes from the first letter in the French word for error.

Moving from the internal history of mathematics into its cultural setting, in "The Centrality of Mathematics in the History of Western Thought" I observe that, although modern mathematicians may find it hard to explain to outsiders what they do, this has not always been the case. For two thousand years mathematics played a key role in philosophy. Whether it was proof in geometry or the algorithmic power of notation in algebra, mathematics provided a non-material "reality" for idealists, empiricists, and realists to dispute about, as well as a model for methods of demonstration and discovery. Examples abound of mathematics influencing philosophy, from Plato to Spinoza's *Ethics* to the *Declaration of Independence* to Condorcet's prediction of universal progress. Examples of hostility to using mathematics as a model of all human achievement abound as well, ranging from Pascal to Wordsworth. "Centrality" may be an awkward word, but it is important to establish and document the phenomenon it describes.

"Descartes and Problem-Solving" presents a special case of some of the generalizations in "Centrality." Although it is common to be told that Descartes' mathematics comes from his general "method," I explore the possibility that it was the other way around. Observing that Descartes did not have Cartesian coordinates helps understand what he actually did. Descartes developed a new method to solve problems in geometry. He translated them into algebra, then used the algorithmic power of algebra to transform the original question into something more tractable, and finally translated that new "something" back into a geometric construction. His focus on problems and the methods of solving them has influenced mathematicians from Newton to Pólya. And the power of Descartes' new problem-solving method reinforced his philosophical commitment to renewing all of science by finding the one true method, and to portraying that method as inspired by mathematics.

In my three articles on Colin Maclaurin, I turn to "rehabilitating" a mathematician of lesser reputation than Descartes, Lagrange, and Cauchy. In "Was Newton's Calculus a Dead End? The Continental Influence of Maclaurin's *Treatise of Fluxions*" I ask why Maclaurin's contributions have so often been undervalued—this is social history—and I describe his considerable impact on 18th- and 19th-century Continental analysis.

It is often considered, especially by the French school exemplified by Lagrange, that old-fashioned geometry hindered progress of mathematics. In "The Calculus as Algebra, the Calculus as Geometry," I argue that Maclaurin's geometric mind-set helped him formulate results in analysis, and that the decline of such geometric insight was a loss, however much the new analysis was a gain. By contrast, Lagrange's achievements came from his rejection of geometric intuition in favor of a commitment to the general and to his appreciation of abstract algebraic structures. The success of these two different approaches suggests that a diversity of approaches will best promote mathematical progress, as is clear also from works on the psychology of invention, like that of Jacques Hadamard [**2**]. We ought to recognize and nourish this variety of approaches as we teach our subject.

In "Newton, Maclaurin, and the Authority of Mathematics," I show how Maclaurin internalized the way of doing mathematical physics exemplified in Newton's *Principia*, and then applied it to solve problems ranging from the shape of the earth to the computation of annuities. This article also illustrates the authority mathematicians and mathematics were thought to have in the eighteenth century. The universal agreement and clear reasoning of the mathematical sciences were exploited to promote political and social consensus on controversial issues. This provides an instructive example of the deference to quantitative measures as "objective" and therefore beyond politics, which is, for good or ill, alive and well today.

"Why Should Historical Truth Matter to Mathematicians? Dispelling Myths while Promoting Mathematics" began its life as a 20-minute talk at the 2004 Joint Meetings in a session entitled "Truth in Using the History of Mathematics in Teaching Mathematics." I expanded this talk into a colloquium lecture, and after I gave it to several mathematical audiences, with different examples each time, two conclusions became more and more apparent. First, misconceptions about the history of the calculus were not the only important ones for me to address. Second, mathematicians are especially interested in seeing the record set straight when doing this can promote mathematics and improve its teaching. In the current version of the paper, I address myths like "all important mathematics is done by men and within the European tradition," "there was no European mathematics in the Middle Ages," and "the mathematical approach can be applied to settle any major question," as well as "Newton invented the calculus in order to do physics" and various long-held but inadequate views about Maclaurin and Lagrange. I hope to show how knowing what actually happened is more valuable in understanding the mathematical experience than any plausible but erroneous story could be.

Finally, in "How Did Lagrange 'Prove' the Parallel Postulate?" I return to Lagrange, this time via an unpublished manuscript. I cannot rehabilitate Lagrange's "proof" of Euclid's Fifth Postulate. And, unlike Lagrange's power-series definition of the derivative, no "success" arose from Lagrange's approach to this problem. The episode, though, further illuminates Lagrange's approach to science and mathematics, as well as casting light on the broader intellectual and social concerns of Continental society.

Although of course society cannot call mathematical results into being at will, society can and does influence a great deal: for instance, the choice of problems, the applications considered most useful, and how mathematics is taught. Mathematics is part of human culture. And mathematics answers questions of interest and importance to the society around it. In the case of Lagrange and Euclid's postulate, we have the mutual influences of geometry and the arts, the professionalization of mathematics in the nineteenth century, and the increasing need for mathematicians to teach, especially from the time of the French Revolution onward. For Enlightenment thinkers, Euclidean geometry was thought to embody the true nature of space, and space was the really existing framework for Newtonian mechanics. The universal agreement claimed for these ideas served as a model for the philosophy of the Enlightenment, and for its search for universally-accepted truths by reason.

The history of mathematics, then, has a lot to teach us. Mathematicians' creative ideas can come from within mathematics or from the wider society. Mathematical progress cannot be programmed. That's why the history of mathematics is interesting. Mathematics progresses, not only by using and extending the prevailing ideas and techniques, but also by transforming and transcending them. Progress depends on people being willing to use, discover, explore, and develop concepts and techniques—to risk being incomplete or even wrong. Appreciating all of this can attract students, help teachers, and inspire researchers. I hope the present volume can contribute to these goals.

References

1. Hadamard, Jacques. *The Psychology of Invention in the Mathematical Field*, Princeton University Press, 1945. Dover reprint, 1954.

2. Hartley, L. P., *The Go-Between*, London, 1953.

3. Parshall, Karen Hunger, *James Joseph Sylvester: Jewish Mathematician in a Victorian World*, Johns Hopkins, Baltimore, 2006.

I

The Calculus as Algebra

J.-L. Lagrange, 1736–1813

Preface to the Garland Edition

J.-L. Lagrange has long been viewed as a seminal figure in the history of physics, of algebra, of number theory, and of the calculus of variations. But to modern mathematicians and, until recently, to most historians, he was also the man who assumed that every function had a power-series expansion, and who defined the derivative of a function as the coefficient of the linear term in its Taylor series. Accordingly, Lagrange's foundation for the calculus has been viewed as a step backward. But this dissertation, the work of various scholars,[1] and my 1981 book *The Origins of Cauchy's Rigorous Calculus,* have made clear that the older view of Lagrange's calculus is wrong. Lagrange was the key predecessor of men like Bolzano, Cauchy, and Weierstrass in creating modern analysis. Lagrange believed that the calculus of Newton and Leibniz could be defended from logical criticisms, and rigorously founded, by being reduced to algebra. His nineteenth-century successors agreed. Though they rejected the algebra of power series Lagrange had proposed as the foundation for the calculus, they did adopt much of his general approach, his terminology and notation, and—above all—many of his methods and proof-techniques.

While *The Origins of Cauchy's Rigorous Calculus* focuses on Cauchy's accomplishments and treats Lagrange insofar as he contributed to them, the present volume concentrates on Lagrange's work on the foundations of calculus in its own right: on its motivation, its sources, and its content, with only some hints of its influence. The introduction to this version (pp. 9–15 below) places Lagrange in his Enlightenment setting, and explains the key features of his foundation for the calculus. This new preface will sketch the importance of Lagrange's work for the subsequent history of the calculus.

As is well known, Augustin-Louis Cauchy, in his great pedagogical works, the *Cours d'analyse* (1821) and the *Résumé des leçons données à l'école royale polytechnique sur le calcul infinitésimal* (1823), gave the first reasonably sound foundation for the calculus. Similarly, scholars have made clear that Bernhard Bolzano had achieved comparable rigor, though in a narrower realm, in his 1817 paper whose purpose was to prove the intermediate-value theorem for continuous functions. Both Cauchy and Bolzano championed algebraic reasoning over geometric intuition in proofs, gave essentially the modern definition of continuous function, and carried out rigorous proofs about the concepts of the calculus, including their quite different proofs of the intermediate-value theorem.

The similarity of these accomplishments of Bolzano and Cauchy is very striking. Although, in the early nineteenth century, Bolzano's work was less well known than Cauchy's, more recent

[1] A new bibliography, updating that in the 1966 version, follows this preface. On the point just made, see especially the works listed there by Bottazzini, Dauben (1982), Dugac, Flett, Fraser (1989), Goldstine, Grabiner, and Struik.

scholarship has made its quality clear. In 1970, Ivor Grattan-Guinness found the coincidence between the work of these two men so impressive that he claimed that Cauchy had read Bolzano's 1817 paper and borrowed the results without acknowledging their author.[2] In the spirited debate which followed this proposal, Hans Freudenthal vigorously, and I think convincingly, defended Cauchy's originality.[3] But, although the work of Cauchy and of Bolzano are not identical, the resemblances remain striking, and Grattan-Guinness was right in saying that they require an explanation. But Cauchy's and Bolzano's work were independent; the similarity therefore results from a common mathematical heritage, and the principal figure in that common heritage was Lagrange. I have discussed these similarities, and documented Lagrange's influence—which Bolzano and Cauchy each explicitly acknowledged—at some length elsewhere.[4] The key point for our present purpose is that *both* of the successful early-nineteenth-century attempts to rigorize analysis depend upon Lagrange, a fact which magnifies his importance. Thus one cannot understand the rigorization of the calculus in the nineteenth century without understanding Lagrange's contributions to it. Since Cauchy was much more influential than Bolzano, however, in what follows we will emphasize Lagrange's influence on Cauchy.

Of the general influence of Lagrange's calculus on Cauchy's there can be no doubt. Cauchy's books abound with references to Lagrange's. Cauchy's biographer Valson tells us that when Cauchy went on his first engineering job, Lagrange's *Théorie des fonctions analytiques* was one of the four books he took with him. The subtitle of Cauchy's *Cours d'analyse*, "Analyse algébrique," comes from the full title of Lagrange's book. Cauchy adopted Lagrange's notation $f'(x)$ for the derivative, called it by Lagrange's term "fonction derivée" (the origin of our term "derivative"), and often used the Lagrange remainder for the Taylor series. But more importantly, Cauchy's overall program, and some of his crucial definitions and proof techniques, also have a Lagrangian origin.

The introduction to Cauchy's *Cours d'analyse* includes a clear statement of his desire to set a new standard for rigor in analysis: "As for methods, I have sought to give them all the rigor which exists in geometry." Besides such general statements, the *Cours d'analyse* puts these standards into practice, with Cauchy's careful investigation of convergence of series, proofs of criteria for convergence, proofs about limits using algebraic inequalities, and his demonstration of the intermediate-value theorem. Two years later in his *Leçons sur le calcul infinitésimal*, Cauchy showed how the wealth of results of the differential and integral calculus could be firmly based on his definition of limit and on his limit-based definitions of derivative and integral. Similarly, Lagrange's major purpose in his *Fonctions analytiques* was to give a rigorous basis for the calculus, which meant, for him, showing how the full complement of existing results could be based on a firm, algebraic foundation. In this sense, Lagrange's book, which was based on his lectures at the Ecole polytechnique, was really the first Cours d'analyse, as its title indicates: "Theory of analytic functions, containing the principles of the differential calculus. . . . reduced to the algebraic analysis of finite quantities."

There are two main ways in which Lagrange's ideas were adopted, and sometimes transformed, in the work of his successors. The first way is for a Lagrangian approximation technique

---

[2] [Grattan-Guinness 1970a].

[3] [Freudenthal 1971].

[4] [Grabiner 1981] and [Grabiner 1984].

to be turned into an existence proof. (In fact, many techniques used in approximations in the eighteenth century can be turned into proofs of convergence, where the "error"—epsilon in Cauchy's notation—in the approximation process can be shown to be less than any given quantity.) The second way is that Lagrange develops what for Cauchy turns out to be an essential defining property or a key method, although for Lagrange it was a useful, but incidental property, or an ad hoc technique. We will give only a few examples here, but refer the reader to the works already cited for more detail.

Both these ways of transforming Lagrangian ideas can be found in Cauchy's definition of continuous function and subsequent proof of the intermediate-value theorem. Cauchy and Bolzano both have essentially our definition of continuous function. Lagrange seems first to have picked out and used the key property in his work on approximations. For instance, he said of a curve continuous at the origin, "We can always find an abscissa . . . corresponding to an ordinate less than any given quantity, and then all smaller values of [the abscissa] correspond also to ordinates less than any given quantity."[5] The relationship between Lagrange's characterization of continuity and the later definitions by Cauchy and Bolzano was pointed out by Bolzano himself in 1830.[6] Cauchy, using his definition of continuity, proved the intermediate-value theorem for continuous functions by a method closely resembling a technique Lagrange had used to approximate the real solutions of polynomial equations. Lagrange divided an interval into an arbitrary number of subintervals, and then looked at the sign of the polynomial at the endpoint of each subinterval, in order to identify the subintervals which contained a solution.[7] Cauchy's proof used this technique of interval division, repeating it over and over for successive subintervals, in order to obtain the intermediate value as a limit.[8]

The most important contribution Lagrange made to Cauchy's foundation for the calculus was another "incidental" result, which I have elsewhere called the Lagrange property of the derivative:[9] Given $f(x)$ and its derivative $f'(x)$, let i be a real-valued increment of x. Then, Lagrange says, $f(x+i) = f(x) + i[f'(x) + V]$, where V goes to zero with i. This means, he continues, that, given any D, i can be chosen sufficiently small so that V lies between D and $-D$, that is, so that $f(x+i) - f(x)$ lies between $i[f'(x) - D]$ and $i[f'(x) + D]$.[10]

The kinship between the Lagrange property of the derivative and modern delta-epsilon treatments should be obvious. Lagrange obtained this property from the Taylor-series expansion of $f(x+i)$. He was then able to use this property to prove, among other things, that a function with positive derivative on an interval is increasing there (see Appendix, pp. 103–104 below), to prove the mean-value theorem for derivatives, to obtain the Lagrange remainder of the Taylor series (itself motivated by the study of error-bounds in approximations), and to deduce his version of the Fundamental Theorem of Calculus. Cauchy, instead, obtained the Lagrange property from

---

[5] Lagrange, *Fonctions analytiques, in Oeuvres* IX, p. 28.

[6] Bolzano, B., *Functionenlehre*, in B. Bolzano, *Schriften*, Prague: Königlichen Böhmischen Gesellschaft der Wissenschaften, 1930, p. 16.

[7] Lagrange, *Traité de la résolution des équations numériques de tous les degrés*, 2d. ed., Paris: Courcier, 1808, in *Oeuvres* VIII, pp. 22–23.

[8] Cauchy, *Cours d'analyse*, in *Oeuvres*, Ser. 2, vol. III, pp. 378–380. A translation of Cauchy's proof appears in [Grabiner 1981, pp. 167–168].

[9] [Grabiner 1981, p. 116].

[10] Lagrange, *Leçons sur le calcul des fonctions,* new edition, Paris: Courcier, 1806, in *Oeuvres* XI, pp. 86–87.

his own definition of the derivative as the limit of the quotient of differences. Once Cauchy had done this, he could adopt—and improve—Lagrange's proof techniques.[11] Although Cauchy's treatment of the integral owes less to Lagrange than does his treatment of limits, continuity, and derivatives, Cauchy's proof of the Fundamental Theorem is also an adaptation of Lagrange's proof.[12]

Lagrange's work on the foundations of the calculus was essential to the rigorization of the calculus in many ways. It was Lagrange who convinced the mathematical world that the calculus could be made rigorous only by being reduced to algebra. It was he who established that the derivative was neither the quotient of infinitesimals nor a kind of motion, but a function. Finally, it was he who, by studying functions by their Taylor series with remainder, first used inequalities fruitfully in the calculus, both to prove as theorems facts previously considered obvious and to work out many results which his successors needed. It is thus important to know how and why Lagrange conceived the calculus as algebra, how his work developed, and why he wanted to show that the whole body of results of calculus could be given a firm foundation. It is also important to investigate his work on approximations and to understand his pioneering proof methods in calculus. These are the goals this edition of *The Calculus as Algebra* hopes to accomplish.

---

[11] In, for instance, Cauchy's proof of the mean-value theorem for derivatives, *Leçons sur le calcul infinitésimal,* in *Oeuvres*, Ser. 2, vol. IV, pp. 44–45. A translation of this proof appears in [Grabiner 1981, pp. 168–170].

[12] Lagrange, *Fonctions analytiques,* in *Oeuvres* IX, pp. 238–239, and compare p. 81 for the mean-value theorem for integrals; Cauchy, *Calcul infinitésimal, Oeuvres*, Ser. 2, vol. IV, pp. 151–152. A translation of Cauchy's proof appears in [Grabiner 1981, pp. 174–175].

Acknowledgements

Much of the research for this work was supported by the National Science Foundation. I should also like to record my gratitude to the many people with whom I have discussed topics in the history of mathematics. Uta C. Merzbach of the Smithsonian Institution has greatly deepened my understanding of numerous topics, and suggested that approximation techniques would be an area worth investigating. Professor Dirk J. Struik has made many useful suggestions and has given me the benefit of his vast knowledge of eighteenth century mathematics. Professor Carl B. Boyer of Brooklyn College read Chapter 4, and madc many helpful comments for its improvement. Professor I. Bernard Cohen has been a constant source of encouragement as well as an excellent teacher of all aspects of the history of scientific thought. The Biblioteca Medicea-Laurenziana, Florence, kindly allowed me to see a microfilm of a rare paper of L. F. A. Arbogast. I should like to record my thanks to the Harvard College Library, and to the Bibliothèque de l'Institut de France, Paris. Finally, I would like to thank my husband, Sandy Grabiner, who has read the entire book and whose encouragement and mathematical insights have been indispensable.

Mathematical Conventions

Unless otherwise specified, all notation in this work follows the practice of the paper or book under discussion.

Introduction

The eighteenth century was an age in which the power of the human mind seemed unlimited. Philosophers believed that they had discovered the nature of human understanding; scientists or "natural philosophers," that they had found the basic laws of the cosmos; observers of society thought that they knew the principles on which government was based. Human history was expected to show perpetual progress in the future.

Such predictions were not made merely out of optimistic desire; they were based on the real success that men saw had been achieved by the sciences. The science whose success most captured the imagination was mathematical physics. Newton's *Philosophiae Naturalis Principia Mathematica*[1] had made universal the view that the cosmos is a machine working according to principles understood by the human mind. Mathematics had been established as the appropriate language in which to describe these principles. Newton and Leibniz had invented a new branch of mathematics, the calculus, which was ideal for expressing and extending the Newtonian physics. The calculus itself was developed further, into a collection of techniques and results which staggers the imagination.

The mathematicians of this time were men of the world. From the *Encyclopédie* to the *Lettres à une princesse d'Allemagne* of Leonhard Euler, from Gaspard Monge's educational projects to Pierre Simon Laplace's *Essai philosophique sur les probabilités*, the mathematicians provided essential components of the Enlightenment that went beyond their technical achievements.[2]

By contrast, "Je ne sais pas" was the statement most often attributed to Joseph-Louis Lagrange. How is such a spirit to be reconciled with that of the Enlightenment? If ever an eminent eighteenth century *mathématicien* lived in an ivory tower, it was Lagrange. He wrote no philosophy, and almost never commented on non-mathematical subjects, even in his correspondence with the Encyclopedist Jean-le-Rond D'Alembert. Though we know that he disapproved of Euler's *Lettres à une princesse d'Allemagne*, this seems less a philosophical judgment than a displeasure that Euler would waste his time on non-scientific subjects.[3] Lagrange lived in Paris

---

[1] Isaac Newton, *Philosophiae Naturalis Principia Mathematica* (Londini: J. Streater, 1687); the title means "Mathematical Principles of Natural Philosophy."

[2] *Encyclopédie, ou Dictionnnaire raisonné des sciences, des arts et des métiers*, par une societe de gens de lettres. Mis en ordre et publié par M. Diderot... et quant à la partie mathematique, par M. D'Alembert. (Paris: chez Briasson, David l'aîiné, Le Breton, Durand, 1751–1765); Leonhard Euler, *Lettres à une princesse d'Allemagne sur divers sujets de physique et de philosophie* (St. Petersburg: Mietau, etc., 1770); Pierre Simon Laplace, *Essai philosophique sur les probabilités* [1812] (Reprinted, Paris: Gauthier-Villars, 1921); on Monge, see Louis de Launay, *Un grand Français: Monge, fondateur de l'École Polytechnique* (Paris: P. Roger, 1933).

[3] Letter to D'Alembert of 2 June 1769. He suggests that the work "n'aurait pas dû publier pour son honneur." Lagrange *Oeuvres* XIII, p. 132.

from 1786 to 1813, yet we do not know what he thought about the Revolution, the Terror, or the Empire.[4] And we have no record of Lagrange's expression of confidence in the future of science, let alone of mankind.

Lagrange saved his unifying vision and his strong opinions for the realm of mathematics, eschewing his contemporaries' concern with everything in this world and the next. When Lagrange was dogmatic, it was on a very narrow subject: the nature of mathematics and mechanics. Like his contemporaries, he took his unifying principles from the dominant tendencies of the century. But instead of inferring perpetual social progress from the progress of science, Lagrange limited himself to developing a view of mathematics and physics based on the nature of the Continental mathematics of the eighteenth century, and to some extent on the philosophical statements made then about mathematics.

Mathematics, according to men as widely divergent in their views as Locke, Leibniz, Berkeley, and Hume, was a demonstrative science, whose proofs were chains of "clear and distinct ideas," in which every idea was immediately perceived by the mind to agree with the next.[5] Mathematical reasoning began with definitions and assumptions, the fewer the better,[6] and proceeded to infallible conclusions. Mathematics, to the eighteenth century, was thus the epitome of what human reason could accomplish.

More specifically, an eighteenth-century Continental[7] mathematician could see in mathematics the triumph of analysis.[8] The power of analysis or algebra, based on the symbolic notation pioneered by François Viète (1540–1603), had been exploited in the seventeenth century by René Descartes and Pierre de Fermat in their analytic geometry, by John Wallis in his treatment of infinite processes, and particularly by Leibniz and Newton in the calculus. On the Continent, where most of the important mathematics of the eighteenth century was done, an algebraic view reigned supreme. Treating the symbols of Leibniz's differential calculus as if they were algebraic objects which could be manipulated almost without thinking, and certainly without any appeal to intuition, men like the Bernoullis and Euler pushed the subject to frontiers beyond the dreams of its inventors. At the same time—and usually by the same investigators—the science of Newton's *Principia* was translated into the language of the differential calculus. New problems were attacked and solved, and the powerful methods of differential equations and the calculus of variations were applied to physics.

---

[4] We have only the remark on the execution of Lavoisier: "It took only a moment to make that head fall, and one hundred years perhaps will not suffice to produce another like it." Quoted by Jean-Baptiste Joseph Delambre, "Notice sur la vie et les ouvrages de M. le Comte J.-L. Lagrange," in Lagrange's *Oeuvres* I, p. xl.

[5] See John Locke, *An Essay Concerning Humane Understanding* (London: Printed by Eliz. Holt for Thos. Bassett, (1689); G. W. Leibniz, *Nouveaux essais sur l'entendement humain*, ed. Emile Boutroux (Paris: C. Delagrave, 1899); George Berkeley, "A Treatise Concerning the Principles of Human Knowledge," in Volume I of *The Works of George Berkeley*, ed. A. C. Fraser (Oxford: Clarendon Press, 1871); David Hume, *An Enquiry Concerning the Human Understanding* [1777] (Chicago: Open Court, 1900); for "clear and distinct ideas," see René Descartes, *Discours de la méthode pour bien conduire la raison, et chercher la vérité dans les sciences* (Leyden: I. Maire, 1637).

[6] Newton said in the Preface to his *Principia* that "it is the glory of geometry that from [its] few principles . . . It is able to produce so many things." In Isaac Newton's *Mathematical Principles of Natural Philosophy*, ed. Florian Cajori (Berkeley and Los Angeles: University of California Press, 1934), p. xvii.

[7] Eighteenth-century British mathematics was more geometrically oriented, and in addition was more prone to rely on the notion of "velocity" in the calculus.

[8] By "analysis," I mean the use of algebraic symbols to solve problems. It "triumphed" over geometrical methods as well as over ignorance.

Yet the very success of physics distressed more than one man of the Enlightenment. As Laplace once said, "There is but one law of the cosmos, and Newton has discovered it."[9] By the end of the eighteenth century, the most important consequences of Newton's physics had been discovered. Denis Diderot, who was hostile to the mathematical method in science, nevertheless caught an important eighteenth-century feeling when he said that men like the Bernoullis, Euler, and D'Alembert had "erected the pillars of Hercules" beyond which later ages would not be able to pass.[10]

The same feeling prevailed in pure mathematics. Lagrange himself once wrote to D'Alembert, "Does it not seem to you that higher mathematics is becoming rather decadent?"[11] This "'fin de siècle' pessimism"[12] had some basis in the nature of eighteenth-century mathematics. The development of the calculus had been largely formalistic in the eighteenth century. Without careful attention to convergence of series, without knowing under what conditions one might change the order of taking limits or of integration, the number and type of results one could obtain were in fact limited. Near the end of the century, the limit had been reached. What we now call elementary calculus, most elementary techniques for solving differential equations, and the formal properties of power series had been worked out. The only major task remaining appeared to be that of creating a synthesis of results.

The power of algebraic or analytic methods in mathematics convinced Lagrange that the most desirable synthesis of results in mechanics or mathematics would be wholly analytic. Thus he composed his *Mechanique analitique*[13] without a single diagram. The point was not that illustrative diagrams were an unmitigated evil, but that geometric intuition had no place in a truly rigorous subject. Similarly, Lagrange rejected all intuitive ideas as foundations for the calculus.[14] This meant that Newtonian fluxions, which Lagrange believed to rest on the mechanical idea of velocity, were not acceptable. He was also led to reject the concept of limit which had been championed by his friend D'Alembert, and which when properly defined was to provide a truly rigorous foundation for the calculus.[15] Lagrange instead turned to algebra—specifically, the algebra of power series—for his foundation of the calculus.[16]

There is more to the relationship between Lagrange's mechanics and his calculus than his belief that the only important task remaining in either was to put the completed results in the most

---

[9] Pierre Simon Laplace, *Précis de l'histoire de l'astronomie* (Paris: Courcier, 1821). Compare the similar remark of Lagrange, that Newton was the greatest genius that ever existed, "and the most fortunate, for we cannot find more than once a system of the world to establish." Attributed to him by Delambre, *op. cit.*, p. xx.

[10] Denis Diderot, *De l'interprétation de la nature*, section IV. In Denis Diderot, *Oeuvres Philosophiques*, ed. Paul Vernière (Paris: Editions Garnier Frères, 1961), pp. 180–181.

[11] Letter of 24 February 1772, in *Oeuvres* XIII, p. 229.

[12] This highly descriptive phrase is used by D. J. Struik, *A Concise History of Mathematics* (New York: Dover, 1948), p. 199.

[13] The first edition was *Mechanique analitique* (Paris: Desaint, 1788); the second *Mécanique analytique* (Paris: Courcier, 1811–1815). Note that Euler had also written an "analytical mechanics," and also avoided diagrams when he could.

[14] The eighteenth century, in its preoccupation with results, had not displayed great interest in the question of foundations, though of course many mathematicians had something to say about the nature of the differential quotient. This point will recur many times in what follows.

[15] For Lagrange's rejection of this concept, see Chapter 3, below. For the development of this concept in the eighteenth century, and its rigorous definition and application by Cauchy, see Chapter 4.

[16] What eighteenth-century algebra consisted of, and how Lagrange was able to use it, is treated in Chapter 2.

elegant form—the algebraic form. The writing of the *Mechanique analitique*, which reduced mechanics entirely to the calculus, made it even more necessary that the calculus itself be given a firm foundation.

"Analytical mechanics" is analogous to "analytic geometry": each for Lagrange is an attempt to discuss an aspect of reality without recourse to intuition, either physical or geometrical.[17] This observation will serve to answer, at least for Lagrange, the philosophical question of why mathematics should be appropriate to describe the physical universe. Copernicus and Galileo had answered this question with their Platonism; subsequent physicists, accepting the appropriateness of mathematics, had been too successful in its use to stop and ask the question. Geometry was universally accepted in the eighteenth century as *the* science of space;[18] its rigorous bases were not questioned, though many people tried to bolster it by proving Euclid's parallel postulate.[19] If algebra describes geometry—and Descartes had shown that it did—it also describes space. But the science of mechanics bears the same relationship to its subject matter that geometry does to space. If mechanics could be reduced to the differential calculus, and the calculus reduced to algebra, the philosophical problem would disappear.

Not only was the *Mechanique analitique* purely analytic, but it was—in Ernst Mach's words—"a stupendous contribution to the economy of thought."[20] Lagrange reduced mechanics to his single principle of the conservation of virtual velocities. This elegance of treatment—reducing an entire subject to an analytically expressed principle—was akin to the approach of his theory of analytic functions.

The Basic Doctrine of the Theory of Analytic Functions

Lagrange's reduction of the calculus to algebra, particularly as expressed in the *Théorie des fonctions analytiques* of 1797, will be the subject of this work. We shall see how Lagrange's ideas on the calculus evolved during his career, and what motivated him to write the *Fonctions analytiques*. We shall examine the way in which the achievements of the *Fonctions analytiques* are related to the algebra of the eighteenth century. We shall discuss in detail the way in which Lagrange derived the results of the calculus from his algebraic foundation. Finally, we shall examine Lagrange's important positive influence on nineteenth-century analysis. This influence came partly through the inequality proof-technique developed in his *Fonctions analytiques* and

[17] Lagrange remarked that "mechanics is a geometry of four dimensions," *FA* (1797), p. 223; *Oeuvres* IX p. 337, meaning apparently that mechanics is the algebra of (x, y, z, t) just as Euclidean geometry is the algebra of (x, y, z). That geometry is algebraic, Lagrange said, was shown by Descartes. See Lagrange's *Lecons élémentaires sur les mathématiques données à l'école normale en 1795*, in Oeuvres VII, pp. 181–288, p. 271.

[18] Voltaire said, "There is but one morality, as there is but one geometry." "Morality," from the *Philosophical Dictionary*, in *The Portable Voltaire*, ed. Ben Ray Redman (New York: The Viking Press, 1961), p. 225.

[19] The idea was to make Euclid's geometry perfect. See, for instance, Gerolamo (or Girolamo) Saccheri, *Euclides ab omni naevo vindicatus* [Euclid freed of all blemishes] ed. and translated by George B. Halsted (Chicago: Open Court, 1920) [1733]. Lagrange himself read a memoir trying to prove the parallel postulate to the Académie des Sciences in Paris, but he apparently thought better of it and never published this. Bibliothèque de l'Institut de France, *MS* 909, folio pages 17–43, "Sur la théorie des parallèles," is this memoir; written on it in Lagrange's hand is "read at the Institut at the meeting of 3 February 1806."

[20] Ernst Mach, *The Science of Mechanics*. Translated from the Ninth German edition of 1933 (Chicago, Sixth American edition: Open Court, 1960).

Leçons sur le Calcul des Fonctions, partly through his establishment of the idea that the calculus must be reduced to algebra.

The *Théorie des fonctions analytiques* is unlike earlier eighteenth-century works on the differential calculus in being primarily concerned with the foundations of analysis. Lagrange was not seeking new results, but trying to consolidate gains already made. The goal of the work is apparent in the full wording of its title: "Theory of analytic functions, containing the principles of the differential calculus, free from any consideration of infinitely small quantities or evanescents, of limits or of fluxions, and reduced to the algebraic analysis of finite quantities."[21]

As was the custom following Leibniz, Lagrange took "differential calculus" to mean "set of rules for operating with differentials," rules such as $d(uv) = v\,du + u\,dv$. Lagrange thought he was supplying a general theory within whose framework the principles and main results of the differential calculus could be included. This theory was, he thought, free from the imperfections of all previous attempts to explain the concepts of the calculus. He rejected the idea that there are special principles relating to infinity or infinitesimals, which principles have to be imported into mathematics in order to justify the calculus. He believed that the principles of the differential calculus must be purely mathematical, reduced to the consideration of finite quantities by mcans of the well-developed science of algebra.

The algebraic foundation was achieved by the use of power series for the definition of the derivative of a function. To eighteenth-century mathematicians, power series were just infinite polynomials, which were manipulated exactly like finite polynomials. The formula for the product of two power series thus needed no more justification than that for the product of polynomials.[22]

To Lagrange, a function of x, $f(x)$, is an analytic or algebraic expression in x: any "expression de calcul" into which x enters in any way.[23] If, in such an expression, $(x + i)$ is substituted for x, he said—and tried to prove—that the corresponding function $f(x + i)$ can be developed into a series of the form

$$f(x + i) = f(x) + pi + qi^2 + ri^3 + \cdots$$

where $p, q, r, \ldots$ are new functions of x which depend on $f(x)$ and are independent of i. The finding of functions like $p, q, r, \ldots$ "derived from the primitive function $f(x)$," Lagrange called the "true object" of the differential or fluxional calculus.[24]

---

[21] Joseph-Louis Lagrange, *Théorie des fonctions analytiques, contenant les principes du calcul différentiel, dégagés de toute considération d'infiniment petits ou d'évanouissans, de limites ou de fluxions, et réduits à l'analyse algébrique des quantités finies* (Paris: Imprimérie de la République, an V [1797]). The second edition was revised and corrected by Lagrange (Paris: Courcier, 1813). This second edition was reprinted in *Oeuvres*, Tom. IX (Paris: Gauthier-Villars, 1881). The *Fonctions analytiques* will hereinafter be called *FA*; *FA* (1797) will mean the first edition.

[22] For what "algebra" meant to Lagrange in 1797, and for details of the manipulations of infinite series, see "The Algebraic Background of the Theory of Analytic Functions," below.

[23] *FA* (1797), p. 1; compare *Oeuvres* IX, p. 15. The history of the function-concept is sketched in "Algebraic Background . . . ," pp. 42–43, 45–46, below.

[24] *FA* (1797), p. 2. Compare *Oeuvres* IX, pp. 21–22. The development was known for all the common functions; Lagrange says it was known "by the theory of series." For what "theory of series" involved, see "The Algebraic Background. . . . " Lagrange, considering only functions which had analytic expressions, gave what he thought was a proof that any function could be so developed at all but a few isolated points; this "proof" is analyzed in "Algebraic Background . . . ," pp. 48–51. Note that i is Lagrange's notation for a *real* increment.

In the development for $f(x+i)$ given above, the coefficient of i, which here is p, was called by Lagrange "the first derived function of $f(x)$."[25] He introduced a new notation $f'(x)$ for the derived function. Thus $f'(x)$ was defined by its position in the power series representation of the function $f(x+i)$ in powers of i.

Preserving a consistent notation, he designated by $f''(x)$, the first derived function of $f'(x)$; that is, the function derived from $f'(x)$ just as $f'(x)$ is derived from $f(x)$. $f^k(x)$ was recursively defined as a function derived from $f^{k-1}(x)$ just as $f'(x)$ is derived from $f(x)$. His term "fonction derivée," which he chose to stress the functional nature of $f'(x)$, $f''(x), \ldots$, and their dependence on $f(x)$, is the origin of our "derivative." Using these definitions, he proved that the coefficients $q, r, \ldots$ are $f''(x)/2!$, $f'''(x)/3!, \ldots$.[26] This he viewed as an algebraic proof of Taylor's theorem.

Lagrange devoted the remainder of his *FA* to showing how the received results of the differential calculus can be deduced from his power series definition of the derivatives of a function. In deriving many of these results, he made use of the Taylor formula with "Lagrange remainder":

$$\begin{aligned} f(x+i) = {} & f(x) + if'(x) + \cdots + i^n/n!f^n(x) \\ & + i^{n+1}/(n+1)!\left[f^{(n+1)}(x+j)\right] \end{aligned}$$

for some j between 0 and i.[27]

There are two major mathematical drawbacks to Lagrange's theory of analytic functions. First, it is not true that any function given by an analytic expression can be expressed as the sum of a convergent Taylor series about any arbitrary point. (It was not sufficient for Lagrange to except a finite number of isolated points at which the function or its derivatives cease to exist.) Related to this is the fact that two functions, all of whose derivatives agree, need not be identical, even though one of them may be the sum of their common Taylor series. This was pointed out by Augustin-Louis Cauchy, in the case of the two infinitely differentiable functions e^{-x^2} and $e^{-x^2} + e^{-1/x^2}$ expanded about $x = 0$.[28] Lagrange's experience with functions was limited to those which are well-behaved; greater experience with functions with anomalous properties, in the nineteenth century, was to make Lagrange's conception look naïve. Yet in the eighteenth century, with the known functions and canons of proof, the assertion that every function has a Taylor series appeared obvious, and Lagrange's attempted proof might have seemed supererogatory to his intended readers.

Second, since there was yet no theory of convergence, he was not able to justify his manipulations with series. For functions of a real variable, with which *FA* deals exclusively, Lagrange's

---

[25] *FA* (1797), p. 14; compare *Oeuvres* IX, p. 32.

[26] The proof appears in *FA* (1797), pp. 13–14. Compare *Oeuvres* IX, pp. 32–33. This is described and analyzed below, in the form in which it first appeared in Lagrange's work, in his 1772 paper. Lagrange does not use the factorial notation, but writes 1.2, 1.2.3, and so on. I shall use the notation 2!, 3!, n!, from here on, without apology. The notation f', $f'', \ldots$ he had used as early as 1761. See page 20, below.

[27] See the discussion in Chapter 3, below, for some examples. The reminder term is derived in *FA* (1797), p. 41ff; compare *Oeuvres* IX, Chapter VI, pp. 71–84.

[28] Augustin-Louis Cauchy, *Résumé des leçons données à l'école royale polytechnique sur le Calcul Infinitésimal* (Paris: de l'Imprimérie Royale, 1823), in *Oeuvres complètes d'Augustin Cauchy*, IIe Série, Tome IV (Paris: Gauthier-Villars, 1899), pp. 229–230.

manipulations are generally acceptable as long as the functions have convergent power series developments. For functions of a complex variable, for which being analytic (having a convergent Taylor series expansion) is equivalent to being differentiable, many of the results and techniques of *FA* are perfectly valid. But Lagrange himself had no idea of this, and it was not until the time of Weierstrass that the theory of analytic functions of a complex variable was rigorously founded on convergent power series.

1

The Development of Lagrange's Ideas on the Calculus: 1754–1797

General Sketch of the Development of Lagrange's Ideas

The evolution of Lagrange's ideas on the calculus can be traced through his writings from 1754—the year of his first paper—to 1797, when he published the first edition of *FA*. It is not generally appreciated how much Lagrange's views on the calculus changed throughout his career. In particular, the point of view of his brief early writings has usually—and, in my view, mistakenly—been identified with the conclusions of his more mature thought.[1]

Lagrange's early work on the calculus (1754–1761) presented investigations of formal relationships which hold between derivatives, differentials, and integrals, but these were not intended to give a formalistic or algebraic foundation for the calculus. It is nevertheless often considered that Lagrange's foundation for the calculus was formalistic,[2] partly because of the formal nature of this early work. Also, his most enthusiastic disciples stressed the formalistic aspects of *FA* and of the Taylor series definition of the derivative.[3]

Yet Lagrange, in 1760, accepted as rigorous the Newtonian doctrine of first and last ratios. His later rejection of these, and of limits and fluxions, in favor of his "algebraic" method was preceded by a period of dissatisfaction with all the received bases for the calculus. I have found

---

[1] See, for instance, P. E. B. Jourdain, "The ideas of the 'fonctions analytiques' in Lagrange's early work," *Proceedings, International Congress of Mathematicians*, Volume II (Cambridge: University Press, 1912), pp. 540–541. Or see Carl B. Boyer, *The History of the Calculus and its Conceptual Development* (New York: Dover, 1959), whose generalizations about Lagrange's views on the calculus, pp. 251–252, encompass writings from 1759 to 1799. See, however, n. 15, p. 20 below.

[2] As suggested, for instance, by Eric Temple Bell, *The Development of Mathematics* (New York and London: McGraw-Hill, 1945), p. 289; compare Boyer, *op. cit.*, p. 253.

[3] The reception of *FA*, and its "disciples," are discussed in S. Dickstein, "Zur Geschichte der Prinzipien der Infinitesimalrechnung. Die Kritiker der 'Théorie des Fonctions Analytiques' von Lagrange." *Abhandlungen zur Geschichte der Mathematik 9* (1899), pp. 65–79, or, *Zeitschrift für Mathematik und Physik*, Supplement to *44* (1899), pp. 65–79. One should also note that the Analytical Society of Cambridge has views stemming from *FA*, with a formalistic flavor. See especially their Notes in the translation: Silvestre François Lacroix, *An Elementary Treatise on the Differential and Integral Calculus* (Cambridge: J. Smith for J. Deighton & Sons *et al*, 1816). Translation, Appendix, and Notes, by Charles Babbage, George Peacock, and John Frederick William Herschel. Cp. pp. 79–80, below.

no evidence that he ever held that infinitesimals were rigorous,[4] though he did grant that they could be of heuristic value once the rules for operating with them had been rigorously justified.[5]

In 1772, he briefly suggested identifying the task of finding the differential quotient of $f(x)$ with that of finding the coefficient of i in the power series development of $f(x+i)$. However, this did not satisfy him. It was not until 1797 that he published a "proof" that every function has a Taylor series, and actually based the calculus on the definition of $f'(x)$ as the coefficient of i in the expansion of $f(x+i)$. Only in 1797 did he work out his "theory of analytic functions" in detail, together with applications to geometry—tangents, areas, arc-lengths—and to mechanics. After 1797, his basic view of the concepts of the calculus did not change, though his standards of rigor and clarity improved.[6]

Infinite Series and the Calculus: From 1754 up to 1772

Lagrange's first paper, written at age 18, presented the purely formalistic discovery that the formula for the differential of a product

$$d^m(xy) = (d^m x)y + md^{m-1}x\,dy + \frac{m(m-1)}{2!}d^{m-2}x\,d^2y + \cdots \tag{1.1}$$

gives the same coefficients as the binomial expansion

$$(a+b)^m = a^m b^0 + ma^{m-1}b^1 + \frac{m(m-1)}{2!}a^{m-2}b^2 + \cdots . \tag{1.2}$$

[7]

Therefore, there is a correspondence between the "calcolo delle infinite" and that "delle finite grandezze."[8] The well-known differential formula (1.1) was not derived here by Lagrange, but easily follows from simple manipulations with finite differences.[9] Since the binomial theorem (1.2) is true for negative m, it was natural for Lagrange to extend the formula (1.1) to negative m also. He interpreted $(dx)^{-1}$ to be the integral of dx. Thus, in the formula (1.1), if $m=-1$ and $x = dx$, the series obtained is

$$\begin{aligned} d^{-1}(y\,dx) &= \int y\,dx \\ &= xy - x^2/2!\,dy/dx + x^3/3!\,d^2y/dx^2 + \cdots \end{aligned}$$

---

[4] As asserted by Jourdain, *op. cit.*

[5] E.g., in the "Avertissement," *Mécanique analytique*, 2d. edition (Paris: Courcier, 1811–1815), in *Oeuvres* XI (Paris: Gauthier-Villars, 1888), p. xiii.

[6] For the applications to geometry, see Chapter 3, below. For one example of greater rigor, see the proof-technique in Chapter 4, pp. 84–88, below.

[7] "Lettera di Luigi de la Grange Tournier, Torinese, all' illustrissimo Signor Conte Giulio Carlo da Fagnano, contenente una nuova serie per i differenziali et integrali di qualsivoglia grado, corrispondente alla Newtoniana per le potesta e le radici," in *Oeuvres* VII, pp. 583–588. The letter was published separately (Torino: Nella Stamperia Reale, 1754). The equations cited appear in *Oeuvres* VII, p. 584. To further the analogy, Lagrange used the notation $(xy)^m$ instead of $d^m(xy)$, where $x^m = d^m x$ and $x^0 = x$. I have used the more familiar differential notation for clarity.

[8] Ibid., p. 583.

[9] As, for instance, by Leonhard Euler, in *Institutiones Calculi Differentialis* (Petropolitanae: Impensis Academiae Imperialis Scientiarum, 1755) in *Opera Omnia*, Series Prima, Vol. X (Leipzig and Berlin: B. G. Teubner, 1913), Section 163.

which is, Lagrange noted, the "Bernoulli series," published by Johann Bernoulli in 1694.[10] The derivation of this well-known result from his formal analogy must have convinced the young Lagrange of the value of his discovery. He sent this discovery to Euler, in a letter of 1754.[11] Euler's reaction consisted, sadly for the young Lagrange, of the statement that the analogy had already been noted by Leibniz.[12]

Lagrange never referred to his first work again, no doubt because it was not new. It should be noted, however, that the formal correspondence he had found between the analysis of the infinite and of the finite, between the calculus and algebra, was closely connected with a power series. It seems likely that this first discovery of his youth directed his attention to the possibility of relating the calculus to algebra by means of power series. Often a youthful discovery so strikes a mathematician that much of his later career involves variations on the same theme: thus Gauss continually returned to the fundamental theorem of algebra, Cauchy to criteria for convergence. Lagrange was to return again and again to the theme that there is a correspondence between the objects of the calculus and those of algebra. Nevertheless, the foundations of the calculus were not mentioned in Lagrange's 1754 paper.

The first occurrence of the subject of the "true metaphysic" of the calculus in Lagrange's published writings was in a letter to Euler of 24 November 1759.[13] Lagrange expressed his pleasure in hearing that Euler was "continuing to enrich the Republic of Letters" with important works on the differential and integral calculus, and on mechanics. He added that he himself had worked out the elements of these subjects for the use of his pupils in Turin, and he believed himself "to have developed the true metaphysic of their principles, as far as this is possible."[14] Lagrange's phrase "autant qu'il est possible" could be taken to mean that he had worked out the "true metaphysic" so thoroughly that nothing further could be desired. But I think it more likely that Lagrange considered the subject of the foundations of the calculus, long in dispute, a rather unsatisfactory one, and meant that his presentation was the best possible under these circumstances. However this may be, Lagrange's comment reflects the common eighteenth-century view of the foundations as essentially a thing apart from the calculus—he did not even bother to relate the details to Euler. They were worked out merely as an explanation for his students. In Lagrange's own contributions to mathematics in the 1750's and 1760's, he did not bother with the "true metaphysic" of the objects of the calculus.

In the absence of a set of notes from Lagrange's course at Turin, we can only speculate about what he thought the "true metaphysic" was in 1759. His isolated and somewhat cryptic utterance, quoted above, has been taken to mean that he had worked out some form of his "theory

---

[10] Lagrange, *op. cit.*, p. 587. See Johann Bernoulli, "Additamentum effectionis omnium quadraturarum et rectificationum curvarum per seriem quandam generalissimam," *Acta Eruditorum* 1694, pp. 437–441. This formula can be obtained by term-by-term integration of the Taylor series $y(0) = y(x) - x\,dy/dx + x^2/2!\,d^2y/dx^2 - \cdots$.

[11] Lagrange's *Oeuvres* XIV, pp. 135–138. The date is established by editor Ludovic Lalanne by Lagrange's reference to the death of J.-C. Wolf (1679–April 1754). This letter remained unpublished until B. Boncompagni published it in *Lettres Inédites de Joseph-Louis Lagrange à Léonard Euler* (St. Pétersbourg: Expedition pour la Confection des Papiers de l'état: 1877), pp. 5–8.

[12] Euler's letter is dated Berlin: 6 September, 1755. In *Oeuvres* (of Lagrange) XIV, pp. 144–146. Leibniz's discovery was first published as "Symbolismus memorabilis calculi algebraici et infinitesimales in comparatione potentiarum et differentiarum. . . . " *Miscellanea Berolinensia*, 1710.

[13] *Oeuvres* XIV, pp. 170–174.

[14] *Loc. cit.*, p. 173.

of analytic functions" in 1759.[15] But the fact that, many years later, Lagrange gave what he considered to be the true metaphysic of the calculus does not prove that this was what he had in mind in 1759. We shall presently see that there is evidence to the contrary.

P. E. B. Jourdain's suggestion that Lagrange defined the derivative by its position in the Taylor series in 1759 is based on Lagrange's introduction of the notation $\xi'(x)$, $\xi''(x), \ldots$ for the differential quotients of $\xi(x)$ in 1761. This was in the section on the lever in a treatise on mechanics by Daviet de Foncenex.[16] Here Lagrange expanded the functions $\xi(x+z)$ and $\xi(x-z)$ into power series in z—unfortunately incorrectly, using the expansion

$$\xi(x+z) = \xi(x) + z\,\xi'(x)/2! + z^2\,\xi''(x)/3! + \cdots;$$

however, the error does not affect the argument for which he used the series. The notation $\xi'(x)$, $\xi''(x)\ldots$ he explains as denoting "successive differences of $\xi(x)$ divided by dx."[17]

The reform in this paper thus appears to be entirely one of notation. The explanation is vague: it is not clear whether the "successive differences" of $\xi(x)$ are finite or infinitesimal. In the Taylor series, the coefficients of z^k are divided by $(k+l)!$ instead of $k!$ There is a misunderstanding here of either the differential calculus or the Taylor series. It is hard to believe that the error was due to Lagrange, who had used the correct Taylor series in 1754; we may blame Foncenex or perhaps the printer.

In any case, the statement identifying $\xi'(x)$ with $d\xi(x)/dx$ was not treated as a new foundation for the calculus; it is merely a new and more convenient notation. This treatment certainly underlines Lagrange's early interest in the Taylor series, but after all, everyone knew by 1760 that $f(x+i) = f(x) + i\,df(x)/dx + i^2/2!\,d^2f(x)/dx^2 + \cdots$. Using another notation for the differential quotient does not in itself provide any new knowledge, either about the differential quotient or about the Taylor series. The notation used here by Lagrange did call attention to the dependence of the successive differential quotients on $\xi(x)$. However, the use of the notation z, z', $z'', \ldots$ for recursively defined quantities, which the successive differential quotients clearly are, is common enough in the eighteenth century. Moreover, the $\xi'(x)$ notation is close to the dot-notation used by Newton for fluxions. We need not assume, as Jourdain has done, that Lagrange had any new concept in mind in introducing it here.

In 1760–61, Lagrange in fact seems to have accepted a Newtonian metaphysic for the calculus. He wrote, in a note to a paper of Hyacinth Sigismund Gerdil which rejected the notion of the actual infinite, that the method of infinitesimals produced correct conclusions only by a compensation of errors.[18] Lagrange's note gives the following example. In finding a tangent to a

---

[15] E.g., by Jourdain, *op. cit.*, p. 540; by Boyer, *op. cit.*, p. 252. Since a set of notes on Lagrange's Turin course have now been published by Maria Teresa Borgato [Lagrange 1987], we can see that, in the 1750's, he based the calculus on "ultimate ratios."

[16] Daviet de Foncenex, "Sur les principes fondamentaux de la mechanique," in Mechanics Section, *Miscellanea Taurinensia*, Tomus Alter. *Mélanges de philosophie et de mathématique de la Societé Royale de Turin*, 1760–1761, pp. 299–322. Jean-Baptiste Joseph Delambre, "Notice sur la vie et les ouvrages de M. le Comte J.-L. Lagrange," in Lagrange's *Oeuvres* I (Paris: Gauthier-Villars, 1867), pp. viii–li, relates that Lagrange told him "à ses derniers instants" that he was the author, p. xi. It is not really clear whether the chapter was actually written by Lagrange, or composed by Foncenex on lines suggested by Lagrange. Foncenex wrote ξx instead of $\xi(x)$; I have modernized this for the sake of clarity.

[17] Foncenex, *op. cit.*, p. 321.

[18] Joseph-Louis Lagrange, "Note sur la métaphysique du calcul infinitésimal," *Miscellanea Taurinensia* II (1760–61), pp. 17–18, in *Oeuvres* VII, pp. 597–599. The basic idea of this is probably, though not certainly, based on the discussion

curve by the method of infinitesimals, the assumption is made that the curve is a polygon with an infinite number of infinitely small sides. This is not true. Later in the computation, some quantities are neglected as if they were zero when they are not zero—only infinitely small. This is an error also. The fact that the correct tangent is found must therefore be explained by the fact that the errors cancel each other out. But the compensation of errors is surely *not* a "metaphysic" for the calculus![19]

On the other hand, Newton's method is, according to Lagrange, "entirely rigorous, in the assumptions and in the procedures of computation [*calcul*]; for [Newton] conceives that a secant becomes a tangent only when the two points of intersection come to fall on each other, and then he rejects from his formulas all the quantities which this condition makes actually zero."[20] Newton's method thus requires that "the quantities whose first or last ratios we seek be regarded as evanescent, that is as zero."[21]

At this time, Lagrange appears to have known of only two competing "metaphysics" for the calculus: that of Newton and the infinitesimals of Leibniz. Lagrange held the concept of infinitely small quantities to be non-rigorous. A quantity is either something or nothing. Perhaps infinitely small quantities could be regarded as useful fictions to shorten arguments and solve problems, but the validity of the results had to be established without them. The idea of first and last ratios, or limits, as expressed by Newton, appeared rigorous to Lagrange in 1760. He had not yet begun to criticize the notions "secant *becomes* a tangent," points "*come to fall* on each other," or "evanescent," which carry with them the intuitive idea of motion.

Thus, Lagrange may well have explained the differential quotient to his students in 1759 as the limit of a ratio, after the manner of Newton. His statements about the calculus in 1760 support this conjecture; they do not suggest that he even considered defining the derivative by its position in the Taylor series.

His change in views from 1760 to 1772, in which year he did so define the derivative, I believe to have been related to his mathematical activity during that period. From 1761 to 1772, Lagrange published nothing on the foundations of the calculus, but he was actively engaged in other kinds of mathematical research. This period marked in particular Lagrange's most important work on algebra.[22] A reading of his papers in this period shows him to have been an algebraist of consummate skill. His manipulative facility was great; his feeling for the general shines through

---

of compensation of errors in George Berkeley, "The Analyst," first published in 1734. See *The Works of George Berkeley*, edited A. C. Fraser (Oxford: Clarendon Press, 1901), Volume III. I do not know at what point in his career Lagrange became acquainted with Berkeley's ideas; it must have been before 1797, since he then knew Colin Maclaurin's rebuttal of the "Analyst," the *Treatise of Fluxions*.

[19] As Carl Boyer seems to imply, *op. cit.*, pp. 257–258. If one claimed to prove that

$$(3+4)+5=12$$

by saying that $(3+4)=8$, and that $8+5=12$, the right answer would be obtained by this compensation of errors, but the proposition could not be considered rigorously established!

[20] Lagrange, *op. cit.*, p. 598.

[21] *Ibid.* He apparently had in mind Newton's *Philosophia Naturalis Principia Mathematica*, Book I, Section 1, Lemma XI, Scholium. The possible influence of Euler's zeros also comes to mind; Lagrange does not mention them here. Euler's *Institutiones Calculi Differentialis* (1755) treated dx as actually zero, but occasionally used limit-arguments. See A. P. Juschkewitsch, "Euler und Lagrange Uber die Grundlagen der Analysis," in *Sammelband der zu ehren des 250 Geburtstages Leonhard Eulers* (Berlin: Akademie-Verlag, 1954), pp. 224–244.

[22] Especially "Sur la résolution des équations numériques" and "Additions au Mémoire sur la résolution des équations numériques," *Mémoires de l'Academie royale des Sciences et Belles-Lettres de Berlin*, 1767, pp. 311–352; 1768,

everywhere. He holds that any particular result of any interest must be a special case of some more general principle. This is the view of modern mathematics; Lagrange is one of its first proponents.[23]

Infinite Series and the Calculus: The 1772 Paper

In 1772, Lagrange published "Sur une nouvelle espèce de calcul rélatif à la différentiation et a l'intégration des quantités variables."[24] In this paper, both the supremacy of algebra in Lagrange's thinking and his generalizing view are illustrated. The paper presents an operational calculus for the differential operator d and related operators like d/dx. The operational calculus views the derivative itself as an object of algebraic manipulation. The particular example for $d^m(xy)$ had been the subject of Lagrange's 1754 paper; here he correctly attributes the result to Leibniz.[25] The 1772 paper extended the operational point of view, exploiting the calculus for d to yield results in finite differences. He not only gave meaning to expressions like d^{-1}, but to those like $e^{d/dx}$.[26]

For,

$$u(x+h) - u(x) = \Delta u,$$

and since $u(x+h) = u(x) + h\, du/dx + h^2/2!\, d^2u/dx^2 + \cdots$,

$$\Delta u = h\, du/dx + h^2/2!\, d^2u/dx^2 + h^3/3!\, d^3u/dx^3 + \cdots .$$

If $d^k u/dx^k$ is treated as the operator $(d/dx)^k$ applied to u, Δu will be given by

$$\Delta u = e^{h(du/dx)} - 1. \tag{1.3}$$

The need for the Taylor series in this particular result of his operational calculus led Lagrange to make some remarks about this series. These remarks are not relevant to the subject of the paper, but they do show how he viewed the Taylor series—and the calculus—in 1772. It was not necessary for him to prove Taylor's theorem in this paper, nor was it necessary to make any remark about the real nature of the function du/dx. That he did both these things shows that the Taylor series was already an object of close scrutiny, and that the objects of the differential calculus were no longer, in his mind, to be best explained by the method of first and last ratios.

---

pp. 111–180; these are in Lagrange's *Oeuvres* II, pp. 539–580, 581–654. Also, "Réflexions sur la résolution algébrique des équations," *Nouveaux Mémoires de l'Académie royale des sciences et Belles-Lettres de Berlin*, 1770, pp. 134–215; 1771, pp. 138–253; in *Oeuvres* III, pp. 205–424. This paper, which investigated why the quintic was unsolvable, initiated the investigations of both Niels Henrik Abel and Evariste Galois.

[23] See especially the "Réflexions sur la résolution algébrique des equations." Compare E. T. Bell, *op. cit.*, passim.

[24] "Sur une nouvelle espèce de calcul rélatif a la différentiation et a l'intégration des quantités variables," *Nouveaux Mémoires de l'Académie Royale des Sciences et Belles-Lettres de Berlin*, "Classe de mathématiques" pp. 185–221, in *Oeuvres* III, pp. 439–476.

[25] The *Miscellanea Berolinensia*, 1710. Lagrange, *op. cit.*, p. 441.

[26] *Ibid.*, p. 450. Lagrange's result is more general than the one I give, below, as (1.3), being derived for a function u of any number of variables. But the one-variable example shows what he had in mind. Note that du/dx in (1.3) is an operator, not the function $u'(x)$. I have used h for the increment for which Lagrange used ξ.

His remarks on the nature of du/dx begin with his introduction of the infinite series for $u(x+h)$. The "known theory of series"[27] gives $u(x+h)$ as a power series in h:

$$u(x) + ph + p'h^2 + p''h^3 + \cdots$$

where p, p', p'', ... are new functions of x, "derived" in a certain way from the function u.[28]

Similarly, if u is a function of x and y, and if we write $u(x, y)$ as u,

$$u(x+h, y+k) = u + ph + qk + p'h^2 + q'hk + r'k^2 + p''h^3 + \cdots$$

for some p, q, r, p', q', r', ... "The differential calculus, considered in all its generality, consists in finding directly, and by easy and simple procedures, the derived functions $p, p', p'', \ldots q, q', q'', \ldots r, r', r'', \ldots$ of u."[29] The integral calculus, he added, consists in finding u by means of the functions p, p', p''....

Lagrange concluded that "this notion of the differential and integral calculus seems to me the clearest and simplest yet given," since it is "independent of all metaphysics and all theory of infinitely small or evanescent quantities."[30] This statement foreshadows the point of view of *FA*, and Lagrange's break with the methods of the past.

Lagrange then used these power series considerations to give an algebraic demonstration for Taylor's theorem.[31] This demonstration begins with the formula:

$$u(x+h) = u + ph + p_1h^2 + p_2h^3 + \cdots \tag{1.4}$$

where the functions u, p, p_1, ... are understood to be evaluated at x. In the proof, Lagrange made use of the similar power series expansions

$$p(x+w) = p + qw + rw^2 + \cdots, \tag{1.5}$$

$$p_1(x+w) = p_1 + q_1w + r_1w^2 + \cdots, \tag{1.6}$$

$$p_2(x+w) = p_2 + q_2w + \cdots, \tag{1.7}$$

and so on. Since the series expansion (1.4) holds for any x and h, Lagrange could develop $u[(x+w)+h]$ about $(x+w)$ in powers of h, and set this equal to $u[x+(h+w)]$, developed about x in powers of $(h+w)$.[32]

---

[27] See notes 22 and 24 of Introduction.

[28] Lagrange, *op. cit.*, p. 442. Note that in general p^k is *not* d^kp/dx^k. Of course it does not matter whether we focus on d^ky/dx^k or on $k!\ d^ky/dx^k$. Unfortunately, Lagrange introduces the prime notation for derivatives later in this paper. Thus, a cursory reading of the passage just cited might suggest that in the sequence $p, p', p'' \ldots$ each "primed" function is the derivative, in the modern sense, of its predecessor. This is not Lagrange's meaning; he says only that they are derived from—i.e., depend upon—their predecessors.

[29] *Ibid.*, p. 443.

[30] *Ibid.*, p. 443.

[31] *Ibid.*, p. 444–447. In 1772 Lagrange said that he thought it was the simplest of possible demonstrations, p. 447. Later, in the *FA* of 1797, p. 5, in which a more elegant form of the proof was given, he called it an algebraic demonstration of Taylor's theorem. In 1797, he realized that he did not have to assume the existence of the series in (1.5), (1.6), (1.7), etc.

[32] I have substituted subscripted letters for the primed letters Lagrange used for the Taylor series coefficients, to prevent confusion with our notation for derivatives. Where he used Greek letters, I have substituted Latin ones. Thus, where Lagrange wrote $p', \omega', p', \ldots, p'', \omega, p'' \ldots$ I write $p_1, q_1, r_1, \ldots, p_2, q_2, r_2 \ldots$

Developing about $(x+w)$, and using the series expansions for $p(x+w)$, $p_1(x+w), \ldots$

$$\begin{aligned} u[(x+w)+h] &= u(x+w) + p(x+w)h + p_1(x+w)h^2 + p_2(x+w)h^3 + \cdots \\ &= u + pw + p_1 w^2 + p_2 w^3 + \cdots + (p + qw + rw^2 + \cdots)h \\ &\quad + (p_1 + q_1 w + r_1 w^2 + \cdots)h^2 + \cdots. \end{aligned}$$

But $u[x+(h+w)] = u + p(h+w) + p_1(h+w)^2 + p_2(h+w)^3 + \cdots$.

Expanding and collecting powers of h, the expression for $[x+(h+w)]$

$$\begin{aligned} &= (u + pw + p_1 w^2 + \cdots) + h(p + 2p_1 w + 3p_2 w^2 + 4p_3 w^3 + \cdots) \\ &\quad + h^2(p_1 + 3p_2 w + 6p_3 w^2 + 10 p_4 w^3 + \cdots) \\ &\quad + h^3(p_2 + 4p_3 w + 10 p_4 w^2 + \cdots) + \cdots \end{aligned}$$

Now he equated the coefficients of $h, h^2, \ldots$ in the expressions for the equal quantities $u[(x+w)+h]$, $u[x+(h+w)]$. For instance, in the expressions for the coefficients of h,

$$p + 2p_1 w + 3p_2 w^2 + 4p_3 w^3 + \cdots = p + qw + rw^2 + \cdots.$$

Equating the coefficients of w, $w^2, \ldots$

$$q = 2p_1, \quad r = 3p_2, \ldots.$$

Lagrange then noted that q is derived from p just as p is derived from u (See (1.4) and (1.5)), and, similarly, just as q_1 is derived from p_1, etc. To express this relationship, Lagrange introduced the prime notation. That is, he set

$$p = u', \quad \text{and} \quad q = p', q_1 = p_1', \ldots.$$

He then let u'' designate a function derived from u' just as u' is derived from u. Thus,

$$q = p' = u''.$$

But since $q = 2p_1$,

$$p_1 = u''/2.$$

Similarly, let u''' be derived from u'' just as u'' is derived from u'; then, since $3p_2 = q_1$ (shown by equating the coefficients for h^2 in the expressions $u[(x+w)+h]$, $u[x+(h+w)]$),

$$p_2 = q_1/3 = p_1'/3 = u'''/2.3.$$

Similarly, $p_3 = u''''/4!, \ldots$. Thus, he concluded

$$u(x+h) = u + u'h + u''/2!\,h^2 + u'''/3!\,h^3 + \cdots.$$[33]

Only now did he show the equivalence of these concepts to those of the well-known differential calculus, by treating h as infinitely small. The change in u is then $u'h$. If $u'h = du$, $h = dx$,

$$u' = du/dx.$$

[33] *Ibid.*, p. 445.

Thus to find the function u', said Lagrange, we need only find the differential du by the rules of the "calcul des infiniment petits," and divide it by the differential dx. Then, u'' being derived from u' just as u' is from u, we have

$$u'' = d(du/dx)/dx = d^2u/dx^2.$$ [34]

Finally, $u(x+h) = u + du/dxh + d^2u/dx^2h^2/2! + \cdots$, Taylor's theorem, where du, d^2u, $d^3u \ldots$ he took to designate the first, second, third… differences of u when x varies through the "infinitely small difference dx."[35] Lagrange in 1772 apparently saw Taylor's theorem as having as its subject an infinite series of quotients of infinitely small differences. Thus, writing $f(x+h) = f(x) + f'(x)h + f''(x)/2!\,h^2 + \cdots$ is not a proof; the coefficients must be identified with $\frac{d^k f(x)/dx^k}{k!}$. That is, $f'(x)$ is not conceived to be equal to $df(x)/dx$; this must be shown.

In this paper, Lagrange exhibited the equivalence of his conception to the differential calculus, and also indicated how the functions u', u'', … might be found by methods already in use by mathematicians. Nevertheless, this paper is far from having the outlook of *FA*. The series development $u(x+h) = u + ph + p_1h^2 + \cdots$ is taken completely for granted; the algorithmic rules of the differential calculus are not derived from the new definition of u'; and the expressions d^ku/dx^k are treated as quotients of infinitely small differences. Further, the rest of the paper uses the Leibnizian notation du/dx, d^2u/dx^2…. In 1772, Lagrange was interested principally in his new operational calculus, which gave him many results as special cases of general formulas like

$$\Delta^\lambda u = (e^{h\,du/dx + k\,du/dy + \cdots} - 1)^\lambda$$ [36]

To this end, he was willing to speak of infinitely small differences without apology.

His desire to find a basis for the calculus independent of infinitely small and evanescent quantities, and the fact that he did give an algebraic derivation for the Taylor series coefficients, show the importance he attached to finding an algebraic foundation for the calculus in 1772. His remarks about the true foundation for the calculus were to become of importance in his later work, but they were irrelevant to the main goals of the 1772 paper. Like so many other treatments of the "true metaphysic of the differential calculus" by eighteenth-century analysts, Lagrange's remarks were incidental. It is clear that in 1772 Lagrange thought that infinite series would provide the clearest foundation for the calculus, but this was an embryonic idea, not a realized goal.

The Period of Indecision: The Berlin Prize

Lagrange did not feel he had a satisfactory foundation for the calculus after 1772, until at least 1789.[37] We may conclude this partly from the fact that he published nothing with any claim to solve the problem between 1772 and 1797. In addition, we possess a document which testifies

---

[34] *Ibid.*, pp. 446–447. It will be recalled that Lagrange had used the prime notation in Foncenex's memoir of 1759. But it was there a notational convenience; the recursive properties of the function u' were not pointed out. He also used the prime-notation in "Nouvelle méthode pour résoudre les équations littérales par le moyen des séries," *Mémoires de l'Académie Royale des Sciences et Belles-Lettres de Berlin*, 1768, pp. 251–326, in Oeuvres III, pp. 3–73. Here, too, it was just a notational convenience; in fact, following the introduction of the prime-notation, the differential calculus notation for differential quotients of $\psi'(x)$ occured many times.

[35] "Sur une nouvelle espèce de calcul…," p. 447.

[36] *Ibid.*, p. 450.

[37] When he heard a memoir of L. F. A. Arbogast on this subject; Lagrange praised this memoir in *FA*. See page 28, below.

to Lagrange's continuing dissatisfaction with the state of the foundations of the calculus, dating from the middle of the period 1772–1797. This document makes clear that Lagrange did not regard his 1772 work as definitive.

In 1784, the Berlin Academy's "Classe de Mathématique," of which Lagrange was President, recognized the unsatisfactory state of the foundations of analysis and offered a prize for "a clear and precise theory of what is called 'Infinity' in mathematics."[38] The resident members of the mathematics section at this time were Lagrange, Johann III Bernoulli (1744–1807), and Johann Karl Gottlieb Schulze (1749–1790).[39] Lagrange, as the director and leading talent, no doubt proposed the subject, and probably was responsible for the language in which the details and motivation were described. The pre-eminence of Lagrange among the academicians, the history of his concern with this question, and the close resemblance of the language of the prize-proposal to that later used by Lagrange to justify the ideas of *FA* all support this conclusion. The obvious discrepancy between the standards of rigor in geometry and arithmetic, and those of the calculus, was responsible for the attacks which had been made against the subject by Bishop Berkeley and Bernhard Nieuwentijdt.[40] These objectors had provoked spirited—if inadaquate—defenders. Lagrange saw the attacks as pointing out a serious problem which needed to be solved.[41]

The introductory statement of the proposal attributed the usefulness of mathematics, the great respect in which it was held, and the "honorable denomination of 'exact science,' *par excellence*" to the "clarity of its principles, the rigor of its demonstrations, and the precision of its theorems."[42] But, it continued, "to assure to this beautiful part of our knowledge the continuation of these precious advantages, we ask a clear and precise theory of what is called 'infinity' in mathematics."

The writers of the proposal plainly were disturbed by the fact that such a vast structure of analysis had been erected on what was apparently an inadequate foundation. How, they hoped the competitors would explain, had so many theorems "been deduced from a contradictory supposition"—for "infinite" and "magnitude," they said, appeared contradictory to many. It is simply false to say that not until the work of Cauchy and Niels Henrik Abel was anyone uneasy about the insufficiency of the foundations of analysis for the size of the structure which had been

---

[38] *Nouveaux Mémoires de l'Academie Royale des Sciences et Belles-Lettres, Année 1784, Avec l'Histoire pour la même année*, "Histoire," pp. 12–14, "Prix proposés par l'Académie Royale des Sciences et Belles-Lettres pour l'année 1784," p. 12.

[39] A list of the members in 1786, the year of the death of Frederick the Great, is given by Adolf Harnack, *Geschichte der Königlich Preussischen Akademie der Wissenschaften zu Berlin*, 3 volumes, (Berlin: Reichsdruckerei, 1900), Volume I, Section 1, p. 480. In addition to Bernoulli (who became a member in 1764), and Schulze (1777), Friedrich Adolf Maximilian Gustav von Castillon (1747–1814) became a member in 1786. Lagrange himself became both a resident member and director in 1766, replacing Euler who had gone to St. Petersburg. Harnack, p. 467. A biographical index of the members of the Berlin Academy may be consulted in either *Deutsche Akademie der Wissenschaften zu Berlin: Biographischer Index der Mitglieder*, ed. Kurt-R. Biermann, Gerhard Dunken (Berlin: Akademie-Verlag, 1960), or, a list of members with capsule biographies in Erik Amburger, *Die Mitglieder der Deutschen Akademie der Wissenschaften zu Berlin: 1700–1950* (Berlin: Akademie-Verlag, 1950). Incidentally, the best-known Berlin academician in the Mathematics section, besides Lagrange, in the late eighteenth century, was Johann Heinrich Lambert, who died in 1777.

[40] See, on Berkeley, G. A. Gibson, "Berkeley's Analyst and its critics: an episode in the development of the doctrine of limits," *Bibliotheca Mathematica*, N. S., XIII (1899), pp. 65–70. On Nieuwentijdt, see Moritz Cantor, *Vorlesungen über Geschichte der Mathematik* (4 vols. Leipzig: B. G. Teubner, 1900–1908), Volume 3, passim.

[41] See, especially, the 1760 discussion of the doctrine of compensation of errors, and the critique of previous attempts to found the calculus in *FA*, discussed in Chapter 3, below.

[42] "Histoire," 1784, p. 12.

erected. But was it possible to do the calculus without using this troublesome, if fruitful, notion of "infinity"?

The Academy voiced the hope that another principle could be substituted for "the infinite": a principle which would be "certain, clear—in a word, truly mathematical." The new principle should not, furthermore, make the use of the calculus too difficult or tedious.

Thus Lagrange, presumably along with his fellow-academicians Bernoulli and Schulze, did not think that the limit-ideas of Newton and D'Alembert had provided an adequate basis for the calculus. Lagrange probably would have said, if asked, that notions like "limit" and "evanescent" had not been formulated in a manner "certain, clear... truly mathematical": he was to say this explicitly in his *FA* of 1797. The "Method of Exhaustion," on the other hand, while considered rigorous in the eighteenth century, was widely recognized as much too unwieldy for actual use. The Academy showed its dissatisfaction with these methods not by explicitly attacking them, but by asking for a "new principle," one to be presented "with all possible generality, and with all possible rigor, clarity, and simplicity."[43] The Academy thus recognized the problem and appealed to the entire learned world to solve it.

Unfortunately, the results were not quite what had been hoped for. The prize was awarded in 1786, by unanimous decision, to the entry of Simon L'Huilier.[44] But the report of the prize committee makes clear that the unanimity did not mean wild enthusiasm. The Academicians sounded like professors handing back a disappointing group of essays. Though many contributions had been submitted, they said, the authors had all forgotten "to explain 'how so many true theorems have been deduced from a contradictory supposition.'" The contributions had lacked "clarity, simplicity, and especially rigor." Finally, most of the contributors had not even seen that the principle desired had to be "not limited to the infinitesimal calculus, but extended to Algebra and to Geometry treated in the manner of the Ancients"; it seems from this that contributors had tried to find an *ad hoc* principle, avoiding the reduction of the calculus to basic concepts of mathematics. The prize question, the Academy felt, had "received no complete answer."[45]

L'Huilier's answer had come the closest, and—committed to give a prize—the Academicians awarded it to him. But from the tone of the report we may see that his essay was regarded as the best of a bad lot, and that the problem of finding a satisfactory foundation for the calculus was still unsolved.

L'Huilier had based the calculus on the verbally-expressed limit-concept common in the eighteenth century, stating that "the axiom... that what is true up to the limit is true at the limit, is involved in the very conception of a limit."[46] His work did make extensive use of the infinite-series development of functions—which is probably what made it attractive to

---

[43] *Ibid.*, p. 13

[44] "Histoire de l'Académie..." 1786, "Prix proposés par l'Académie Royale des Sciences et Belles-Lettres pour l'Année 1788," p. 7. L'Huilier published his essay later as *Exposition élementaire des calculs supérieurs*, (Berlin, [1787]), and, translated into Latin and expanded, as *Principiorum calculi differentialis et integralis expositio elementaris* (Tubingae: J. G. Cottam, 1795). A brief account of the 1787 version appears in Boyer, *op. cit.*, pp. 255–257.

[45] "Histoire," 1786, p. 8.

[46] *Exposition élémentaire*, p. 167; quoted by Boyer, *op. cit.*, p. 256. This is "intuitively obvious," but false; Boyer suggests that we consider the case of an irrational defined as a limit of a sequence of rationals. L'Huilier's definition of "limit" is as follows: "Let there be a variable quantity, always smaller or larger than a proposed constant quantity, but which can differ from the latter by less than any proposed quantity, no matter how small: this constant *quantity* is called the *limit* above or below [more literally, in greatness or in smallness] of the variable quantity." *Exposition élémentaire*, p. 7.

Lagrange. But L'Huilier's attempt to avoid infinite series in favor of the limit-concept whenever possible would not have been to Lagrange's liking.

The episode of the Berlin Academy prize makes it clear that in 1784–1786 Lagrange no longer thought the doctrine of first and last ratios to be clear and rigorous. Thus, there could be no way of showing what the "errors" in the use of infinitesimals were—and, therefore, no way of explaining the wealth of true results obtained from infinitesimals by the doctrine of compensation of errors. The failure of the competition to elicit a solution to the problem meant that the Academy's appeal to the whole learned world had failed. And if Lagrange at this time believed that he had an acceptable answer to the question, he kept it well hidden. Members of the Academy were of course ineligible for the prize;[47] but certainly if Lagrange had worked out a version of the "theory of analytic functions" satisfactory to him, he would have wanted to publish it in the Academy's *Mémoires*.

It has been suggested that Lagrange, dissatisfied with the prize-winning paper, returned to work on the possibility of a power series foundation for the calculus, briefly suggested in his 1772 paper.[48] But this does not seem to have been what made him write *FA*. Lagrange did not have a word to say on the subject in 1786—nor had he published anything on it ten years later.[49]

The Period of Indecision: Arbogast's Sketch of a Solution

The last important event in the development of Lagrange's ideas on the calculus before the writing of *FA* occurred in 1789, though it did not bear fruit until much later. This was his acquaintance with a memoir on the foundations of the calculus by L. F. A. Arbogast: "Essai sur de nouveaux principes de calcul différentiel et intégral, indépendans de la théorie des infiniment-petits et celle des limites."[50] This essay, presented to the *Académie des Sciences* of Paris, and read before that body in 1789, shared some of the goals outlined in the 1784 Berlin prize-proposal. Lagrange must have been very impressed with Arbogast's effort, since he said in 1797 that Arbogast's "beau mémoire" ought to leave nothing to be desired—except that it had not been published.[51] It will therefore be desirable to consider Arbogast's essay in some detail.

The "Essai" was in fact never published; though the manuscript still exists. My discussion will be based chiefly upon Arbogast's own discussion of his "Essai" and a twentieth-century dissertation by Karl Zimmerman.[52]

---

[47] "Histoire," 1784, p. 13.

[48] By G. Vivanti, in "Infinitesimalrechnung," Abschnitt XVII (pp. 639–869) in Moritz Cantor, *Vorlesungen über Geschichte der Mathematik*, Volume 4 (Leipzig: B. G. Teubner, 1908), p. 645.

[49] In fact his stay in Paris began in 1786 with a long period of depression, during which he did no mathematics. Delambre relates that this "répos philosophique" lasted until Lagrange was pressed into service during the Revolution on the Weights and Measures Commission. Delambre, *op. cit.*, p. xxxix.

[50] The MS is in the Bibliotheca Medicea-Laurenziana, Florence, Codex Ashburnham Appendix, sign. 1840.

[51] *FA* 1797, p. 5.

[52] Karl Zimmermann, *Arbogast als Mathematiker und Historiker der Mathematik* (Inaugural-Dissertation zur Erlangung der Doktorwürde der Hohen Naturwissenschaftlich-Mathematischen Falkutät der Ruprecht-Karls-Universität zu Heidelberg, 1934) [abridged]; L. F. A. Arbogast, *Du Calcul des Dérivations*, (Strasbourg: Levrault Frères, An. VIII, 1800). Through the kindness of the Bibliotheca Medicea-Laurenziana, Florence, I have been able to examine the MS in microfilm and verify that Zimmermann's account is accurate, though not complete.

The purpose of the paper, according to Arbogast, was "to bring to the higher calculus the same evident quality which reigns in ordinary algebra."[53] He defined the differential quotient by its occurrence as the coefficient of Δx in the power series expansion of $\phi(x + \Delta x)$. He thus felt he was "showing the link between the differential calculus and the general method of series, and making visible that the former is nothing but a particular case of the latter."[54] These statements would surely have impressed Lagrange, since they are reminescent of his 1772 paper.[55]

It will be easier to present the essential points of Arbogast's paper if we follow his own summary. Arbogast's summary, published in his book on the operational calculus, was intended to insure that his unpublished memoir of 1789 would not be confused with his book of 1800.[56] He listed, in 1800, "six principles" on which the 1789 paper rested, but gave no proofs or details as to how the principles were applied. Since the summary was written after Lagrange's *FA* of 1797, it stressed aspects of the "Essai" which appeared important in the light of *FA*, not necessarily those which played the greatest role in the "Essai" itself. Nevertheless, the principles indicate the major applications of the theory of infinite series to the calculus, and therefore provide a convenient framework in which to discuss the "Essai" of 1789.

The first principle he called "the principle of separation of independent quantities." Though he enunciated it in a very general and somewhat obscure manner, it meant in practice that if

$$\sum_{k=0}^{\infty} a_k x^k = \sum_{k=0}^{\infty} b_k x^k \quad \text{for all } x \text{, then} \quad a_k = b_k \quad \text{for all } k.$$[57]

It was very useful in comparing the coefficients of two equal power series; we have seen Lagrange use it in 1772. The second principle stated that a function $\phi(x)$, evaluated at $(x + \Delta x)$, gave a series in integral powers of Δx:

$$\phi(x + \Delta x) = \phi(x) + p\Delta x + q/2!\,(\Delta x)^2 + r/3!\,(\Delta x)^3 + \cdots$$

where $p, q, r, \ldots$ are functions of x, each of which "derives from" its predecessor by the same law by which p "derives from" ϕ. Arbogast attributed this principle to Lagrange's 1772 memoir.[58]

The third principle suggests for the first time that Arbogast, in 1789, may have been doing something new: any function of $x + \Delta x$ can always be developed in a series which proceeds according to integral powers of Δx, for all but particular isolated values of x.[59] Arbogast did more than state this; he tried to prove it.[60] His "proof" begins with the assumption that any given

---

[53] Quoted by Zimmermann, *op. cit.*, pp. 44–45.

[54] Quoted by Zimmermann, p. 45.

[55] In 1800, Arbogast acknowledged the influence of Lagrange's 1772 paper on his 1789 "Essai," see *Calcul des Dérivations*, p. xii, n.

[56] See *Calcul des Dérivations*, 1800, p. xii, and pp. xii–xiv, notes, for the discussion. Boyer, *op. cit.*, p. 203, seems to have been led by Arbogast's title into believing the 1800 book to be a published version of the 1789 "Essai."

[57] He said that, given an equation containing both quantities depending on a variable and quantities independent of it, if the equation holds for all values of the variable, the terms dependent on, and independent of, the variable, will be, respectively, equal, p. xii, n.

[58] *Ibid.* Compare the discussion of Lagrange's memoir, pp. 22–25, above.

[59] *Ibid.*

[60] Zimmermann, p. 46, quotes both Arbogast's theorem and his proof. The theorem states: "If y is any function of x and x is increased by Δx, we can always express what y becomes by this increase by a series which proceeds according to integral powers of Δx."

function y can be written in the following form:

$$y = Ax^{\alpha} + Bx^{\beta} + Cx^{\gamma} + \cdots. \tag{1.8}$$

This assumption, with identical notation except for the use of z instead of x for the independent variable, had been given by Euler in his *Introductio in analysin infinitorum* where "function" had been defined as "analytic expression." Here $\alpha, \beta, \gamma, \ldots$ were any real numbers, and the form given in (1.8) seems to have been viewed by Euler as the most general analytic expression.[61] If, said Arbogast, x is increased by Δx, then y becomes $y + \Delta y$ and

$$y + \Delta y = A(x + \Delta x)^{\alpha} + B(x + \Delta x)^{\beta} + \cdots. \tag{1.9}$$

By the binomial theorem, Arbogast expanded $(x + \Delta x)^{\alpha}, (x + \Delta x)^{\beta}, \ldots,$ rearranged the resulting series, and concluded

$$\begin{aligned} y + \Delta y = {} & [Ax^{\alpha} + Bx^{\beta} + Cx^{\gamma} + \cdots] \\ & + (\Delta x)[A\alpha x^{\alpha-1} + B\beta x^{\beta-1} + C\gamma x^{\gamma-1} + \cdots] \\ & + \frac{(\Delta x)^2}{2!}[A\alpha(\alpha-1)x^{\alpha-2} + B\beta(\beta-1)x^{\beta-2} + C\gamma(\gamma-1)x^{\gamma-2} + \cdots] + \cdots \end{aligned} \tag{1.10}$$

or, calling the bracketed coefficients $p, q, r, \ldots$

$$y + \Delta y = +p\Delta x + q/2!\,(\Delta x)^2 = r/3!\,(\Delta x)^3 + \cdots \tag{1.11}$$

It is evident from the form of the coefficients $p, q, r, \ldots$ given in (1.10) that r is derived from q, and q from p, just as p was derived from y.[62] If $\alpha = 0, \beta = 1, \gamma = 2, \ldots,$ this proof is identical with Lagrange's 1772 algebraic derivation of the Taylor series coefficients.

But Arbogast claimed to have done more. He started with non-integral exponents in the Euler form (1.8). Thus he made it appear that he had proved that *any* arbitrary function must have a power series expansion. Of course, he had assumed the binomial theorem for non-integral powers, and the possibility of reversing the order of summation when series more complicated than power series were involved. Lagrange himself, in 1797, would try to prove that every function had a power series expansion (except at a few special points) without these two assumptions, though he too would assume the Euler form (1.8). Elsewhere in *FA*, Lagrange used the binomial theorem and rearranged power series, but he never rearranged—nor owing to the theorem that every function has a power series, did he need to rearrange—a non-power series.

Arbogast's attempted proof that any arbitrary function has a Taylor series may well have made Lagrange believe that such a proof was necessary. Arbogast's beginning with the general analytic expression (1.8) may have led Lagrange to begin there too; the influence of Euler's *Introductio* on Lagrange, however, might by itself suffice to explain Lagrange's return to this source.[63]

---

[61] *Introductio*... (Lausanne: Bousquet, 1748), in *Opera*, Series I, Volumes 8–9, Section 59.

[62] Zimmermann does not quote Arbogast as making this observation, but then his presentation of Arbogast's work does not claim to be a complete transcription. In any case, it is obvious. Also, Arbogast pointed it out in the discussion of his second "principle" in 1800, as mentioned above.

[63] An extensive critical discussion of Lagrange's attempted proof, and its dependence on the work of Euler, is given in "Algebraic Background," below.

Beginning with the formula (1.11), Arbogast defined the *finite* quantity $p(\Delta x)$ as the first differential of y, dy; $q(\Delta x)^2$ as the second differential of y, $d^2y, \ldots$. He then substituted dx for Δx, so that $dy = pdx$, $d^2y = q\,dx^2, \ldots$. Finally,

$$y + \Delta y = y + dy/dx(\Delta x) + 1/2!\,d^2y/dx^2(\Delta x)^2 + \cdots, \tag{1.12}$$

Taylor's theorem.[64] From the power series definition of the quantity dy/dx, Arbogast deduced the rules for finding differentials of particular functions, of a product, and so on.[65] Such a deduction is of course necessary if the definition of dy/dx is to be a foundation for the calculus. Arbogast resembles Leibniz and Euler, and differs from Lagrange, in focussing on the differential, not the derivative, as the basic concept of the calculus. But to Arbogast, the differential is a finite quantity, not infinitesimal or zero.

Arbogast's fourth principle had the greatest influence on Lagrange: "In the development of $\phi(x + \Delta x)$ one can always take for Δx a value finite and assignable, small enough so one of the terms of the series will be greater than the sum of all those which follow." He added that he never allowed any differential or term of a series to vanish; sometimes he had stopped the series development of a function after only a few terms, "but always taking account of the remainder of it."[66] The possibility of choosing (Δx) so that a fixed non-zero term of the series for $\phi(x + \Delta x)$ would exceed, in absolute value, the sum of all the remining terms was to become basic to the application of Lagrange's theory.

Since in the power series expansion (1.12) Δx could be taken "as small as we wish... it is... evident that each of the terms containing a power of Δx will be larger than the sum of all those which follow it, if we take Δx such that each term $1/n!\ d^ny/dx^n\ (\Delta x)^n$ is greater than twice the following term $1/(n + 1)!\ d^{n+1}y/dx^{n+1}\ (\Delta x)^{n+1}$, which is always possible."[67] What Arbogast must have had in mind is a comparison with the geometric series $\sum_{k=1}^{\infty} 1/2^k$.[68] It is of course always possible, for a *given n*, to choose (Δx) sufficiently small so that the size of the nth term will be greater than twice the $(n + 1)$st. Arbogast does not seem to have been aware that this was not enough, that he had to show that (Δx) could be chosen sufficiently small for such inequalities to be simultaneously true for *all n*.[69] This result—that a term could be made to exceed the sum of all those which follow it in a power series—was well known.[70] Arbogast, however, was the first to recognize that it could be used in the deduction of the received results of the calculus from a power series definition of the differential quotient.[71] This result was to

---

[64] Zimmermann, p. 47.

[65] Compare Zimmermann, p. 48.

[66] Arbogast, *op. cit.*, p. xiii, n. Presumably, he meant to exclude zero terms from the general statement of principle four.

[67] Quoted by Zimmermann, pp. 47–48.

[68] If, for a series $\sum_{n=0}^{\infty} a_n$, $1/2|a_{n-1}| \geq |a_n|$ for all n, then, since $1/2^n = \sum_{k=n+1}^{\infty} 1/2^k$,

$$|a_n| \geq \left| \sum_{n+1}^{\infty} a_k \right|.$$

[69] In a power series $\sum c_k(\Delta x)^k$ with no $c_k = 0$, a necessary and sufficient condition for choosing such a (Δx) is for $|c_{k+1}/c_k|$ to be bounded.

[70] For instance, the result appears in L'Huilier's *Principiorum calculi differentialis...*, pp. 24–26, with a similar proof. It is unlikely that L'Huilier knew Arbogast's unpublished memoir.

[71] Zimmermann, p. 49–50, gives an instructive example: Arbogast's derivation of the rule for finding maxima and minima. For a minimum, $y(x + \Delta x)$ and $y(x - \Delta x)$ must both exceed y. That is, $y' = y + dy/dx(\Delta x) + d^2y/dx^2(\Delta x)^2 + \cdots$

play the same central role in the later work of Lagrange. Instead of always using this result directly, Lagrange usually employed a finite Taylor formula with the "Lagrange Remainder;" then he calculated the value of (Δx) for which that remainder would be less than a given quantity. Lagrange's derivation of his remainder term, however, required itself the use of the principle under discussion.[72]

A feature of Arbogast's work which appears to have impressed Lagrange was the application of Arbogast's definitions of the concepts of the calculus to geometry. S. F. Lacroix, in an influential textbook, later credited Arbogast with being the first to use a "purely analytic" theory in applying the calculus to the theory of curves,[73] i.e., using nothing but the algebra of polynomials, infinite as well as finite. Lacroix then noted the resemblance of Lagrange's theory of curves in *FA* to Arbogast's treatment, saying that Lagrange's point of view led to the same result. What Arbogast did in studying tangents was to characterize two curves as tangent at a point when the first differential quotients of their equations were equal; from this, he easily found the straight line tangent to a given curve.[74] Arbogast's fifth principle characterizes orders of contact between curves.[75] It implies in particular that no straight line passing through the common point can be placed between a curve and its tangent. Lagrange used this geometric "no other line" criterion as his definition of the tangent, and proved it equivalent to the equality of the derivatives.

Arbogast did not state the "no other line" property, but did argue that two curves with equal first derivatives would touch each other without cutting.[76] He showed this by arguing that the sign of the difference between the $(\Delta x)^2$ terms in the Taylor Series for the two curves would be the same, for small Δx, on both sides of the common point.

---

and $y_1 = y - dy/dx(\Delta x) + d^2y/dx^2(\Delta x)^2 - \cdots$ must both exceed y. Suppose $dy/dx \neq 0$. Since "Δx can always be taken small enough for $dy/dx(\Delta x)$ to exceed all the terms that follow, taken together, if $y' > y$, then $y_1 < y$, since the term in (Δx) dominates all the others. Therefore, to have both $y' > y$ and $y_1 > y$, dy/dx must equal zero.

72 See pp. 67–69, below.

73 Silvestre François Lacroix, *Traité du Calcul Différential et du Calcul Intégral* (Paris: 3 volumes, J. B. M. Duprat, 1797–1800). Volume I, p. 370.

74 Zimmermann, pp. 51–52.

75 Given two curves, $y(x)$, $u(t)$, such that

$$y(x + \Delta x) = y + dy/dx(\Delta x) + \frac{d^2y/dx^2}{2!}(\Delta x)^2 + \cdots$$

and, if $\Delta x = \Delta t$,

$$u(t + \Delta t) = u + \frac{du}{dt}(\Delta x) + \frac{d^2u/dt^2}{2!}(\Delta x)^2 + \cdots$$

"if the n constants of the latter curve as determined by the equations $u = y$, $du/dt = dy/dx$, $d^2u/dt^2 = d^2y/dx^2$, etc., $d^{n-1}u/dt^{n-1} = d^{n-1}y/dx^{n-1}$; this curve so determined will be, of all curves of the same nature, the one whose course will approach closest to the course of the first, so that it will be impossible to make another curve of of the same nature as the second pass between these two curves." In particular, if $u = y$, the two curves will have a common point; "making $u = y$, and $du/dt = dy/dx$ [he has dy/dt instead of dy/dx, a printer's error] the second curve will be tangent to the first." *Calcul des Dérivations*, pp. xii–xiv, notes.

76 MS, Part 2, Section 50. "Touching without cutting" is the definition used by Euclid, *Elements* III, Def. 2. The "no other line" property, for circles, is a consequence of "touching without cutting." See *Elements* III, 16. The importance of this property for Lagrange was that it made it possible for him to characterize the tangent as the "closest" straight line to a curve, without using the limit-concept. See p. 75 ff, below.

Arbogast's sixth and final principle is really a special case of the fourth: given three expressions ordered according to the powers of Δx:

$$U = a + b(\Delta x) + c(\Delta x)^2 + \cdots$$
$$V = a_1 + b_1(\Delta x) + c_1(\Delta x)^2 + \cdots$$
$$W = a_2 + b_2(\Delta x) + c_2(\Delta x)^2 + \cdots$$

so that either $U \leq V \leq W$ or $U \geq V \geq W$, "if the two extreme series have a certain number of their first terms equal between them, the series for V will have the same number of its terms equal to the corresponding terms of the other series. If we have, for example, $a = a_2, b = b_2$, we have also $a_1 = a, b_1 = b$."[77] He declares, after stating this principle in his 1800 book, that it has given him "in a rigorous way the differential of the area . . . of any curve." But this description is more appropriate to Lagrange's area-computation, which we shall discuss in connection with *FA*; Arbogast's area-computation in 1789, as quoted by Zimmermann, seems to have been differently conceived.[78]

How are we to assess Arbogast's accomplishment? Arbogast took up Lagrange's suggestion of 1772, together with algebraic techniques for manipulating series and inequalities. Using these, he worked out in some detail how the series definitions of the concepts of the calculus might be used to give some of the important known results, while avoiding the drawbacks of evanescents, infinitesimals, and limits. Lagrange, hearing Arbogast's "Essai" read in 1789, must have been pleased to see his idea carried so far. No doubt this made it even more likely that any calculus

[77] *Calcul des Dérivations*, p. xiv, n. His notation is $a^I, a^{II}, \ldots$. I have substituted subscripted Roman numerals. Arbogast gives no proof of this principle. It could be proved by remarking that principle 4 (letting us choose Δx small enough for any term of the series to exceed the sum of the remaining terms) implies that any series will have the sign of the first non-vanishing term for sufficiently small Δx; then applying this remark to the series for $V - U$, $W - V$. Lagrange uses a set of inequalities like $U \leq V \leq W$ in his area-computation, but U, V, and W are finite Taylor series with Lagrange remainders. In any case, by 1800 Arbogast had read the *FA*; it is not clear whether he knew this sixth principle *explicitly* in 1789.

[78]

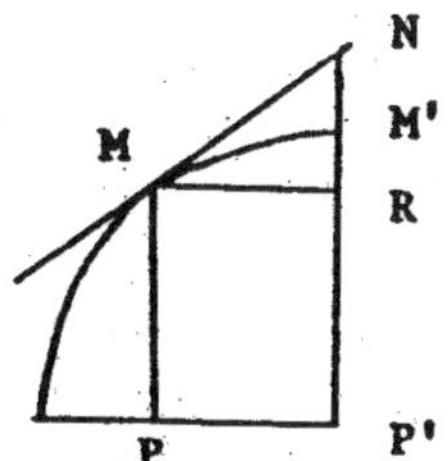

Suppose z is the area to be found, as a function of x. If $\Delta x = \overline{PP'}$, then $\Delta z =$ area $PMM'P'$, $\Delta y = \overline{P'M'} - \overline{PM}$. MN is tangent to the curve at M. The Taylor series for $z(x + \Delta x)$ gives

$$z = p\Delta x + q/2!\,(\Delta x)^2 + \cdots$$

Then, using a geometric argument based on the contribution to the area Δz by the individual terms of the power series expansion for Δy, Arbogast gives area $PMRP' = y\Delta x$, area $RMN = 1/2\,dy/dx(\Delta x)^2$. Arguing that $1/2(dy/dx)(\Delta x)^2$ "belongs already to the second term of the series for Δz," he concludes that $p = y$. See Zimmermann, p. 55. Arbogast *could* have argued that area $PMRP' \leq$ area $PMM'P' \leq$ area $PMNP'$, i.e.,

$$y(\Delta x) \leq p(\Delta x) + q/2!\,(\Delta x)^2 + \cdots \leq y(\Delta x) + \frac{1}{2}\frac{dy}{dx}(\Delta x)^2;$$

and then applied something like his sixth principle to conclude that $p = y$. He did not do this, through his procedure may have suggested doing this to Lagrange.

course Lagrange might be asked to teach in the future would begin from his 1772 suggestion. However, it should be noted that Lagrange's *FA* went far, far beyond Arbogast's "Essai." This is especially evident, as we shall see, in arguments involving inequalities, in relating concepts defined by infinite series coefficients to their classical definitions, and through the use of the remainder term of the Taylor series.

Conclusion

Lagrange's long and deep concern with the foundations of the differential calculus cannot be understood in terms only of the internal development of his mathematical ideas. One must consider also the wider intellectual and social background of eighteenth-century mathematics.

Lagrange's early interest in algebra would not, by itself, have produced *FA*. To earn a living, he found himself a teacher of elementary calculus at Turin in 1759. At this time, his attempts to explain the calculus to beginners called his attention to the inadequacy of the existing foundations.

Frederick the Great, as a respected patron of science and learning in the Age of Reason, gave Lagrange the position of director of the mathematics section of the Berlin Academy in 1766. The importance of academy prizes as a stimulus to mathematical work in the eighteenth century made Lagrange believe that a solution to the problem of the foundations of the calculus might result from setting this question as a prize competition. No doubt the nonsense he had to read in judging the contest increased his dissatisfaction with the approaches taken to the foundations of the calculus by his contemporaries. The limited use made of infinite series in L'Huilier's prize-winning work was no doubt one of the things that impressed Lagrange about it; yet this, too, was not wholly satisfactory. The prize competition surely convinced Lagrange—if he was not already convinced—that the state of the foundations of the calculus was a serious problem. The problem had not been solved by the masters of the past, even Leibniz, Newton, Euler, and D'Alembert. And nobody in the learned world who thought he understood the basis for the calculus was able to explain it on paper.

Lagrange's writings contain no metaphysics. He was a mathematician, not a philosopher like his friend D'Alembert. Yet he was aware of philosophical issues in mathematics in an especially active way. His own reading as well as his long friendship with D'Alembert would have contributed to this. Berkeley had pointed out in his *Analyst* that the calculus as then practiced did not resemble the kind of rigorous mathematics believed in by philosophers; thus Berkeley attributed the correct results obtained to a compensation of errors. Lagrange, as the language of the Berlin prize-proposal shows, took such critiques seriously. He realized that his task was not to prove that Berkeley was a fool, but that the calculus could be made to conform to expectations about mathematics in general.

Finally, the French Revolution, and the subsequent pressing into service of great mathematicians as teachers at the Ecole Polytechnique, made Lagrange consider anew the problem of the foundations of the calculus. Clearly the best course in analysis would not only include the standard results, but would derive them from a solid foundation. This is the view of a modern pure mathematician, not one concerned solely with methods of problem-solving; this was Lagrange's view. Yet the more practically oriented founders of the Ecole Polytechnique were responsible for Lagrange's putting this view into the form of the published *FA*.

It was in 1795 that Lagrange was asked to teach mathematics at the newly founded Ecole Polytechnique.[79] While developing an explanation of the calculus for his students, Lagrange would naturally have undertaken a deep examination of the basic ideas of the subject—as he had once tried to do when teaching at Turin in 1759. Lagrange himself said he had been "engaged by particular circumstances to develop the general principles of analysis," clearly meaning the "circumstances" to be his teaching position at the Ecole Polytechnique.

His earlier recognition of the inadequacy of infinitesimals, limits, first and last ratios, was not enough for this task. He had to have a positive doctrine: one that had to be, as he had recognized in 1784, clear, general, simple—and, above all, rigorous. Accordingly, he tells us that he "recalled [his] old ideas" on the principles of the differential calculus and worked them our further.[80] After he had developed the theory of analytic functions, he relates, he decided to publish it, because it would be very useful for those studying the differential calculus.[81] Thus the *Théorie des fonctions analytiques*, based on his lectures at the Ecole Polytechnique, was presented to the general public in 1797.

---

[79] The Ecole Polytechnique, at which Lagrange taught the calculus, began in 1795. Lagrange had also taught briefly at the shortlived original Ecole Normale (1794–5); according to Delambre, he lectured there only on arithmetic, and on algebra and its applications to geometry. Delambre, *op. cit.*, p. xli.

[80] *FA* 1797, p. 5; compare *Oeuvres* IX, p. 19. The "old ideas" were those indicated in 1772, reducing the concepts of the calculus to those of the algebra of power series.

[81] *Ibid.*

2

The Algebraic Background of the Theory of Analytic Functions

Introduction

The *Théorie des fonctions analytiques* of Joseph-Louis Lagrange has always appeared puzzling to historians of the calculus. Viewing the work from the point of view of the history of the concepts of the calculus, historians have had difficulty in understanding why Lagrange should have begun with the assertion that all functions had Taylor series; they have attributed it to a formalist tendency. Looking at Lagrange's definition of the nth derivative as the nth coefficient in the Taylor series multiplied by $n!$, it has been hard to see why the work commanded any attention at all since it defines a relatively simple concept by means of a complex one; it is puzzling that a man of Lagrange's stature should have so trivially begged the question.[1] Finally, it has not been explained why, in this apparently formalist work, Lagrange gave the first expressions for the remainder term of the Taylor series. The book seems to be outside the main stream of development in the story of the foundations of the calculus, and its major achievements—an algebra of power series and the first expression for the remainder term in the Taylor series—seem unrelated to each other as well as to that main stream.

One may ask whether viewing the *Fonctions analytiques* from the standpoint afforded by the history of algebra would shed any light on the structure of the work or on the historical antecedents of its lasting achievements. Since eighteenth-century algebra and calculus share a common interest in infinite series and approximations, and since the work itself is avowedly an attempt to reduce the calculus to algebra, the question is a natural one. In fact such a view can answer many of the questions raised about the *Fonctions analytiques*, and may lead to a juster estimate of its importance and influence.

Many of the techniques developed in algebra for treating polynomial equations were extended to the study of infinite series in the seventeenth and eighteenth centuries. Lagrange's interest in this algebraic work on series made it possible for him to derive most of the received results of

---

[1] See Carl B. Boyer, *The History of the Calculus and its Conceptual Development*, (New York: Dover, 1959) pp. 251–254, 260–263, and the sources he cites. Compare Eric Temple Bell, *The Development of Mathematics*, (New York and London: McGraw-Hill, 1945) pp. 289–290, and Nicolas Bourbaki, *Eléments d'histoire des mathématiques* (Paris: Hermann, 1960), pp. 217–218.

calculus by manipulation of power series. This meant to Lagrange that the calculus could be given a firm foundation in concepts purely mathematical, without needing vague or intuitive notions like "limit" or "infinitesimal."

Convinced by the actual series developments of many well-known functions that any analytic or algebraic expression could be given as an infinite power series, Lagrange attempted to give a proof (of course, fallacious) that this is always so. An examination of the proof and the algebraic results it uses will show both from where the proof-procedure came and how it was possible for Lagrange so to delude himself.

Lagrange's calculuation of the remainder term in the Taylor series does not rely in any way on Lagrange's "proof" that any analytic function can be so represented. Rather, it recognizes that a Taylor series expansion is likely to be used as an approximation, either in algebra or in physics, and that a precise error-estimate is therefore desirable. The first systematic study of estimates of error arose in Lagrange's work on approximation techniques in algebra.[2] Thus, the best-known achievement of the *Fonctions analytiques* has its roots in algebra. The remainder term was derived through a skillful use of algebraic inequalities, just as were the error-estimates in the *Résolution des équations numériques*; these manipulations are direct ancestors of those today termed "epsilonics."[3]

Lagrange had a generalizing approach to mathematics. He tried to found each subject with which he dealt on the most economical possible basis: the differential calculus, the analytical mechanics, the calculus of variations. The differential calculus seemed to him a set of powerful methods used to solve particular classes of problems, but insufficiently justified. He tried, therefore, to supply general, abstract foundations for these algorithms—unexceptionable first principles from which the particular results already developed could all be deduced. The source of these first principles had to be a part of mathematics which appeared in no need of propping up, which seemed to the eighteenth century to have the certainty, universality, and self-sufficiency expected of a rigorous subject—algebra.

The Attractiveness of Algebra: Certainty

Eighteenth-century mathematicians stressed the generality and certainty of algebra through the idea of "universal arithmetic." The heuristic power of symbolic notation—elevated almost into a philosophy of knowledge by Leibniz—goes hand in hand with the universality of its expressions.

The description of eighteenth-century algebra by its practitioners is most often as a "universal arithmetic," in which the operations of ordinary arithmetic are applied to letters. The letters are understood to represent any numbers whatsoever. Thus, the algebraist could obtain complicated relations which yielded valid arithmetical results when numbers were substituted for the letters.

---

[2] In papers published in the *Mémoires* of the Berlin Academy, 1767–1768. These were published, together with extensive notes which tripled the length of the work, as *De la résolution des équations numériques de tous les degrés*, (Paris: chez Duprat, An VI [1798]). A second edition (Paris: chez Courcier, 1808) is included in *Oeuvres de Lagrange*, ed. J.-A. Serret, Volume VIII (Paris: Gauthier-Villars, 1879); the title of this second edition was *Traité de la résolution des équations numériques de tous les degrés.* All subsequent citations of this work will be from the *Oeuvres*, Volume VIII, and I shall refer to the work as *REN.*

[3] That is, delta-epsilon proof methods. See below, "From Proof-technique to Definition."

The authority for this view was Isaac Newton, whose lectures on algebra at Cambridge (1673–1683) were published as *Arithmetica Universalis*:

> Computation is done either by numbers as in common arithmetic, or by species as is the custom in analysis. Both are founded on the same principles, and lead to the same result: Arithmetic in a definite and particular manner, Algebra in an indefinite and universal manner. But in the latter, almost all the statements, and especially the conclusions, are [true] Theorems.[4]

The view was echoed by, among others, Leonhard Euler, who called this branch of mathematics "die Analytik oder Algebra,"[5] and cited by D'Alembert in the mathematical dictionary.[6]

This view takes for granted the laws and definitions of arithmetic. Some writers inquired into these as well, notably Newton: "By number we understand . . . an abstract ratio of any quantity to another of the same kind which is regarded as unity."[7] But the discussion of arithmetic and number as such usually held little interest in an algebraic work. After an initial definition of number, sometimes as "plurality" and sometimes following Newton, arithmetic was considered well enough founded to be left behind while algebra was pursued. And since algebra was just a generalized arithmetic, the truth of its conclusions was believed equally well founded.[8]

The "universal arithmetic" description of algebra is not, however, an adequate account of what algebraists actually did. No eighteenth-century algebraist was interested in the mere proclamation of general rules for arithmetic operations, even in symbolic language. But being able to treat unknown or undetermined quantities just as known ones are treated—by symbolic representation—gives algebra the power to solve verbal or geometrical problems. The subject treated at greatest length in eighteenth-century algebra was finding exact or approximate procedures for solving equations, where the equations were taken as independent objects of study, not merely as translations of verbal or geometrical problems, though such translation was in part their origin.

Lagrange took this wider view, that algebra includes more than universal arithmetic. It is not just supposed to give the individual values of the quantities sought, but to find "the system of operations," either arithmetical or geometrical, which are to be performed on the given [known] quantities to find the unknowns.[9] Instead of terming algebra an analytic (i.e., problem-solving)

---

[4] Isaac Newton, *Arithmetica universalis, sive de compositione et resolutione arithmetica liber* (Cantabrigiae: Typis Academicis, & Londini: Benjamin Tooke, 1707), p. 1.

[5] Leonhard Euler, *Vollständige Anleitung zur Algebra* (Stuttgart: Reclam-Verlag Stuttgart, 1959), p. 42. This is a reprint of the first German edition, St. Petersburg, 1770.

[6] *Dictionnaire encyclopédique des mathématiques*, par d'Alembert, l'abbé Bossut, de la Lande, le Marquis de Condorcet, &c (Paris: Hotel de Thou, 1789), Article "Algèbre." The *Dictionnaire* often provides useful summaries of eighteenth-century points of view, though its French origin led to some neglect of the work of Euler and the Bernoullis.

[7] Newton, *op. cit.*, p. 2. This account of number was repeated by Euler, *op. cit.*, p. 42. An analysis of Newton's view is given by Gottlob Frege in his *Grundlagen der Arithmetik*, translated by J. L. Austin as *Foundations of Arithmetic* (New York: Harper and Brothers, 1960), in Section 19, pp. 25–26.

[8] It was not until the nineteenth century that acceptable definitions of real number were sought and given, by such men as Richard Dedekind (1831–1916).

[9] Lagrange, *REN*, pp. 14–15. In the *Leçons sur le calcul des fonctions*, (Paris: chez Courcier, 1806) in *Oeuvres* X, p. 10, he added that development of an analytic expression in an infinite series is also an "operation." The *Calcul des fonctions* is hereinafter cited as *CF*.

art, he said that "algebra taken in the widest sense is the art of determining unknowns by the functions of known quantities, or quantities regarded as known."[10]

Lagrange's view of algebra as essentially the study of functions includes at once the generality of the subject and its actual chief activity: it is at once a universal study of systems of operations and an equation-solving art.

Thus if Lagrange said he would reduce the calculus to algebra, he meant that its subject matter would be systems of operations which are expressible by symbolic formulas. Algebra was as firmly based for him as it was for his predecessors; if he could carry out the reduction, the calculus would be rigorously founded. And the analogies between algebra and the calculus gave reason to believe that the reduction was in fact possible.

The Attractiveness of Algebra: Methods

Eighteenth-century algebraic theory is essentially the theory of polynomial equations. The approach used is firmly based in the philosophy of algebra worked out in the previous century. The seventeenth century, conscious of the observable features of algebraic equations—the use of letters and the consequent visibility of the operations which had produced an algebraic quantity—based its equation theory in those features of equations observable when the equation was generated by the ordinary operations of arithmetic.

An equation was considered as built up by the multiplication of factors. For instance, if a and b are the roots of an equation, we know that $(x-a)=0$ and $(x-b)=0$, and therefore that their product $(x-a)(x-b)=0$; hence, $x^2-(a+b)x+ab=0$. The final form was taken to represent any quadratic equation, and exhibits clearly the relation between the coefficients and the roots of such an equation.[11] This procedure makes it easy to construct an equation which has a particular set of quantities as its roots, and hence makes natural the conclusion that any equation can be built up as a product of factors.[12]

The fundamental root-coefficient relationships of the general nth degree equation were given, although without explicit derivation, as early as 1629, together with a statement of the Fundamental Theorem of Algebra.[13] These were repeated and enlarged upon in the seventeenth and eighteenth centuries. Typical statements would take the form which follows: if an equation is written

$$x^n + px^{n-1} + qx^{n-2} + \cdots + s = 0,$$

---

[10] With a definition of function like many given in the eighteenth century: "when the way a quantity depends on other quantities can be expressed by a formula which contains those quantities, we say it is a function of these quantities." *REN*, p. 15.

[11] This process of equation-building is owed to Viète; the first systematic exposition was given by Thomas Harriot in 1631, in *Artis analyticae praxis ad aequationem algebraicas nova, expedita, et generali methodo, resolvendas* (Londini: apud Robertum Barker . . . et Haered. Io. Billii, 1631).

[12] This conclusion was called the "Fundamental Theorem of Algebra" by Carl Friedrich Gauss, who in 1799 gave the first acceptable proof of it, together with a critique of previous attempts at proof. A convenient edition of Gauss' work on this theorem is in *Die Vier Gauss' schen Beweise fur die Zerlegung ganzer algebraischer Functionen in reele Factoren ersten oder zweiten Grades* (*1799–1849*), ed. E. Netto (Leipzig & Berlin: Wilhelm Engelmann, 1913).

[13] By Albert Girard, in *Invention nouvelle en l'algèbre* (Amsterdam: chez Guillaume Iansson Blaeuw, 1629) reprinted by D. Bierens de Haan, (Leiden: chez Muré Fréres, 1884), pages $E_4R - E_4V$.

then the coefficient of the second term with opposite sign $(-p)$ is equal to the sum of all the roots; the coefficient of the third term (q) is equal to the sum of the products of the roots taken two by two; the last term (s) is the product of all the roots having their sign changed. Further, $p^2 - q$ is the sum of the squares of the roots. Similar results can be derived to almost any desired complexity.[14]

Looking at a polynomial and deducing from it the relations between its coefficients and its roots is a very impressive achievement of seventeenth-century algebra. If one begins with a factored polynomial, the results appear trivial; but if one begins with the polynomial form, the fact that methods exist for obtaining the relations between roots and coefficients appears nothing short of miraculous.

With these relations, solutions for polynomial equations could be found. A number of techniques had been devised for manipulating polynomials to make them more amenable to solution. I shall mention the few most frequently employed, which were applied to infinite series by Newton, Euler, and Lagrange. First, a polynomial $P(x) = 0$, can be transformed into another one $P(X + h)$ by making the substitution $x = X + h$. Usually h was chosen to alter the coefficients in some way which made the polynomial easier to work with. The earliest important example is that used by Jerome Cardan in 1545 in treating the cubic equation $x^3 + ax^2 + bx + c = 0$. Letting $y = x - (a/3)$ and setting $P(y) = 0$, we obtain an equation which has the form $x^3 + mx + n = 0$. In general, as was realized by Cardan and François Viète, an nth degree polynomial can be divested of its $n - 1$th degree term ax^{n-1} by making the substitution for x of $x - a/n$.

A similar substitution can produce an equation all of whose coefficients are positive, and therefore which has no positive root; this makes it possible to find an upper bound for the roots of the original equation.[15] In general, if $0 = P(x) = \sum_{k=0}^{n} a_k x^k$, then the auxiliary equation formed from setting $x = y + h$ is $0 = P(y) + \sum_{k=0}^{n} \frac{p^{(k)}(h) \cdot y^k}{k!}$, by Taylor's theorem. We find, though, that such men as Newton and Maclaurin did not identify the quantities in the auxiliary equations with the calculus, apparently because they preferred not to bother with something which introduced extraneous notions where they were not necessary—compare Newton's eschewing of fluxions in the *Principia*. This is another example of the way in which the eighteenth century viewed algebra as simpler than, logically prior to, or better founded than the calculus.

Another technique used which is closely related to the calculus considers, again, the polynomial $P(x)$. Subtracting $P(x + h) - P(x)$ gives a polynomial in h, $Q(h)$, of degree $n - 1$. The solution of $Q(h) = 0$ provides $n - 1$ different numbers such that, in general, any one of them lies between two roots. These $n - 1$ numbers are used to serve as first approximations to solutions in the many iterative processes for approximating solutions not capable of being found exactly. In fact, solving $P(x + h) - P(x) = 0$ is equivalent to finding the points for which $dP/dx = 0$, and therefore the turning points of the graph $y = P(x)$. Pierre de Fermat had used precisely this technique—stated algebraically since the calculus had not yet come to be regarded as a separate study—to find maxima and minima.[16]

---

[14] See, for instance, Colin Maclaurin, *A Treatise of Algebra in Three Parts* (2d edition; London: Printed for A. Millar & J. Nourse, 1756), p. 141; Alexis Claude Clairaut, *Elemens d'Algébre* (Paris: chez Courcier, An X (1801)), Vol. 2, pp. 38–39, p. 149.

[15] See Maclaurin, *op. cit.*, where this is done for a cubic equation.

[16] See Boyer, *op. cit.*, page 155 ff.

These examples of setting $x + h$ for x and seeing what happens to the coefficients of $P(x)$ testify to how entrenched this technique was: both in approximations and in exact procedures, two apparently dissimilar parts of algebra which have a great deal in common in the eighteenth century.

In dealing with infinite instead of finite polynomials, many of the same procedures seemed appropriate. In the eighteenth century, the study of infinite polynomials belonged just as properly to algebra as the study of finite ones. It seemed natural to inquire into the relations between the coefficients and the roots in the equations involving infinite polynomials. For instance, take the equation $f(x+h) = f(x) + ph + qh^2 + \cdots$ where x is a fixed quantity. Then $f(x+h)$ can be written $f(x) + P(h)$, where $P(b)$ is an infinite polynomial in h with fixed coefficients for a given x. When the equality holds, there will clearly be definite relations between x and the coefficients in $P(h)$. To investigate the possibility of satisfying this equality, it is thus necessary to investigate what these relations are—just as $x^2 + px + q = 0$ can be solved by investigating the relations which hold between x and values of p and q. The particular relations found will be different, but the same techniques in handling polynomials may well be appropriate.[17] Thus, the theory of power series appeared to Euler and Lagrange as quite naturally borrowing many of its key questions from the algebra of polynomials, and therefore borrowing its techniques from the algebra of finite polynomials as well. Techniques of polynomial-manipulation had been extended to infinite series in the seventeenth century;[18] Euler and especially Lagrange added to the common stock of procedures the systematic study of the coefficients in general infinite power series and the laws governing their formation.

The most impressive eighteenth-century study of infinite series is Euler's *Introductio in analysin infinitorum* of 1748,[19] a direct predecessor of Lagrange's *FA*. The subject of the *Introductio* is the analysis of the infinite; that is, the book was intended to give an account of infinite analytic expressions, just as theories of equations have given an account of finite ones.[20] It is not a book on the calculus, but a study of infinite series, infinite products, and infinite continued fractions—"algebraic" objects.

Euler saw as the general subject of the analysis of the infinite the function: "A function of a variable quantity is an analytic expression composed, in any manner, of that same quantity and of numbers, or of constant quantities."[21] This definition is obviously equivalent to that given by Lagrange: a function, or analytic function, of a quantity is any "expression de calcul" (finite or infinite) into which the quantity enters in any way.[22]

---

[17] Compare Joseph-Louis Lagrange, *Théorie des fonctions analytiques*, (Paris: Imprimérie de la Republique, An V [1797]), *passim*. The second edition (Paris: chez Courcier, 1813), is reprinted as Volume IX in *Oeuvres de Lagrange*; all further citations in this chapter will be from this edition, hereinafter called *FA*.

[18] A masterly treatment is to be found in Isaac Newton, *A Treatise of the Method of Fluxions and Infinite Series* (London: T. Woodman & J. Millan, 1737).

[19] (Lausanne: Bousquet, 1748). This is republished in Euler's *Opera Omnia*, Series I, Volumes 8–9.

[20] "Analytic" and "algebraic" are used interchangeably by Euler; see *Algebra*, p. 42. "Algebra of the infinite" or "infinite algebraic expression" would serve just as well here.

[21] *Introductio*, in *Opera*, Series I, Vol. 8; p. 18, Section 4. Because of the many editions of the works of Euler, I shall cite them henceforth by section numbers. It is worth mentioning that, in a paper published in 1748 on the vibrating string, Euler was willing to consider piecewise continuous functions for which he thought no analytic expression could be given. See C. Truesdell, "The Rational Mechanics of Flexible or Elastic Bodies, 1638–1788," in *Leonhardi Euleri Opera Omnia*, Vol. X & XI, Seriei Secundae (Turici: Orell Füssli Turici, 1960), pp. 246–248.

[22] *FA, Oeuvres* IX, p. 16.

Making the function-concept primary in analysis may thus appear as Euler's achievement, not Lagrange's. However, Euler's text of the differential calculus is not the *Introductio*, but the *Institutiones Calculi Differentialis.*[23] Here the differential quotient becomes a major object of study. The *Introductio* is a work narrower in scope than the calculus in general; it is a study of analytic expressions for their own sake.

FA attempts to deduce the laws of the differential calculus from those governing analytic expressions—in particular, power series. A function which has no analytic expression can thus not properly be studied by the methods of *FA* and therefore does not belong in rigorous analysis, in Lagrange's view. Since he had rejected previous methods of defining the derivative, his definition of function had to conform to the methods of *FA*. Lagrange, through his algebraic approach to the calculus, made the function-concept primary *to the calculus.* It has remained primary, although not in its common eighteenth-century sense.

Euler's definition of a variable as an "indeterminate" quantity[24] is also algebraic. "Variable" or "varying quantity" suggests a kinematic image just as "fluxion" did, while the idea of an indeterminate or general symbol which could take on any particular determined value is derived from the idea of universal arithmetic. The universal arithmetic framework, distinguishing only between particular and general, determined and undetermined, helps to obscure the distinction between dependent and independent variable; nevertheless, it provides an algebraic sense of "variable" to supplant the kinematic one.[25] Lagrange followed this algebraic usage in his *FA*, in fact preferring the term "indeterminate" to "variable."[26]

Euler used, on infinite series of a single variable, all the well-known algebraic techniques for manipulating finite polynomials. For instance, implicit in all Eulerian derivations of infinite series by means of undetermined coefficients is the theorem that, if $\sum_{k=0}^{\infty} a_k x^k = \sum_{k=0}^{\infty} b_k x^k$ for all x, then, for all k, $a_k = b_k$. The legitimacy of this procedure for finite polynomials can be clearly seen in the light of the fundamental theorem of algebra; in infinite series it is taken on faith, an obvious formal extension of reasoning from the finite to the infinite. In fact, it is valid if the series agree on an interval—or on any infinite set, containing a limit point, within the interval of convergence. Euler considered infinite polynomials $P(z)$ as functions of z; he spoke of "factoring" them;[27] and, in Chapter X, used methods of root-coefficient relations to convert infinite sums to infinite products and vice versa.[28] If

$$1 + Az + Bz^2 + \cdots = (1 + az)(1 + bz)(1 + cz)\ldots,$$

then, whether the expressions are finite or infinite, we have the results

$$A = a + b + c\ldots, \qquad B = ab + ac + ad \cdots + bc + bd + \cdots.$$

[23] (Petropolitanae: impensis Academiae imperialis scientarum, 1755). This is Volume 10 of *Opera Omnia, Series Prima* (Leipzig & Berlin: G. Teubner, 1913). This work will henceforth be called *ICD.*

[24] *Introductio*, Section 2.

[25] Variable, or flowing quantities, were most often used by fluxionists. See Florian Cajori, *A History of the Conceptions of Limits and Fluxions in Great Britain from Newton to Woodhouse* (Chicago: Open Court, 1931).

[26] *Passim.* Actually, Lagrange used both terms. Euler occasionally used the letter i for an indeterminate or variable quantity: e.g., *Introductio*, Section 115. Lagrange followed this natural, yet to us somewhat unusual notation, using i where we would write h or Δx in defining the derivative. For instance, see *CF*, pp. 10–15.

[27] For instance, in *Introductio*, Section 34.

[28] *Op. cit.*, Section 165.

The spirit of these extensions of arguments from finite to infinite is apparent in Lagrange's work on infinite series also.

The *Introductio* gave infinite series developments for more functions than had any previous work: functions as apparently different as quotients of polynomials and exponentials, as cosines and logarithms. It appears from the *Introductio* that all the functions commonly studied in the calculus, even those defined geometrically, could be represented by infinite series.

The *Introductio* was, as we have observed, conceived as a study of the algebra of the infinite, and not as a work on the conceptual basis of the calculus. Lagrange decided that the concept of function used there could be a bridge between the calculus and algebra. Perhaps the same description—a bridge—will serve to assess the importance of the *Introductio* in conditioning Lagrange's work. He was able to take the formal achievements of the *Introductio*, and, stressing their algebraic nature, use them as a basis for the calculus. They seemed to him to partake of the certainty of algebra and yet embody theorems of the calculus.

According to Lagrange, however, the *Residual Analysis* of John Landen (1719–1790), not the *Introductio*, was close to his ideas on the foundations of the calculus.[29] Landen himself, best known for a theorem about elliptic integrals, was a working surveyor who did pure mathematics in his leisure time.[30] Little is known about his education; from references in his work it is clear that he read and esteemed the work of his friend Thomas Simpson, and that of James Stirling and Roger Cotes. He knew some Continental writers through a few of their books, he said, but added that he did not read the *Mémoires, Histoire,* or *Commentaries* of "foreign academies."[31]

The purpose of the Residual Analysis is to provide a "natural" basis for the calculus. This basis is to be found in algebra. Infinitesimals are not to be considered, and no principles are to be borrowed from the doctrine of motion. Whatever can be done "by the method of computation which is founded on these borrowed principles, may be done, as well, by another method founded on the anciently-received principles of algebra."[32] These goals, with which Lagrange thoroughly agreed, were to be accomplished by Landen through use of the "residual analysis." A quantity like $x_0 - x_1$ is a "residual," formed by subtraction; the calculus contains countless considerations of such quantities.

By use of simple techniques for the manipulation of literal quantities, Landen was able to determine derivatives, and apply them to the usual geometrical examples.[33] A fair sample of his

---

[29] *FA, Oeuvres* IX, p. 18.

[30] On Landen, see H. G. Green and H. J. J. Winter, "John Landen, F.R.S. (1719–1790), Mathematician," in *Isis 35* (pp. 6–10), 1944.

[31] John Landen, "Postscript" to "Supplement" to *Observations on Converging series, occasioned by Mr. Clarke's translation of Mr. Lorgna's Treatise on the same subject* (London: printed for the author, 1781), "Supplement," p. 31. He also stated here that he had not read the *Acta Eruditorum*, though he cited a 1704 article by John Bernoulli in that journal in his *Discourse concerning the Residual Analysis: A new branch of the algebraic art, of very extensive use, both in Pure Mathematics and Natural Philosophy* (London: J. Nourse, 1758), p. 9. It should be noted that these statements were made in a priority controversy, where it was to Landen's advantage to claim no acquaintance with work which had preceded his.

[32] *Discourse*, pp. 4–5. The full programme is executed in *The Residual Analysis* (London: printed for the author, 1764), hereinafter called *RA*.

[33] In what follows I have modernized Landen's notation. He writes x and $x,$ for what I shall write x_0 and x_1; he wrote $[x/y]$ as abbreviation for $\frac{y_0 - y_1}{x_0 - x_1}$; I will not use the abbreviation. The value of that ratio when x_0 "is equal to" x_1 he wrote as $[x - y]$, which I shall write as dy/dx. See *RA*, p. 3.

procedure is the following:

$$\text{Clearly } u_0w_0 - u_1w_1 = w_0(u_0 - u_1) + u_1(w_0 - w_1).$$
$$\frac{u_0w_0 - u_1w_1}{v_0 - v_1} = w_0\frac{u_0 - u_1}{v_0 - v_1} + u_1\frac{w_0 - w_1}{v_0 - v_1}.$$

When we set $v_0 = v_1$, $u_0 = u_1$, and $w_0 = w_1$, therefore,

$$d(uw)/dx = w(du/dx) + u(dw/dv).^{34}$$

This is simply an example of assuming that what is true for all distinct v_0, v_1 remains true when they are equal.

The most frequent task set in *RA* is to evaluate the quotient $\frac{F_0-F_1}{x_0-x_1}$ when $x_0 = x_1$, and where F is a function of x. Landen's definition of a function of a variable quantity is "an algebraic expression composed, in any manner, of any power or powers of [it], with any invariable coefficients."[35] The task is thus reduced to computing quantities like $\frac{v^{m/r}-w^{m/r}}{v-w}$ as manageable expressions—usually infinite series—and then setting $v = w$. Functions like n^x and $\log x$ are dealt with by setting them equal to a power series in x with undetermined coefficients, which are then determined in the same formal manner used by Euler in his *Introductio* (although Euler's treatment is much more elegant).[36]

RA obviously has some striking similarities to Euler's *Introductio*, but seems to me to be independent of it. It is possible that Landen never read the *Introductio*; there is nothing in *RA* that requires us to believe that he had read it before composing his work. The similarities can be attributed to the common predecessors of both Euler and Landen; notably, Isaac Newton's *Method of Fluxions*, the work of Cotes on infinite series for trigonometric functions, and Johann Bernoulli's definition of function.[37]

Lagrange found in *RA* the announcement of a programme which struck a sympathetic chord; the calculus was to be founded on algebra. But he also found an inelegant execution of it, with a confusing notation and a neglect of important properties of the series obtained—such as their convergence and their kinship with the Taylor series. The stress on "residuals" like $(x_0 - x_1)$ when $x_0 = x_1$ is closer in spirit to the "quantities absolutely zero" of Euler's *ICD* than to the Taylor series treatment of Lagrange.

Lagrange and Landen agreed that the calculus should be based on algebra, which meant that it would be freed from the consideration of motion or of the infinitely small. But Landen's conception of algebra is narrower than Lagrange's. Landen seems to have viewed algebra as a way of computing with literal quantities, Lagrange as a general study of operations. Landen viewed the calculus as a set of special cases of literal formulas involving ratios of residuals; the

---

[34] *RA*, pp. 11–12.

[35] This is similar to Euler's definition. The probable source of both is Johann Bernoulli, "Rémarques sur ce qu'on a donné jusqu'ici de solutions des problèmes sur les isoperimetres," in *Opera Omnia*, Tomus Secundus, pp. 235–269. (Lausannae et Genevae: Bousquet, 1742); the Latin original was published in the *Acta Eruditorum* 1718, pp. 15–31, 74–88. The definition is given on p. 241 of Volume II of the *Opera*. "We here call a *Function* of a variable quantity, a quantity composed in any manner of that variable quantity and constants."

[36] See, especially, *RA* pp. 28–30.

[37] Bernoulli's name appears on a list in the "Postscript," cited in Note 31, along with those of Euler, D'Alembert, Fontenelle, and Huygens, as men whose work Landen knew, at least in part. There are occasional citations of the work of other Continental mathematicians in *RA*—e.g., of L'Hospital's classification of points of inflexion in curves, *RA*, p. 97. All this makes clear that Landen did not have a reluctance ever to credit Continental work.

formulas are then applicable to geometry, summing series, etc. Lagrange made the calculus and the solving of equations both cases of his theory of analytic functions.

Lagrange's emphasis on the function-concept led him away from the concentration on quotients. Landen may have wanted to eliminate this too, yet began his work with a basic theorem for computing $\frac{v^{m/r}-w^{m/r}}{v-w}$. Lagrange saw the derivative as a function in its own right, not a quotient.

Finally, understanding of the importance of the Taylor series is entirely absent from Landen's work. He does not seem to have recognized the Taylor series of differential quotients as identical with the algebraically generated infinite series obtained in the *Residual Analysis*.

The Algebraic Character of the Taylor Series

What made it possible for Lagrange to make the Taylor series the foundation of the calculus was his algebraic view of that series—not the Taylor series expansions of *ICD*, which are based on what Lagrange regarded as the wrong metaphysic of the calculus and a shaky definition of the derivative. It is the sort of work with infinite series done in the *Introductio* which attracted Lagrange to them, while knowing the *fact* that the derivatives play the role that they do in the Taylor series may be owed to *ICD*. Taylor series developments—i.e., with the infinite series coefficients explicitly identified with differential quotients—for the common functions of analysis are worked out in *ICD* with Euler's usual elegance.[38]

The *Introductio* contains the actual development, without appeal to the calculus, of many known functions in infinite series: we find the obvious algebraic ones, and the series known since Newton and Leibniz for sine and cosine; there is also the elegant—and purely formalistic—development of the exponential, logarithmic, and trigonometric series and their mutual relations.[39] Use of the definitions of the relevant functions, together with arguments on very large or very small quantities, shows explicitly that $\sin x$, $\cos x$, $\log x$, a^x are analytic expressions, that is, have infinite series representations.[40] That this could be done with no mention at all of differentials, limits, or fluxions set a powerful example for Lagrange.

---

[38] An excellent study of the relation between *ICD* and *FA* is given by A. P. Juschkewitsch, "Euler und Lagrange über die Grundlagen der Analysis," in *Sammelband der zu ehren des 250 Geburtstages Leonhard Eulers*, Deutschen Akademie der Wissenschaften zu Berlin (Berlin: Akademie-Verlag, 1954), pp. 224–244.

[39] Chapters 7–8, Sections 114–143.

[40] Perhaps an example will lend definiteness to this description of Euler's procedure. Consider the trigonometric functions sine and cosine, defined in the now standard way as lines in the unit circle. Euler derived the formulas for $\cos(y+z)$ and $\sin(y+z)$ from the geometric definitions (see *Introductio*, Sections 126–132). Then he computed (where $i^2=-1$) $(\cos z + i\sin z)^2 = \cos^2 z - \sin^2 z + 2i\cos z\sin z = \cos 2z + i\sin 2z$ and, in general, $(\cos z \pm i\sin z)^n = \cos nz \pm i\sin nz$ (Sections 132–133). If z is "an infinitely small arc," (Section 134) we have $\sin z = z$, and, if n is infinitely large, Euler added, $u = nz$ is finite. This technique—allowing n to become infinite and z to go to zero simultaneously, preserving a finite product, was often used by Lagrange. See, for example, the Lemma used to derive the Remainder Term in the Taylor series, discussed in Chapter 4, below.

From the general equation, expanding the left hand side as a binomial and taking the real part:

$$\cos nz = (\cos z)^n - n(n-1)/2!\,(\cos z)^{n-2}(\sin z)^2 + \cdots$$

Recalling that $nz = u$ and that $\sin z = z$, we have, as n becomes infinite, that $n(n-1)(n-2)\ldots(n-(k-1)) = n^k$ and that, therefore,

$$\cos u = 1 - u^2/2! + u^4/4! \text{ \&c.}$$

To be sure, the vital question of convergence is obscured by this sort of stress on formulas, but it should be noted that the Taylor series representations for $\sin x$, e^x, and similar functions used by Euler converge for all x. The ease of handling and freedom of application of infinite series in eighteenth-century work horrified men like Niels Hendrik Abel and Augustin-Louis Cauchy, who pursued systematic investigations of convergence; yet most eighteenth-century results, with appropriate saving clauses here and there, are still valid. This was only in part due to the "luck" in selecting convergent series; the divergence of the harmonic series was, after all, well known. The ideas of convergence in the eighteenth century were not quite as primitive as is sometimes alleged.[41]

For most of the infinite series dealt with in the *Introductio*, knowing the first few terms enables one to determine those which follow. For these series, a "general term" can be given, which is a simple analytic expression. For instance, the general term of the cosine series can be written $\frac{u^{2k}}{(2k!)}(-1)^k$. For all functions considered, the series developments are recursive and this fact is stressed.

Lagrange, because of his algebraic approach to the calculus, made much of the recursive features of the Taylor series. Just as a_{k+1} is simply determined from a_k, and is the same sort of quantity, $f^{(k+1)}(x)$ is simply determined from $f^{(k)}(x)$, and is the same sort of function; in fact, for Lagrange, the relationship is exactly the same since $f^{(k)}(x)$ is just $k!$ times the kth co-efficient in the Taylor series expansion of $f(x+i)$ in powers of i. In particular, $f'(x)$ is a function derived from $f(x)$ in exactly this way—hence the name derivative, from *fonction derivée*, the term introduced by Lagrange. It is just another function, obviously subject to the laws of analysis or algebra. Lagrange in fact *defined* the derivative by its position in the Taylor series.[42]

In writing the general term of the Taylor series for $f(x+i)$ as $\frac{f^{(k)}(x)i^k}{k!}$ he found an extraordinarily expressive notation $f', f'', \ldots, f^{(k)}(x)$, which stresses the fact that the derivative is not to be broken up into numerator and denominator. Only an algebraist, who was willing to stop thinking about differential quotients for a time, could have devised the prime notation for derivatives; but the essential step suggested by the notation is conceptual: to consider the *functions* f, f', f'', etc., instead of successive ratios of fixed quantities.[43]

---

[41] Some of the confusion on this point is due to the fact that "converge" was used in the eighteenth century to mean merely that the nth term of a series went to zero, and *not* in the modern sense as meaning that the series had a finite sum. One could say that a series "converged" and yet had no finite sum in the case of the harmonic series $\sum 1/k$. For this use of "convergence," see the article on it in the *Dictionnaire*, cited in Note 6. Since power series in one variable with bounded coefficients converge (in the modern sense) for the absolute value of the variable less than 1, as do alternating series whose terms go to zero, eighteenth-century mathematicians often treated such series as having finite sums on the sole grounds that the nth term approached zero; nevertheless they knew that this was not a sufficient criterion for all series. Jakob Bernoulli had investigated the divergence of the harmonic series as early as 1689; this example is discussed in the *Dictionnaire*, article "Série ou Suite."

[42] *FA, Oeuvres* IX, p. 33. As will be recalled, from his definitions of $f', f'', \ldots$ Lagrange gave an "algebraic proof of Taylor's theorem." See Chapter 1, above.

[43] The use of "prime" notation for successively defined quantities occurs often in the eighteenth century: e.g., *Introductio*, Section 213. Simon L'Huilier had said that the derivative was not to be broken up as a quotient, but the temptation was great to think of it as a quotient of differentials when it was written dy/dx. See Boyer, *op. cit.*, page 255.

Origins of the "Proof" that Every Function has a Taylor Series

It is sometimes asserted that Lagrange assumed that any function had a Taylor series expansion.[44] In fact he gave an algebraic "proof" of this. An analysis of this proof will show it to be based on material in Euler's *Introductio*, and will underline again the algebraic nature of *FA*.

Euler himself, stating that functions of z can always be represented by expressions of the form $A + Bz + Cz^2 + \cdots$, said that if the possibility of such a development is doubted, the actual development of each function could leave no doubt.[45] But since he did not have a proof that any function could be so expressed, he admitted that non-integral powers of z might be used as well: "There remains no doubt that *any* function of z can be transformed into an infinite series of form

$Az^\alpha + Bz^\beta + Cz^\gamma + Dz^\delta + \cdots$, where $\alpha, \beta, \gamma, \delta \ldots$ express any numbers."[46] But he made no further use of this more general series.

Lagrange's false proof[47] starts from this doubtful point. The proof contains many Eulerian elements, though the attempt at a proof was new. Lagrange observed, echoing Euler, that the theorem was "verified in fact" by the development of the different known functions; however, he added that nobody had tried to "demonstrate it *a priori*." Furthermore, he added that the differential calculus rested on the assumption that any function could be so developed—this is indeed the whole point of *FA*. Surely a fact so fundamental should be proved. And since, in some cases, the series development is not valid[48] a proof is all the more necessary to see what is really happening. In this we see the transitional character of Lagrange's work: a man of the eighteenth century in his certainty that, if there are exceptions, they are not very important; a man of the nineteenth century in demanding proofs whenever possible to insure precision of application. Shoddy proofs may be less attractive than either intuitive certainty or valid arguments, but they are nevertheless signposts on the road to rigor.

In going through Lagrange's proof it must be kept in mind that a "function" is an analytic expression, as Euler had defined it in the *Introductio*.[49] Lagrange began by substituting $(x + i)$ for x in the function $f(x)$; this yields a new analytic expression involving x and i. The new expression will contain terms of the form pi^r (my notation), where p is a function of x and r is some real number. This follows because Lagrange implicitly assumed a sum of terms like $\sum_{k=0}^{\infty} p_k\, i^{r_k}$ to be the most general possible analytic expression in i. This view is exactly that of the *Introductio*, though Lagrange did not cite this work anywhere in the course of his proof.

The question now becomes, what values can the exponents of i have in the development

$$f(x + i) = f(x) + pi + qi^2 + \cdots + ui^r + \cdots?$$

(r-notation is mine; Lagrange asked the question verbally.) In answering the question Lagrange used some remarks Euler had made about the values of many-valued functions. In general, modern

---

44 E.g., by Carl Boyer, *op. cit.*, pp. 252–253.

45 *Introductio*, Section 59. Recall that "function" here meant "analytic expression."

46 *Ibid.* (my italics) Euler understood that, if m, n, p, q are integers and α was irrational, $1 \le m/n < \alpha < p/q$ implies $z^{m/n} < z^\alpha < z^{p/q}$. See *Introductio*, Section 97.

47 *FA* Section 2, *Oeuvres* IX, pp. 22–23. In what follows, I have supplied all the detailed justifications not directly attributed to Lagrange.

48 But, Lagrange said, only at isolated points. See *FA, Oeuvres* IX, Chapter 5.

49 Section 4. Lagrange's definition in *FA, Oeuvres* IX, p. 16.

definitions of function require that they be single-valued; Euler studied many-valued functions also, calling them "multiform."[50] He gave as an example: if $y = z^n - Pz^{n-1} + Qz^{n-2}\ldots$, a polynomial in z, z is a multiform function of y with as many values for each y as there are units in the exponent n. In this case, $P, Q, \ldots$ must be uniform (one-valued) functions of z (including constants); if not, z will have, for a given value of y, more values than there are units in n.[51]

Lagrange used an exactly analogous argument on the expression given above for the development of $f(x+i)$. The radicals of i (that is, cases where the exponent r is a fraction) can only come from the radicals included in the original function $f(x)$. (If $f(x)$ were rational always—for all x—clearly $f(x+i)$ must be too—except, he explicitly noted, for particular values of x and i.) Now, he said, the substitution of $x+i$ for x can neither increase nor diminish the number of these radicals, nor change their nature. He meant that, for instance, as long as x and i are indeterminate (variable), something like $ax^{m/n}$ will become $a(x+i)^{m/n}$; we still have *one* irrational quantity, and it is still an nth root.

"By the theory of equations . . . any radical has as many different values as there are units in its exponent,"[52] so that "any irrational function has . . . as many different values as one can make combinations of the different values of the radicals which it includes." What he meant, then, is that something like $a^{1/n} + b^{1/m}$ has $n \times m$ different possible values, real and complex. He did not mention complex numbers here, but they are needed to make the statement true; in fact, if n is odd, $a^{1/n}$ has only one real value, as Euler remarked in Section 15 of the *Introductio*.

Having made these preliminary remarks, Lagrange struck out on his own. Suppose the development of $f(x+i)$ contains a term of the form $ui^{m/n}$. "The function $f(x)$ will be necessarily irrational." Thus it will have some fixed (presumably finite) number of different values, "which will be the same for the function $f(x+i)$ as for its development." He meant that the *analytic expression* (such as $(x+i)^{1/2}$) and its *development as a series in i* (such as $x^{1/2} + i/2x^{1/2} - i^2/8x^{3/2} + \cdots$) will have the same number of possible values.

But if the development is given by the series (his notation)

$$f(x+i) = f(x) + pi + qi^2 + \cdots + ui^{m/n} + \cdots,$$

"each value of $f(x)$ will be combined with each of the n values of the radical $\sqrt[n]{i^m}$, so that the function $f(x+i)$ developed will have more different values than the same function not developed, which is absurd." He meant that, if $f(x)$ has, say, k values whatever x may be, then $f(x+i)$ has k values. But $f(x) + ui^{m/n}$ has $k+n$ possible values; thus the right-hand side of the above equation has more possible values than the left-hand side, which violates what is meant when we write the equal sign.[53]

Going back now to the expression for $f(x+i)$ on page 48, Lagrange added that the quantity we have called r cannot be negative. If it were, he pointed out, setting $i = 0$ would make $f(x+i) = f(x)$, but the right-hand side would then be infinite, a contradiction. Thus, he had "proved" that all values of r must be positive integers.

---

[50] *Introductio*, Section 10.

[51] *Op. cit.*, Section 14.

[52] By the "Fundamental Theorem of Algebra." Compare *Introductio*, Section 14.

[53] Leibniz, for instance, had said that quantities were equal when they could be substituted for one another *salva veritate*. Quoted by Gottlob Frege, "On Sense and Reference," in *Philosophical Writings* (Oxford: Basil Blackwell, 1960), p. 64.

What are we to think of this proof? The algebraic ingenuity of dealing with the number of values which $f(x)$ can take on is apparent; arguments of this sort work for finite expressions and can even be used to check finite expansions. But Lagrange's proof will not do in the infinite case. First of all, there may be infinitely many terms of the form $i^{m/n}$; both sides of the equation on page 48 may have infinitely many possible values—as does, for instance, $\sin^{-1} x$, or $\log x$. Also, each $i^{m/n}$ may itself be representable as an infinite power series in i with integral exponents; a possibility which he did not consider, but which surely ought to have occurred to him since he was thinking about the possibility of power series expansions for all functions. Further, he never considered the possibility of the quantity r being irrational—although such r are just as conceivable as rational ones and were at least implicit in Euler's treatment. In addition, Lagrange was dealing with real, not complex-valued functions; the number of imaginary values may indeed differ for two functions equivalent for real numbers. For instance, $(\sqrt{x})^2$ and $|x|$ coincide for x real, but not for x complex.[54] In fact, the number of real positive values for $x^{1/2n}$, or the number of real values for $x^{1/2n+1}$, is uniquely definable—and it is only in this context that we can understand the setting equal of two analytic expressions involving a real variable; the equivalence of the formulas must mean that they give the same values for the same arguments. Lagrange had the germ of this idea here in his attempt to deal with two equivalent formulas' getting the same results, but he made the number of values, not some uniquely defined value, his major concern. He did not appreciate the distinction between a function representable by an analytic expression and the expression itself. It is one thing to say that functions must possess analytic expressions—a narrow definition for the sake of having something that can be worked with—and another to identify the two.[55]

Lagrange's proof may reveal the algebraic background of *FA*, but the lacunae in it are so gaping that one hesitates to take it seriously. Did he? To be sure he called it a proof, placed it proudly at the first of both editions of *FA*—and of the *Leçons sur Calcul des Fonctions*—which appeared in his lifetime. André-Marie Ampère and Augustin-Louis Cauchy, who dispensed with the Taylor series in their algebraic definitions of $f'(x)$,[56] did not bother to discuss Lagrange's proof, indicating that *they* did not take it very seriously.

I think that Lagrange did believe in the proof, but not very deeply; he did not discuss it very much. He certainly believed the theorem to be true; he gave a proof—as he gave a few others—to fill a gap in Euler's work. The proof is basic to *FA*, but its omission would not alter the rest of the book, which would then apply to those real-valued functions which have Taylor series expansions—i.e., those functions we today term "analytic functions."[57] And Lagrange apparently knew no analytic expressions which did not.

---

[54] Another possible difficulty: if

$$(x+i)^{1/2} = x^{1/2} + 1/2x^{1/2}\,i - 1/8x^{3/2}\,i^2 \ldots,$$

the left-hand side has two possible values; each term on the right-hand side has two possible values. If each is taken separately the right-hand side can have infinitely many values. Presumably Lagrange would argue that the right-hand side really has only two values, since it can be written $x^{1/2}(1 + 1/2xi - 1/8x^2i^2 + \cdots)$, but he did not make this clear.

[55] The first to really make something of this distinction was Dirichlet. The idea of general dependence relations—in fact first expressed by Euler in 1748—was not the only insight Dirichlet brought to function theory. See Bell, *op. cit.*, p. 293.

[56] See below, Chapter 4.

[57] Indeed, Lagrange said as much in the second edition of *FA*: "The method . . . only applies, in general, to the development of a function of x and i insofar as this function can be reduced to a series in positive integral powers of i; for the reasoning of Section 2 [the proof under discussion], by which we have proved that all functions of $x+i$ are, generally speaking, reducible to this form, cannot be applied to any function of x and i." *Oeuvres* IX, p. 29. "But, in the case in which this

In any case, the proof illustrates that Lagrange was not a formalist who assumed that the Taylor series must exist for all functions. He was not being a formalist, but an algebraist; his sin lay not in assuming that any process representable on paper could unambiguously and without contradiction be carried out, but in assuming too close a kinship between finite and infinite polynomials in the course of his proof, and neglecting subtleties concerning rational, real, and complex numbers.

Yet this ill-fated argument illustrates again the spirit with which Lagrange attacked the calculus, and the spirit in which Cauchy and Karl Weierstrass were ultimately to master it: it was an attempt to make the basic theorems of the calculus rest on a purely algebraic proof. This proof, however fallacious, can be credited with encouraging Lagrange to erect the impressive structure of the theory of analytic functions.

The Algebraic Background of the Lagrange Remainder: Approximations

"Why is the remainder term of the Taylor series treated in *FA?*" may not at first seem a difficult question. After all, the book is really about the Taylor series. But what has been said so far about eighteenth-century algebra may have suggested that it was largely concerned with exact or formalistic procedures, or with the theory of equations; the calculation of an error-estimate does not seem to belong to the subject at all. Nevertheless, in Lagrange's broader view, error-estimates have a natural place in the study of approximation techniques, which are properly part of the determination of roots as functions of their coefficients—the resolution of algebraic equations.

The solution of equations that do not yield to exact procedures—including that of the general equation of degree five or higher—occupied a major part of algebraists' time throughout the eighteenth century. In keeping with the general eighteenth-century drive for results and lack of interest in foundations, the emphasis is on finding new and easier methods of approximating to the solutions of equations, not in calculations of the precision of these methods. Most algebraists before Lagrange were not interested in estimating bounds on errors.

Approximations in the eighteenth century were carried out by a wide variety of methods: by use of the binomial theorem;[58] by undetermined coefficients;[59] by the use of substitutions of the form $x \to x + h$;[60] by the calculus;[61] by discussing the limiting cases of particular algebraic expressions;[62] by methods equivalent to the use of the calculus;[63] by easily constructed telescoping

---

reduction is possible," he continued, his results will hold. These remarks are obviously afterthoughts, since he gave the proof itself without comment.

[58] A method based on work done by Isaac Newton.

[59] E.g., by le Marquis de Courtivron, "Sur une manière de resoudre par approximation les équations de tous les degrés," *Mémoires de l'Académie des Sciences*. Paris, 1744, pp. 405–414.

[60] As by Newton in *Method of Fluxions*; compare p. 41, above.

[61] As by Euler in *ICD*, and by Lagrange: see p. 57, below.

[62] See an example of this, p.106ff, below, by Daniel Bernoulli, from "Observationes de seriebus quae formantur ex additione vel subtractione quacunque terminorum se mutuo consequentium, ubi praesertim earundem insignis usus pro inveniendix radicum omnium Aequationum Algebraicarum ostenditur," *Commentarii Academiae Scientiarum Imperialis Petropolitanae*, Tom. III, 1728, pp. 85–100.

[63] Euler, *Algebra*, translator's note on the "rule of double position," p. 296 of John Hewlett translation (London: Longman, Orme, & Co., 1840).

inequalities.[64] With the possible exception of Maclaurin, it seems clear that all these approximators were concerned with deriving, or showing how to derive, infinite analytic expressions for the roots of any polynomial, or, occasionally, transcendental equation. Maclaurin followed a more classical tradition in deriving upper and lower bounds on the sought-for root as statements of approximation. This tendency was largely abandoned by others because of the greater ease of deriving infinite expressions. When inequality-conditions were appealed to, it was to insure that the terms of the series under consideration got smaller and smaller, which generally is sufficient to insure that the value of the successive approximations get closer and closer to the root.

The preference for infinite expressions was so deep-rooted by the mid-eighteenth century that, even in those rare instances when methods were evolved by inequality considerations rather than by algebraic substitutions, they would be immediately converted into infinite analytic expressions and the inequalities discussed no more. Johann Heinrich Lambert[65] found an approximation technique by inequalities—and discarded the inequalities at the end to present a series, making his method look just like the others. It is a method for solving $x^m + px = q$. The series obtained is

$$x = q/p - q^m/p^{m+1} + mq^{2m-1}/p^{2m+1} - \frac{m(3m-2)}{2}(q^{3m-2}/p^{3m+1}).\ldots$$ [66]

The convergence-condition he gave for this series was $(m-1)^{m-1}p^m > m^m q^{m-1}$, the precise justification for which I am unable to supply. Apparently—as the powers of p and q used indicate—a ratio test is being used (in fact, for this series $p^m > q^{m-1}$ would suffice for convergence). In any case, we have an infinite series with a sufficient condition for the successive ratio of the terms to decrease, not an error-estimate. Lambert abandoned his bounding inequalities as soon as he had his result.

The Algebraic Background of the Lagrange Remainder: Error-Estimates Before Lagrange

Error-estimates were not entirely unknown in the eighteenth century. An ingenious approximation method which might have suggested by its nature the calculation of a type of error-estimate had been published in 1728 by Daniel Bernoulli.[67] This method was designed to approximate to

---

[64] Maclaurin, *op. cit* p. 230. This list of methods is not intended to be exhaustive, but does include most of the methods referred to by Lagrange in *REN*.

[65] In "Observationes Variae in mathesin puram," *Acta Helvetica* III (1758), pp. 128–168.

[66] The simplest illustration is for $m = 1$, the case with which he began. Suppose $x + px = q$, where $p, q > 0$ and $x < q/p$. Since $x < q/p$, adding

$$x + px < q/p + px$$

Therefore $q < q/p + px$ and

$$q/p > x > q/p - q/p^2$$

Then $q = x + px > (q/p - q/p^2) + px$ (by adding px)

Therefore, $q > q/p - q/p^2 + px$ and

$$q/p - q/p^2 + q/p^3 > x > q/p - q/p^2.$$

In general, we have the series for x: $x = q/p - q/p^2 + q/p^3 - \cdots$ which converges if $p > 1$.

The case for $x^m + px = q$ was derived in the same way.

[67] See Note 62.

the largest (in fact, largest in absolute value, though he did not say so) real root of a polynomial equation. The approximation generates a sequence of numbers in which the ratio of the last to the next-to-last will approach the root. Bernoulli's method was presented without a precise error-estimate, though he sketched an argument to show that the successive approximating values did approach the root. The method caught the fancy of Leonhard Euler; accordingly, the practice of it for some particular cases is described in Euler's *Algebra*,[68] while its derivation, closely following Bernoulli's treatment, appears in the *Introductio*.[69]

Bernoulli's method is based on the algebraic properties of fractions and on the assumption that polynomials can be broken up into linear factors. Suppose the equation to be solved is

$$x^m - \alpha x^{m-1} - \beta x^{m-2} \cdots = 0$$

and suppose at first that it has m real unequal roots (other cases are dealt with later). If $z = 1/x$, we obtain

$$1 - \alpha z - \beta z^2 \ldots$$

as the same expression in z. We can then consider the fraction

$$\frac{a + bz + \cdots}{1 - \alpha z - \beta z^2 - \cdots - \mu z^m}$$

where the numerator is an arbitrary polynomial. This fraction, as was then well known, can be written as an infinite series in z of the form $\sum_{k=0}^{\infty} a_k z^k$ (my notation). The denominator can be factored into $(1 - pz)(1 - qz)(1 - rz)\ldots$ and the fraction can be written as follows:

$$\frac{a + bz + \cdots}{1 - \alpha z - \beta z^2 - \cdots - \mu z^m} = \frac{A}{1 - pz} + \frac{B}{1 - qz} + \frac{C}{1 - rz} + \cdots$$

for some $A, B, C, \ldots$. Since

$$A/(1 - pz) = A(1 + pz + p^2 z^2 \ldots)$$
$$B/(1 - qz) = B(1 + qz + q^2 z^2 \ldots)$$

the coefficient of z^n in the expression for the original fraction will be

$$(Ap^n + Bq^n + Cr^n + \cdots)$$

But if p is larger than $q, r, \ldots$ (in fact, larger in absolute value), Euler argued, the term Ap^n will exceed $(Bq^n + Cr^n + \cdots)$ for n sufficiently large. Thus, in the notation introduced above,

$$a_n \cong Ap^n$$
$$a_{n+1} \cong Ap^{n+1}$$

---

[68] P. 360ff, with a different rationale.

[69] Sections 333–355. I follow Euler's exposition, below.

and if n is infinite, $a_{n+1}/a_n \cong p$.[70] Euler gave rules to find the a_k's in terms of $\alpha, \beta, \gamma, \ldots$ and $a, b, \ldots$. But if $(1 - pz)$ divides the denominator of the original fraction, then $z = 1/p$ is a root of that denominator. Then $x = p$ will be a root of the original equation, and the root of maximal absolute value.

It should be noted that the whole method rests on the assumption that $(1 - \alpha z - \beta z^2 - \cdots \mu z^m)$ has m factors of the form $(1 - u_k z)$ and that there exists u_0 such that $|u_0| > |u_k|$ for $k \neq 0$.[71]

If we were not to neglect q, presumably the next smaller root, we would have instead:

$$a_{n+1}/a_n \cong \frac{Ap^{n+1} + Bq^{n+1}}{Ap^n + Bq^n} = p - \frac{Bq^n(p-q)}{Ap^n + Bq^n}.$$

The term $Bq^n(p-q)/(Ap^n + Bq^n)$ is included in Euler's presentation only to discuss the case where q is negative. The term shows that the ratio a_{n+1}/a_n (for large n) differs from p by a quantity which is positive or negative depending on whether n is odd or even. The actual calculation of such a term, which approximates the error committed by neglecting q, foreshadows later work on finding upper bounds on errors, but this had to wait for Lagrange. Euler was interested in getting as much information as possible from his formulas—and derived formulas with great facility—but he did not seem interested in a general theory of error-estimates.

Instead of computing general error-estimates, Euler showed the smallness of the error in a particular computation, presumably intended to be typical. For $x^2 - 3x - 1 = 0$, he generated the series 1, 2, 7, 23, 76, 251, 829, 2738....[72] Computing a_7/a_6, he obtained $2738/829 = 3.302774\ldots$, which he subtracted from the true root of this quadratic, $(3 + \sqrt{13})/2 = 3.3027756\ldots$. This sort of error-computation may increase one's confidence in the method, but it does not "prove" the goodness of approximation methods for those cases in which they are really useful—equations whose true solutions have not been found.

Many of the tendencies already discussed appear in the history of the most venerable member of the collection of approximation techniques used in the eighteenth century, the method due to Newton.[73] The method was expounded by him, not in general, but for a simple cubic equation from which the generalization was obvious:[74]

$$y^3 - 2y - 5 = 0.$$

---

[70] Euler wrote the equal sign where we would write $\cong$. I use $\cong$ below, without further apology.

[71] Euler noted in the *Algebra*, p. 362, that the method does not always work. For instance, suppose $x^2 - 2 = 0$ be the equation, and let the fraction be converted into a series. The ratio a_{n+1}/a_n in the case Euler gave there is alternately 1 and 2; this happens, though he did not explicitly say so, because the two roots are equal in absolute value and q^n cannot be neglected with respect to p^n.

The method can be generalized to include cases in which the equation has complex roots, or two or more equal roots which are not maximal. See *Introductio*, Section 346ff.

[72] Section 338, *Introductio*.

[73] Usually credited by eighteenth-century writers to the *Method of Fluxions*.

[74] *Method of Fluxions*, p. 7ff.

As a first approximation to the solution of this equation,[75] 2 is chosen "as differing from the true root less than by a tenth part of itself."[76] Now, setting $y = 2 + p$, Newton obtained, from the original equation, an equation in p

$$p^3 + 6p^2 + 10^p - 1 = 0.$$

But, since p is small, the higher order terms can be neglected, yielding

$$10p - 1 = 0,\ p = .1.$$

The value $p = .1$ is now treated as a first approximation to the solution of

$$p^3 + 6p^2 + 10p - 1 = 0.$$

Setting $p = .1 + q$ and substituting, we obtain

$$q^3 + 6.3q^2 + 11.23q + .061 = 0.$$

Assuming q and q^2 to be small with respect to the other terms, this last equation gives $q = -.061/11.23$. We can continue in this way as long as desired. If preferred, the cubic equation for p could be solved neglecting only the p^3 term, for more accuracy.

Finally, he added that even the third figure (i.e., third significant figure) of p can be found if, in the general equation for p of form

$$p^3 + cp^2 + bp + e = 0$$

b^2 is less than 10ec.[77]

It was usual in eighteenth-century expositions of Newton's method to repeat the stricture that p (actually, $|p|$) be less than 1/10 of the first approximation, justifying the neglect of higher powers of p in procedures like the above. In the absence of an explanation from Newton, the reason for the stricture was not resurrected—other than the general one of insuring quick convergence.

The method of Newton was always cited as one of greatest importance.[78] Newton and Maclaurin had chosen to present this approximation technique in their algebras without reference to the method of fluxions. In a sense this was good mathematical practice—never assuming more than absolutely necessary. But in another sense it was unwise: though they avoided the possibly

---

[75] Finding acceptable first approximations is an art in itself, not here discussed by Newton. A number of considerations from the theory of equations are used in doing this. Most obviously, if $P(y) = 0$ is a polynomial equation and we find by repeated trials z, w such that $P(z)$ is positive, $P(w)$ negative, then there is a root of the equation y between z and w. Now let us consider $P\left(\frac{z+w}{2}\right)$. If it is positive, set $z' = 1/2(z + w)$, $w' = w$. If negative, then set $z' = z$, $w' = 1/2(z + w)$. Now repeat the process again for z' and w'. The quantities z and w, z' and w', etc., can be used as first approximations. See Maclaurin, *op. cit.*, p. 230.

Alternatively, use may be made of the fact that the graph of $P(y)$ has turning points between two real roots, so that if $P(a) = P(b) = 0$, there is a Y between a and b such that $P'(Y) = 0$. This is Rolle's theorem, although in his statement of it P' is not defined by means of the calculus. See Michel Rolle, *Demonstration d'une méthode pour resoudre les égalitez de tous les degrés* (Paris: chez Jean Cusson, 1691). The value of Y found by solving $P'(Y) = 0$ can be used as a first approximation.

Of course, trial-and-error can also be used.

[76] *Method of Fluxions*, p. 7.

[77] His treatment is verbal, rather than with inequalities: *op. cit.*, p. 10.

[78] By Euler, *Algebra*, pp. 356–357—not, however, with Newton's name; by Maclaurin, *op. cit.*, p. 233ff.

controversial use of fluxions, they deprived their readers of the opportunity to see the reasons that their approximation techniques worked, to say nothing of the possibility of proving them. Maclaurin had "generalized" Newton's method by telescoping several steps into one formula, thus: if the equation to be solved is $x^n + px^{n-1} = \cdots + A = 0$, and k is the first approximation,[79] then setting $k + f = x$, we obtain for once and for all (neglecting higher order terms in f)

$$f = \frac{-A - k^n - pk^{n-1} - qk^{n-2}\ldots}{nk^{n-1} + p(n-1)k^{n-2} + \cdots}$$

Euler chose, instead, to generalize these approximation methods by writing them in the language of the differential calculus. One advantage of this was that the methods were applicable to transcendental equations as well as to polynomials; Euler solved $x - n\log x = 0$ in this way,[80] justifying his arguments by the Taylor series, since Taylor had made the substitution $x \to x + h$ for once and for all. Of course Newton knew that his approximation method was related to fluxions, but Euler made explicit and extensive use of the Taylor series in what was unquestionably the study of approximations in algebra.[81] The calculus provides a broader mathematical framework in which to discuss approximation techniques—a framework of which Lagrange quickly availed himself.

Euler gave the following derivation of Newton's method. Suppose, for y any function of x, the equation $y = 0$ is given.[82] Suppose, too, that f is a root, so that $y(f) = 0$. Then

$$y(x + (f - x)) = 0.$$

Expanding in a Taylor series, we obtain

$$y(x) + (f - x)dy/dx + (f - x)^2/2!\, d^2y/dx^2 + \cdots.$$

If $(f - x)$ is very small, all but the first two terms can be neglected and therefore, Euler concluded, $f \cong x - y(dx/dy)$, which is equivalent to the formula given by Maclaurin, above. This procedure can be repeated as often as necessary.

The particular example $y = x^n - a^n - b = 0$ had, in Section 230 of the *Introductio*, been solved by this method, and a remark made about convergence. $f = a + b/na^{n-1}$ is the first approximation. How small must b be to neglect the higher order terms? Euler specified $(a^n + b) < (a+1)^n$, a and b presumably assumed positive. He gave no justifying argument, but it is clear if a and b are in fact positive, since

$$a^n < (a^n + b) < (a + 1)^n \quad \text{and} \quad a^n + b = f^n,$$

then $|a| < |f| < |a + 1|$ and, therefore, $|f - a| < 1$.

Thus Euler's condition is equivalent to the case in which the variable in the power series has absolute value less than 1 when $x = a$. But a series of form $\sum_{k=0}^{\infty} c_k(f - a)^k$ has its terms get smaller when $\text{Lim}\,|c_k(f - a)^k| \to 0$; it is not enough for $|f - a| < 1$. Of course, proving that the terms go to zero would not in itself prove that the series converges.

---

[79] *Op. cit.*, p. 239.

[80] *ICD*, Section 242.

[81] *Op. cit.*, Part II, Chapter 9.

[82] ICD, Section 234. Presumably "analytic expression" is meant for "function," through *ICD* begins with a broader definition.

In fact, Euler may not even have been justified in neglecting the second term, which is $\frac{(f-a)^2 n(n-1)}{2!} a^{n-2}$, which can exceed the first term, $(f-a)na^{n-1}$, unless $1/2(f-a)(n-1) < a$. The particular examples used by Euler always satisfy this inequality, because in illustrating the use of the method, he tried it out on equations for which n is relatively small. If $n = 4$, the highest value he treated, the condition that the first term exceed the second becomes $(f-a) < 2/3a$. Euler's first approximations always are close enough for that inequality to hold—in fact he usually chose them to satisfy Newton's criterion $(f-a) < 1/10a$. Thus it is not surprising that Euler did not meet any obvious counterexamples when he applied this method to particular numerical examples.[83]

The Algebraic Background of the Lagrange Remainder: Lagrange and Bounds on Error

The possibility that the terms of an approximating series would increase for a time before beginning to decrease, in the way discussed above, seemed great enough to Lagrange for him to feel compelled to work out the precise conditions under which the terms would decrease. Lagrange, following Euler,[84] used the Taylor series to give a recursive formula for Newton's approximation method.[85] For instance, if $F(x) = 0$, and $a + p = x$, he wrote

$$F(a) + pF'(a) + \frac{p^2}{2!}F''(a) + \cdots = 0,$$

yielding $p = -F(a)/F'(a)$, again, as a first approximation. He had done, as yet, nothing new here, except for the F' notation which "frees the formulas from this multitude of d's which lengthens and even disfigures them ... and which continually calls to mind the false idea of the infinitely small."[86]

What is new in Lagrange's work with this method is the calculation of precisely what conditions must be satisfied for the first order term to exceed the second order one. In using Newton's method, Lagrange noted that at each step one had to neglect terms whose value was unknown. This meant that the exactness of the next term added to the approximating series could not be judged—a situation intolerable to Lagrange, the pioneer in calculation of error terms. The terms of the series obtained in the Newton approximation might get small very slowly, he observed, or might even increase with n after having decreased for a time.[87] We have noted above how Euler had worked with a series whose terms might decrease only after increasing for a time, without his ever having examined this state of affairs. Lagrange did not mention this, but Euler's work on approximations was known to him; this work may well have suggested the examination of the problem: how exact is Newton's well-known and much-used method?[88]

---

[83] If $(f-a) = a/10$, and the equation $y = x^{21} - a^{21} - b$ be treated by the Euler-Newton method, the second term will be equal to the first.

[84] *ICD*, Section 234.

[85] *REN*, pp. 258–285, of *Oeuvres* VIII.

[86] Lagrange, *Op. cit.*, p. 243. The $f'(x)$ notation had, of course, been used earlier by Lagrange. See p. 20, above.

[87] He spoke of a series which "diverges after having been convergent," using the term "converge" to mean "terms get smaller." *REN*, p. 17.

[88] In Note V, pp. 159–167, of *REN*.

Lagrange attacked the problem in the following way. Consider the equation

$$x^m - Ax^{m-1} + Bx^{m-2} \cdots = 0. \tag{2.1}$$

Suppose α is the desired root, a the first approximation, $(a+p)$ the second. For the second approximation to be closer than the first, we must have

$$|\alpha - (a+p)| < |\alpha - a|, \quad \text{or}$$

$$\left|\frac{1}{\alpha - a - p}\right| > \left|\frac{1}{\alpha - a}\right|.^{89} \tag{2.2}$$

(2.1) can also be written as a product of n linear factors:

$$(x-\alpha)(x-\beta)(x-\gamma)\ldots = 0. \tag{2.3}$$

By the exposition of Newton's method given previously, if $x = a + p$, he obtained

$$p = -X/X',$$

where X is the polynomial in (2.1) evaluated at $x = a$, and X' is the derivative of the polynomial, evaluated at $x = a$. From this, he obtained (differentiating in the form given in (2.3))

$$-X/X' = \frac{1}{\frac{1}{\alpha-a} + \frac{1}{\beta-a} + \frac{1}{\gamma-a} + \cdots} = p.^{90}$$

But if R is defined by $R = \frac{1}{\beta-a} + \frac{1}{\gamma-a} + \frac{1}{\delta-a} + \cdots$ the inequality (2.2) holds if R has the same sign as $(\alpha - a)$.[91] But, suppose R and $(\alpha - a)$ have opposite sign. When, then, is

$$\frac{1}{(\alpha-a-p)^2} > \frac{1}{(\alpha-a)^2}?$$

This requires that

$$2(\alpha - a)R + 1 > 0.^{92} \tag{2.4}$$

---

[89] P. 162. This is Lagrange's notation, except that he did not use the absolute value sign; he did have the idea, saying "abstraction being made of the sign." I use absolute value notation from here on, without further apology.

[90] Because, he argued, $X = (a-\alpha)(a-\beta)(a-\gamma)\ldots$, and X', by the product rule, satisfies $X' = (a-\beta)(a-\gamma)\cdots + (a-\alpha)(a-\gamma)\cdots + (a-\alpha)(a-\beta)\cdots + (a-\alpha)(a-\beta)(a-\gamma)\ldots$, *ibid.*

[91] *Ibid.* For

$$p = \frac{1}{\frac{1}{\alpha-a} + R}$$

Thus,

$$\alpha - a - p = \alpha - a - \frac{1}{\frac{1}{\alpha-a} + R} = \frac{R(\alpha-a)}{\frac{1}{\alpha-a} + R}$$

and therefore

$$\frac{1}{\alpha-a-p} = \frac{1}{\alpha-a} + \frac{1}{(\alpha-a)^2 R}.$$

Clearly the term $\frac{1}{(\alpha-a)^2 R}$ has the same sign as R.

[92] P. 163. For, from the above,

$$\frac{1}{(\alpha-a-p)^2} = \left[\frac{1}{(\alpha-a)} + \frac{1}{(\alpha-a)^2 R}\right]^2 = \left(\frac{1}{(\alpha-a)^2} + \frac{2}{(\alpha-a)^3 R} + \frac{1}{(\alpha-a)^4 R^2}\right).$$

Lagrange pointed out that, if a exceeds all the roots $\alpha, \beta, \dots$ or if a is less than all of them, $(\alpha - a)$ and R will have the same sign and (2.2) will be satisfied, but that it is not clear, in all other cases, that R will satisfy (2.4). It should, he added, be easy to construct a counter-example.

Lagrange did not give conditions for the series to converge. He did not even give conditions for all successive terms to decrease to zero, or even to decrease at all. This is a rather primitive discussion of the precision of an approximation, which only gives conditions for the second approximation to be closer than the first—or, by extension, that any given approximation be closer than the previous one. This is a very early example of the examination of the precision of a complicated approximation by means of careful argument about inequalities. It is the ancestor of complete discussions of both the fact and the rapidity of convergence.

While Lagrange did not here give such a complete discussion, he did do so in other cases. The way was open for him to ask how fast the approximating series approached the root, and within what bounds the difference would lie; the use of inequality-manipulations to estimate the relative sizes of the terms in an approximation procedure is akin to their use in finding estimates of error.

An actual error-estimate was found by Lagrange in conjunction with his "New Method to Approximate Roots of Numerical Equations," given in Chapter 3 of *REN*. The conception of this method resembles somewhat that of Newton's method, but leads to an expression for the root, not as a series, but as a continued fraction.[93]

The method is as follows. Suppose

$$Ax^m + Bx^{m-1} + Cx^{m-2} \cdots + K = 0. \tag{2.5}$$

Suppose that p is a first approximation to a positive root x, such that

$$p < x < p + 1$$

Now let $x = p + 1/y$, where $0 < 1/y < 1$.

I think that Lagrange worked this out by direct analogy with Newton's method—as presented algebraically by Newton: instead of letting $x = p + y$ for $|y| < 1$, he let $x = p + 1/y$ for $|1/y| < 1$.

Substituting this new value of x in (2.5) and multiplying by y^m, we obtain, for some A', B', C', ... an equation of the form

$$A'y^m + B'y^{m-1} + C'y^{m-2} + \cdots + K' = 0. \tag{2.6}$$

By choice of y, (2.6) has a root greater than 1 corresponding to the original root x.

Now let q be the integer such that $y = q + 1/z$ with $0 < 1/z < 1$. Substituting again, multiplying through by z^m, we obtain

$$A''z^m + B''z^{m-1} + \cdots + K'' = 0. \tag{2.7}$$

So, the inequality formed by squaring (2.2) holds when and only when

$$\frac{2}{(\alpha - a)^3 R} + \frac{1}{(\alpha - a)^4 R^2} > 0$$

or, when (2.4) holds.

[93] Euler had given a theory of continued fractions in the *Introductio*, Chapter 18, on which Lagrange drew freely in what follows. The subject had been pioneered by Rafael Bombelli (sixteenth century) and Pietro Cataldi (1548–1626).

Continuing by setting $z = r + 1/u$, etc., Lagrange finally obtained

$$x = p + \cfrac{1}{q + \cfrac{1}{r + \cfrac{1}{u + \cdots}}}$$ [94]

(This was, and still is, standard notation for the sequence p, $p + 1/q$, $p + \frac{1}{q+1/r}$, etc.)

Lagrange asserted here that nobody had yet thought to use continued fractions to solve equations. Euler, however, had considered the reverse problem in the *Introductio*, although Lagrange did not mention this. Euler evaluated several continued fractions by finding solvable equations of which they were solutions.[95] He did not consider the possibility that the fractions he considered would not converge; he just gave numerical values for the first few values of the fraction, and compared them with the numerical solution of the corresponding equation. He was not interested in calculating a general formula for the error committed in breaking off the infinite fraction after a finite number of terms.

The importance of Lagrange's general continued-fraction method, for our purposes, is that it lent itself readily to an estimate of error—the very form of the successive approximations suggested such an estimate. Further, in the derivation of some of these error-estimates, we can see an analogy to the error-estimate applied by Lagrange to the Taylor series.

Returning to Lagrange's exposition: if the successive fractions are $p/1$, $p + 1/q$, $p + \frac{1}{q+1/r}$ etc., he defined the integers $\alpha, \alpha', \beta, \beta', \ldots$ by setting

$$\begin{aligned} \alpha &= p, \quad & \alpha' &= 1 \\ \beta &= q\alpha + 1, \quad & \beta' &= q\alpha' = q \\ \gamma &= r\beta + \alpha, \quad & \gamma' &= r\beta' + \alpha', \text{ etc.} \end{aligned}$$

Then the successive fractions are $\alpha/\alpha', \beta/\beta', \gamma/\gamma', \ldots$ and these are such that the true value of x is always included between two consecutive fractions.[96]

---

[94] *REN*, pp. 41–45. The method was published in the original 1767–68 papers.

[95] See Sections 377–378, especially. Suppose, for instance, the continued fraction $x = \cfrac{1}{a + \cfrac{1}{b + \cfrac{1}{a + \cdots}}}$ be considered. Euler observed that $x = \cfrac{1}{a + \cfrac{1}{b + x}}$ which form, he said, is equivalent to the quadratic equation $x^2 + bx - b/a = 0$.

[96] P. 46ff. This was well known; compare *Introductio*, Section 361. Lagrange did not justify this, but it is not difficult to do so.

By choice of p, $x > p$, $x - p < 1$.

$$P + 1/y = x, \quad \text{or} \quad y = 1/(x - p) \quad \text{and} \quad q < y, \text{ by choice of } q.$$

Therefore, $1/(x - p) > q$.

Since $(x - p)$ and q are positive,

$$x - p < 1/q$$
$$p < x < p + 1/q.$$

Now, since $\beta\alpha' - \alpha\beta' = 1$, $\beta\gamma' - \gamma\beta' = 1$, etc.,[97]

the fractions α/α', β/β', ... are in lowest terms. (He gave no reason, but it was then well known that, if m, n, s, t are integers, $mn \pm st = 1$ means that the greatest common divisor of any two of them is unity.)

But clearly $\beta/\beta' - \alpha/\alpha' = 1/\alpha'\beta'$, $\gamma/\gamma' - \beta/\beta' = 1/\beta'\gamma'$, etc. Thus, the consecutive fractions cannot differ by more than $1/\alpha'\beta'$, $1/\beta'\gamma'$, etc. In fact, $\alpha' < \beta' < \gamma' \ldots$ (for instance, $\beta' < \gamma'$ since, substituting, $q < rq + 1$, since $r \geq 1$). He thus concluded

$$\left|\frac{\beta}{\beta'} - \frac{\alpha}{\alpha'}\right| < \frac{1}{(\alpha')^2}, \text{ etc.,} \tag{2.8}$$

which gives a precise error-estimate.[98]

Euler had given a form of the differences between successive fractions also, but only so that he could convert the fraction into an infinite series.[99] Lagrange wanted to show that the error, which gets smaller and smaller, can be given a precise bound.

His inequalities of the form (2.8) show in particular that, since x lies between any two consecutive fractions, and since the denominators are increasing integers, the difference between x and the kth approximation is certainly less than $1/k^2$. Lagrange did not say this, because it is not the kind of result which interested him; he preferred to find the best possible error-estimate. Given Cauchy's definition of a limit, however, the $1/k^2$ error-estimate would be well suited to prove the convergence of this approximation procedure to x. The existence of results like Lagrange's made the rigorous study of convergence of infinite processes much easier.

Although computation of error-estimates and discussions of convergence often use very different techniques, nevertheless a good treatment of either requires a minimal acquaintance with the algebra of inequalities and facility in handling them. Inequality-techniques originated as a mere shorthand way of expressing simple relations of order. But they contain the germ of the rigorous theory of limits—calculation of the requisite delta from complicated expressions when given an epsilon—when their use is exploited. Lagrange used what was known about the algebra of inequalities as a tool in the investigation of closeness of approximation techniques used to solve polynomial equations in algebra. Elsewhere, we shall see how he extended the technique

---

Since z and r are defined to have the same relation to y and q that y and q have to x and p, we have

$$q + 1/r > y \quad \text{and so}$$

$$p + \frac{1}{q + \frac{1}{r}} < p + 1/y = x < p + 1/q, \text{ etc.}$$

[97] In the first case, for instance, we can see that

$$\beta\alpha' - \alpha\beta' = (q\alpha + 1)(1) - p(q\alpha') = qp + 1 - qp = 1.$$

(He gave no argument.)

[98] *Op. cit.*, p. 47. This convergence holds without using the fact that, say, q is the greatest integer less than y. This is needed when he continued, showing that these are the best approximating rational fractions with the given denominators: the proof uses various number-theoretic techniques.

[99] *Introductio*, Section 363.

of error-estimation to the Taylor series. The estimate of error, the Taylor series remainder term, he derived using a Lemma proved by means of these inequality-manipulations.[100]

Lagrange's attempt to see exactly how good approximations were can be seen as part of his passion for exactness. If a series was broken off at a particular point, there had to be, for him, a method of determining the largest possible error committed in taking the sum of that truncated series for the true value of the sum of the infinite series. The same would have to be true for taking a finite portion of any other infinite expression, such as a continued fraction.

The derivation of the remainder term of the Taylor series is an important contribution to the rigorization of analysis, both because of the proof-technique used[101] and because of the result itself. The remainder term in the Lagrange form enables one to calculate the maximum possible error, and so can be used to investigate the convergence of the Taylor series. But Lagrange wanted the error-estimate for its own sake. In this view, we can call the derivation of the remainder term a fitting capstone to Lagrange's work in the theory of approximations, and therefore a major contribution to algebra.

---

[100] See Chapter 4, below.

[101] See Chapter 4, below.

3

The Contents of the Fonctions Analytiques

Introduction

In 1797, Lagrange published the *Théorie des fonctions analytiques*, the expression of his mature views on the nature of the calculus. These views were, as we have seen, the product of a long period of concern with the foundations of the calculus. He had concluded that algebra would provide the only satisfactory foundation; as we have seen, the algebra of the eighteenth century was sufficiently rich to provide a basis for the many diverse results of the *Fonctions analytiques*.

The *Théorie des fonctions analytiques*, in this first edition, is not an especially attractive work, particularly in comparison with the second edition (1813), which is the one reprinted in Lagrange's *Oeuvres*. The first edition has no division into chapters, and the organization seems arbitrary; Lagrange later said that he had written it "comme d'un seul jet, à mesure qu'il s'imprimait."[1] Nevertheless, it was received with enthusiasm and was widely read.[2] As the word of a master analyst, it would of course be of interest to leading mathematicians. Interest was heightened since the problem to which Lagrange addressed himself was so fundamental.

For our purposes, two features of the *Fonctions analytiques* are most worthy of discussion. The foremost is the way in which Lagrange derived the received results of the calculus from his algebraic foundation. I shall undertake an extensive and technical description of some of the details; one reason for this is to convey the flavor of *FA*. In particular, we shall see the extent to which *FA* is not a collection of computations with formal power series, but instead proceeds by means of manipulation with algebraic inequalities—a method which was to influence Cauchy and Weierstrass. The other feature we shall consider is Lagrange's account of the inadequacy of other foundations for the calculus. This appears at first to be a brief history of the calculus, and thus may remind the reader of the historical introduction to the *Mechanique analitique*; but it differs from the earlier venture in its critical tone. In the *Mechanique analitique*, Lagrange had

---

[1] "At one sitting, for the printer." See "Avertissement" to Lagrange's *Théorie des fonctions analytiques* (Second edition, Paris: Courcier, 1813), reprinted in Lagrange's *Oeuvres*, ed. J.-A. Serret, Volume IX (Paris: Gauthier-Villars, 1881), p. 13.

[2] The immediate reaction to it is summarized in S. Dickstein, "Zur Geschichte der Prinzipien der Infinitesimalrechnung. Die Kritiker der 'Théorie des Fonctions Analytiques' von Lagrange," *Abhandlungen zur Geschichte der Mathematik 9* (1899), pp. 65–79, also in *Zeitschrift fur Mathematik und Physik*, supplement to *44* (1899), pp. 65–79. Compare pp. 79–80, below.

been synthesizing the work whose history he gave; in *FA*, he was superseding the older work on the foundations.

Lagrange's Critique of Earlier Methods

In the opening pages of *FA*, Lagrange gave a history and a critique of previous accounts of the foundations of the calculus. This went beyond the Berlin prize-proposal[3] by specifically pointing out the weaknesses of the most widely used foundations.

The use of infinitesimals, though it led to correct results, was not taken seriously by Lagrange. He held that Leibniz, the Bernoullis, and L'Hospital, "content with reaching exact results by procedures of this [differential] calculus in a prompt and sure way . . . did not occupy themselves with demonstrating its principles."[4] Lagrange did not mean that justifications for the calculus, based on infinitesimals, had not been put forward; but he seems no longer to regard such arguments as worthy of being called "demonstrations"—if indeed he ever did so regard them.

Lagrange described the compensation of errors, considered by some as a justification for infinitesimals, much as he had in 1760.[5] Although it might be easy to see what errors are compensated in a particular example—he probably had in mind the fact that a curve is not a polygonal line, but a small finite line is not zero either—"it would be difficult to give a demonstration" that errors are always compensated.[6]

Newton and the English school had worked out a calculus of fluxions, which Lagrange viewed as leading to the same operations as the differential calculus. But the conceptions were different. Newton "considered mathematical quantities as engendered by motion," and the method of fluxions sought "the ratio of the variable velocities with which the quantities are produced." Lagrange recognized that Newton's view had a deceptive plausibility, saying that "everyone has or believes to have an idea of velocity."[7] But Lagrange held that we do not have a clear enough idea of an instantaneous velocity when that velocity is variable.[8] And he had a more fundamental objection to this view. The calculus has only "algebraic quantities" as its object. Velocity is thus, in Lagrange's view, "a foreign idea," and its introduction into the calculus would force us to regard quantities properly algebraic "as lines covered by a moving body."[9]

Even Newton himself, Lagrange added, preferred the method of "last ratios of evanescent quantities" to that of fluxions. Lagrange believed that Newton's first and last ratios, Euler's method of zeros, and the method of limits were fundamentally alike.[10] For instance, he said that the limit-method was only an algebraic translation of the method of last ratios. And he rejected

---

[3] See pp. 26–27, Chapter 1, for the language of the Berlin prize-proposal.

[4] *Théorie des fonctions analytiques* (Paris: Imprimérie de larépublique, an V [1797]), pp. 2–3; second edition reprinted in *Oeuvres* IX, p. 16. Henceforth, the first edition will be cited as *FA* 1797; the second edition reprinted in the *Oeuvres* as *FA Oeuvres* IX.

[5] Compare pp. 20–21, Chapter 1.

[6] *FA* 1797, p. 3; *Oeuvres* IX, p. 17.

[7] *FA* 1797, p. 4; *Oeuvres* IX, p. 17.

[8] *Ibid.* He noted in this connection the difficulties encountered by Colin Maclaurin in his *Treatise of Fluxions* (Edinburgh: T. W. and T. Ruddimans, 1742).

[9] *Ibid.*

[10] For Newton, see *Mathematical Principles of Natural Philosophy*, edited by Florian Cajori (Berkeley and Los Angeles: University of California Press, 1934), Book I, Section I, especially "Scholium" to Lemma XI, pp. 37–39. For Euler, see

the limit-concept as a foundation for the calculus. He felt that this concept was "not clear enough to serve as the principle of a science whose certainty ought to be founded on what is evident."[11] For the limit-method considers the ratios of "evanescent" quantities; thus it considers quantities "in the state in which they cease, so to speak, to be quantities." Though it is easy to conceive of the ratio of two finite quantities, "this ratio no longer offers a clear and precise idea to the mind, when the terms of the ratio become zero simultaneously."[12]

These assertions by Lagrange rest, I believe, on the unspoken assumption that algebra—the science which deals with quantities—tells us nothing whatever about quantities which are vanishing. Therefore, mathematically speaking, we have no knowledge about quantities in that state. We may well have some intuitive idea about what happens to $\frac{f(x+i)-f(x)}{i}$ as i goes to zero, but this would not be enough to satisfy Lagrange. The calculus could not rest on so flimsy a foundation. For Lagrange was not seeking to render the objects of the calculus intuitively plausible, but to give the subject a firm mathematical basis.

These introductory passages in *FA* show that the best-known attempts to give foundations for the calculus failed to meet the criteria already enunciated in the Berlin prize-proposal. None of them was "clear and precise," motion was not a properly mathematical principle but a "foreign idea," compensation of errors could not be rendered general. The only other possibility Lagrange could conceive was an appeal to the algebra of finite quantities. Lagrange's own systematic treatment of the calculus was intended to avoid the disadvantages of the earlier, non-algebraic methods; he tried to do this by basing the calculus on the algebra of power series.

Any later account of the foundations, even one which strove to replace *FA* as did Cauchy's, would have somehow to come to terms with the criticisms Lagrange had raised. For instance, Bernhard Bolzano, giving the first reasonably acceptable proof of the intermediate-value theorem for continuous functions, repeated Lagrange's attack on ideas of motion in the calculus and his call for strictly algebraic methods.[13] Cauchy rendered the limit-concept itself algebraic, thus making the concept Lagrange rejected suitable to fulfill the goal Lagrange had outlined.[14]

The Results of the Calculus: By Means of Formal Power Series

Lagrange believed—as has been said above—[15] that the required clarity could be provided by basing the calculus on the algebra of power series. As we have seen, Lagrange began with the development

$$f(x+i) = f(x) + ip + i^2q + i^3r + \cdots$$

Institutiones Calculi Differentialis (Petropolitanae: Impensis Academiae Imperialis Scientiarum, 1755), in *Opera Omnia*, Series Prima, Volume 10 (Leipzig & Berlin: B. G. Teubner, 1913). For limits, see pp. 82–83, Chapter 4, below.

[11] *FA* 1797, p. 3; *Oeuvres* IX, p. 16.

[12] *FA* 1797, p. 4; *Oeuvres* IX, p. 18.

[13] Bernhard Bolzano, *Rein analytischer Beweis des Lehrsatzes, dass zwischen je zwey Werthen, die ein entgegengesetztes Resultat gewähren, wenigstens eine reele Wurzel der Gleichung liege* (Prag: 1817), reprinted in Ostwalds Klassiker der Exakten Wissenschaften 153 (Leipzig: Wilhelm Engelmann, 1905), pp. 6–7.

[14] Augustin-Louis Cauchy, *Cours d'analyse de l'Ecole Royale Polytechnique. Ire Partie: Analyse algébrique* (Paris: Imprimérie Royale, 1821), *passim*.

[15] Chapter 2, pp. 39–40.

where $p, q, r, \ldots$ are functions of x. He let p be written $f'(x)$, and defined this as "the first derived function of $f(x)$."[16] He then defined $f''(x)$ as the first derived function of $f'(x)$, $f'''(x)$ as the first derived function of $f''(x)$, etc. From these definitions, he proved Taylor's theorem, i.e., that $q = f''(x)/2!, r = f'''(x)/3!, \ldots$.[17]

Lagrange's major purpose was to show that $f'(x)$ so defined had the properties possessed by the differential quotient, $df(x)/dx$. An attempt to prove the equivalence would have involved Lagrange in the metaphysical difficulties he was trying to avoid, since there was no rigorous definition for a quotient of differentials. But obviously, every formula involving $f'(x)$ would suggest immediately an analogous well-known formula involving the differential quotient. The suggestion that $df(x)/dx$ could be identified with the power series coefficient p had already appeared in Lagrange's 1772 paper. *FA* went beyond this by actually deducing from its basic definitions many of the results of the calculus. Such a derivation of the standard rules had been one of the consequences Lagrange had hoped would come from the 1784 prize competition.

The simplest results obtained in *FA* are those involving the derivative of sums, products, quotients... of functions, which are found by addition, multiplication, division, and substitution of series. The derivation of these results is implicit in the 1772 paper; Taylor's theorem, the hardest of them to derive, already appears there. Similar to these is the finding of derivatives of the common functions x^r, a^x, $\log x$, $\sin x$, $\cos x$, from their known power series developments. Lagrange called these common functions "simple analytic functions of one variable," and stated that all other functions of the same variable are composed of these by addition, subtraction, multiplication, and division.[18]

An example of this procedure is in Lagrange's derivation of the rule for finding the derivative of a product of two functions multiplied by a constant.[19] Suppose

$$y = apq,$$

where p, q, are functions of x, a is a constant. When x becomes $x + i$, y becomes $y + y'i + y''/2!\, i^2 + \cdots$

Then apq becomes $a[(p + ip' + \cdots)(q + iq' + \cdots)]$

$$= a[pq + i(p'q + q'p) + \cdots].$$

Therefore, y', which is defined as the coefficient of i, is given by

$$y' = a[p'q + q'p].^{20}$$

[16] *FA* 1797, p. 14; *Oeuvres* IX, p. 32.

[17] Compare p. 23ff, and Note 31, Chapter 1.

[18] *FA* 1797, pp. 15–28; *Oeuvres* IX, Chapter 4, pp. 45–56. The statement about the simple analytic functions is from 1797, p. 28.

[19] *FA* 1797, p. 29; *Oeuvres* IX, p. 40.

[20] A more complicated example is the "chain rule" for the derivative of a function of a function. Suppose p is a function of x, f a function of p. When x becomes $x + i$, $f(p(x))$ becomes $f(p(x + i))$. Then, since

$$f(p(x+i)) = f(p(x)) + [p(x+i) - p(x)]f'(p) + [p(x+i) - p(x)]^2 \frac{f''(p)}{2!} + \cdots$$

(though Lagrange does not bother to write the above step), by rearranging the terms in the series, and using the series for $p(x + i)$,

$$f(p(x+i)) = f(p) + ip'f'(p) + \frac{i^2}{2!}[(p')^2 f''(p) + p''f'(p)] + \cdots$$

The Results of the Calculus: Those Needed to Derive the Remainder Term of the Taylor Series

Much more significant are those results which are not an obvious consequence of the infinite series formulas. Probably chief among these is the remainder term of the Taylor Series, which is not a "received result of the calculus" at all, but one first obtained in *FA*. In addition, there are several results which are used in Lagrange's derivation of the remainder term, all of which involve some form of what has become, since Cauchy, the concept of continuity.

For instance, consider the theorem, already enunciated by Arbogast, that in the infinite series for $f(x+i)$, $f(x)+ip+i^2q+\cdots$, there is an i for which ip (or any other term) will exceed the sum of the rest of the terms of the series. This theorem says in fact that a power series in i is continuous at $i=0$. This theorem was used, in the first edition of *FA*, to prove the Lemma that a function with a positive derivative on an interval is increasing there.[21] This Lemma was then employed to obtain the Lagrange remainder.

Let us examine some of Lagrange's remarks, which lead up to the proof of this theorem, and which show Lagrange's appreciation of the central importance in the calculus of what is, in effect, the modern concept of continuity. Lagrange considered $f(x+i)$ as the sum of two expressions, one of which depends upon i, and the other of which does not. Since, for $i=0$, $f(x+i)$ reduces to $f(x)$, Lagrange wrote

$$f(x+i)=f(x)+iP,$$

where $f(x)$ is the part of $f(x+i)$ independent of i. That the part dependent on i has i to an integral power as a factor, Lagrange concluded from his "proof" that non-integral powers of i in the series development for $f(x+i)$ are impossible.[22] Thus the part depending on i may be written iP. Actually, the quantity iP is defined without reference to the series development for $f(x+i)$;[23] however, Lagrange seems to imply that we only know that such a P exists because we know that $f(x+i)$ has a power series development. He noted both that

$$P=\frac{f(x+i)-f(x)}{i} \quad \text{and that} \quad P=p+iq+i^2r+\cdots.^{24}$$

He did not say, but evidently expected it to be clear, that when $i=0$, P would become both p (i.e., $f'(x)$ by his definition) and $df(x)/dx$. Lagrange similarly defined $Q, R, \ldots$ by the equations $P=p+iQ$, $Q=q+iR, \ldots$ where p and q are the parts of P and Q independent of i, and iQ and iR are the parts dependent on i.

To prove the theorem that $f(x)$ is greater than $ip+i^2q+i^3r\ldots$, p is greater than $iq+i^2r+\cdots$, and so on, Lagrange thus needed only to show that iP was less than $f(x)$, iQ less

---

Thus, if $y=f(p)$,

$$y'=p'f'(p), \text{ the desired rule.}$$

See *FA* 1797, pp. 29–30; *Oeuvres* IX, pp. 41–42.

[21] The proof from *CF* of an improved version of the Lemma is discussed in Chapter 4, below, as a prototype delta-epsilon argument. The use of the Lemma in deriving the remainder term of the Taylor Series is discussed in pp. 71–73, below.

[22] *FA* 1797, pp. 7–8; *Oeuvres* IX, pp. 23–24. See above, Chapter 2, pp. 48–50, for this proof.

[23] Writing an expression such as $f(x+i)$ as the sum of two other expressions, one of which would be independent of i, was a common technique in eighteenth-century algebra. Compare Arbogast's use of this, cited in Chapter 1, p. 29.

[24] *FA* 1797, pp. 8–9, p. 12; *Oeuvres* IX, p. 24, p. 29. As usual, $f(x+i)=f(x)+ip+i^2q+i^3r+\cdots$.

than $p, \ldots$, rather than argue directly on the power series; thus he translated the infinite series into a finite form for the purposes of the proof. Presumably, he meant the absolute value of the quantities in these inequalities, but he did not say so.

Let us consider Lagrange's proof, beginning with the first equation

$$f(x+i) = f(x) + iP.$$

P is in general not zero when i goes to zero (unless $p = 0$). But iP does vanish with i; similarly, so do iQ, $iR, \ldots$—as long as P, Q, $R \ldots$ exist and remain finite. (This sort of saving clause "as long as they remain finite" is typical of Lagrange, whose concern with rigor is exemplified in his distaste for generalizations to which he knows there may be exceptions. His predecessors were usually less meticulous—and of course, often enough, so was he.)

"By considering the curve with i the abscissa, and one of these functions [i.e., iP, $iQ, \ldots$ for fixed x] the ordinate, this curve will cut the axis at the origin of the abscissas." And, Lagrange specified, as long as the point in question $[(x, f(x))]$ is not a singular point, "the course of the curve will necessarily be continuous from this point; thus it will, little by little, approach the axis before cutting it, and approach it, consequently, within a quantity less than any given quantity."[25]

This characterization of the continuity of iP at $i = 0$ is really no more geometric than Bolzano's or Cauchy's,[26] for Lagrange added, "so that we can always find an abscissa i corresponding to an ordinate less than any given quantity; and then all smaller values of i correspond also to ordinates less than the given quantity." This may call a geometric picture to mind, but we can easily translate his statement into algebra: given any $\varepsilon > 0$, we can find an $i_0 > 0$ such that, if $|i| < i_0$, $|iP(x,i)| < \varepsilon$ (for fixed x).

Returning to Lagrange's own words, "one can thus take i small enough" so that iP is less than $f(x)$, or so that iQ is less than p, and so on. So, he concluded, i may always be found small enough so that any particular term of the series $f(x) + ip + i^2q + i^3r + \cdots$ is greater than the sum of all the following terms.[27] He added that choosing the smallest of this (infinite) set of i's will then insure that any term will exceed the sum of all those remaining. In fact, this is not always possible,[28] but Lagrange never needed this form of the result.

Of course, Lagrange did not prove that $P(x,i) = p + iq + i^2r + \cdots$ is a continuous function of i when $i = 0$. He did show what it means, in terms of inequalities, to say that P is continuous; though he did not, as we do, recognize this as the defining property of continuity.

More important, he realized that this property of P is central to the calculus—not only in his own formulation, but also in the differential and fluxional calculus. Lagrange said that the theorem we have just discussed was tacitly assumed in the differential and fluxional calculus, and was fundamental to these subjects: "one assumes it in the differential calculus and the fluxional calculus, and it is because of this that one can say that these calculuses give the greatest exactness [*donnent le plus de prise sur eux*], especially in their application to problems of geometry and mechanics."[29]

---

[25] *FA* 1797, p. 12; *Oeuvres* IX, p. 28.

[26] Bolzano, *op. cit.*, pp. 11–12; Cauchy, *op. cit.*, reprinted in Oeuvres (2), III, p. 43.

[27] *FA* 1797, p. 12; *Oeuvres* IX, pp. 28–29.

[28] See Note 69, Chapter 1, and p. 31.

[29] *FA* 1797, p. 12; *Oeuvres* IX, p. 29.

Lagrange did not elaborate on this point; let us do so. Consider

$$f(x+i) = f(x) + if'(x) + i^2 Q.$$

The differential calculus, of course, assumes $f(x+i) - f(x)$ and i are infinitely small; the fluxional calculus, that the ratio $\frac{f(x+1)-f(x)}{i}$ is considered when numerator and denominator simultaneously vanish. But whatever else "infinitely small" or "vanishing" may mean, they are at least equivalent to i being a very small finite number—small enough so that $|iQ|$ may be made less than any given quantity.[30] Thus, any result which could be proved by means of Lagrange's theorem that $|iQ|$ is less than $|P|$ would clearly apply to the problems considered by the differential and fluxional calculus.

Results of the Calculus: Derivation of the Remainder Term

The "Lagrange Remainder" for the Taylor series is perhaps the best known achievement of *FA*. Lagrange used his remainder term to justify his methods for finding tangents, areas, and volumes, and to apply the calculus to mechanics. The remainder term allowed him to use finite expressions instead of infinite series in all these applications. His use of finite expressions whenever possible suggests that he realized—at least in part—that infinite series could not be treated exactly as finite expressions are. What might be viewed as his naïve, formalistic faith in the "theory of series," characteristic of his early work, gave way in *FA* to a desire to dispense with that theory whenever he could so so.

There are two forms of the Taylor series remainder given in *FA*. One is essentially an integral form, though Lagrange did not write it as an integral.[31] In fact, Lagrange did not usc the integral sign at all in *FA*, except to indicate the equivalence of one of his results with a received result of the differential calculus. Lagrange did avail himself of the convenience of the integral sign in a manuscript version of the *FA* derivation of the remainder term.[32] Thus it is clear that he banished this notation from *FA* for reasons related to his philosophy of the calculus. In Lagrange's view, the integral notation was, I think, part of the Leibnizian framework. It suggested the metaphysic of the differential calculus, since integration was usually defined as the inverse of differentiation (the eighteenth century did not follow Leibniz in defining the integral as a sum). Properties of the *definite* integral, which might have been of use to Lagrange—as we shall see—were not available to him; since he rejected limits, it was left to Cauchy to derive properties of the definite integral from the definition of it as the limit of a sum.

The second form of the remainder term that he gave is that usually called the "Lagrange Remainder," the expression $i^n/n!\, f^{(n)}(x+j)$ when

$$f(x+i) = f(x) + if'(x) + i^2/2!\, f''(x) + \cdots + i^n/n!\, f^{(n)}(x+j)$$

for some j between 0 and i. In *FA*, this form is derived from the integral form.

---

[30] See Chapter 4, pp. 84–88, for Lagrange's application of essentially this procedure in delta-epsilon style proofs.

[31] The explicit integral form was first published by S. F. Lacroix, *Traité du Calcul Différentiel et du Calcul Intégral*, 3 volumes (Paris: J. B. M. Duprat, 1797–1800), Tom. III, entitled *Traité des differences et des séries*, pp. 397–399.

[32] *Bibliotheque de l'Institut* (Paris), MSS 907, "Sur la série pour le développement des fonctions d'une ou plusieurs variables. Differentes maniéres de avoir la série et ses limites," folio pages 19–32 (this is mis-titled in another hand than Lagrange's in addition to the correct title, given above, as "Sur le dèveloppement des fonctions de plusieurs variables"). See especially f. 26, 29, 30–31. See also "Sur les limites des séries," MS 907, ff. 33–45.

The integral form is derived recursively. Lagrange let the function f be expanded about $f(x - xz)$ in powers of xz:

$$f(x) = f(x - xz) + xzf'(x - xz) + \cdots . \tag{3.1}$$

(He was eventually to be concerned chiefly with the case in which $z = 1$; in this case, (3.1) becomes what is now usually called the Maclaurin series.) Lagrange first calculated the first order remainder term, as follows. Suppose only the first term of the series is to be taken; then

$$f(x) = f(x - xz) + xP, \tag{3.2}$$

where P must be determined from the knowledge that, when $z = 0$, P must become zero also.[33] Taking derivatives of both sides of (3.2) with respect to z,

$$0 = -xf'(x - xz) + xP'.$$

Thus, $P' = f'(x - xz)$.

Similarly, if $f(x) = f(x - xz) + xz\, f'(x - xz) + x^2 Q$,

$$Q' = zf''(x - xz).$$

If R is analogously defined, $R' = z^2/2!\, f'''(x - xz) \ldots$[34] Lagrange did not give the general formula for the nth remainder, but was satisfied with the recursive derivation; this derivation would enable anyone to compute the derivative of the nth remainder as $z^{n-1}/(n-1)!\, f^{(n)}(x - xz)$.[35]

Lagrange noted that setting $z = 1$ yields the series (with, for example, the third order remainder)

$$f(x) = f + xf' + x^2/2!\, f'' + x^3 R$$

where R' is defined as above.[36]

---

[33] *FA* 1797, p. 43; *Oeuvres* IX, p. 72. Because the equation (3.2) must remain true on setting $z = 0$, $f(x) = f(x) + xP(x, 0)$. This argument resembles that given above, p. 67. Note, however, that this is not the same function P.

[34] *FA* 1797, p. 44; *Oeuvres* IX, p. 72.

[35] Lacroix, *op. cit.*, p. 397, considered D'Alembert, in *Recherches sur differens points importans du système du monde* (Paris: chez David, 1754), Section II, p. 50ff, Lagrange's predecessor in using this procedure.

D'Alembert considered the function u such that

$$\phi(z + h) = \phi(z) + u.$$

Differentiating both sides with respect to h, treating z as constant,

$$\phi'(z + h)\, dh = du.$$

(D'Alembert of course did not use the prime-notation, but called ϕ', $\Delta(x)$; also, he wrote ξ for what I have called h.) This yields $u = \int \phi'(z + h)\, dh$.

He continued, defining a new function which I shall call u_1 by $u_1 = \phi'(z + h) - \phi'(z + h) - \phi'(z)$, though he had no special notation for it. A repetition of the above procedure yields for the nth remainder an n-fold iterated integral. To express the remainder by a single integration, it is necessary either to follow Lagrange in using an ingenious change of variable, or to employ integration by parts—as is often done in modern texts which derive the integral remainder.

For an excellent study of the history of the remainder term, the reader may consult Alfred Pringsheim, "Zur Geschichte des Taylorschen Lehrsatzes," *Bibliotheca Mathematica*, III Folge, Erster Band (Leipzig: B. G. Teubner, 1900), pp. 433–479.

[36] *FA* 1797, p. 44; *Oeuvres* IX, p. 73.

It may be of interest to compare this with the modern "integral form" of the remainder term. Since $R'(z) = z^2/2!\, f'''(x - xz)$, integration yields

$$x^3 R(z) = x^3 \int_0^z w^2/2!\, f'''(x - xw)\, dw.$$

Setting $z = 1$ gives

$$x^3 R(1) = x^3 \int_0^1 w^2/2!\, f'''(x - xw)\, dw$$

which, by the simple substitution $t = x - xw$, becomes

$$\int_0^x \frac{(x-t)^2}{2!} f'''(t)\, dt.^{37}$$

In the second edition of *FA*, Lagrange added the computation of remainder terms for several particular functions.[38] However, this was merely for concreteness, and need not concern us here. Lagrange was not principally concerned with the exact value of the remainder for the Taylor series of a given function. To compute with finite portions of infinite expressions in the calculus, it would suffice for him to set bounds on the error committed in taking only the first few terms of an infinite series. This is, as already noted, analogous to computing bounds on the errors in algebraic approximations.[39]

To find these bounds for the remainder of the Taylor series, Lagrange stated and proved the Lemma that a function $f(z)$ with a positive derivative on an interval $[a, b]$ is increasing there.[40] The statement and proof of this lemma was altered in Lagrange's *Leçons sur le Calcul des Fonctions.* In this improved version, it was indirectly to influence Cauchy in the development of the proof-techniques to go with his rigorous definition of the derivative.[41] Accordingly, the proof from the *Calcul des Fonctions* and the way in which it influenced Cauchy will be treated at length in Chapter 4. The proof of the Lemma in the *FA* version need not be treated here; it is given in the Appendix.

Having proved the Lemma, Lagrange was able to find the bounds on the Taylor series remainder. This derivation is close to a standard modern argument, deriving the Lagrange remainder from the integral form by means of what is essentially the theorem: Let Z be a function of z defined on $[a, b]$ where $a \geq 0$, with maximum value M and minimum N. Then

$$N \int_a^b z^m\, dx \leq \int_a^b z^m Z(z)\, dz \leq M \int_a^b z^m\, dz.$$

Lagrange did not use the definite integral. His version of this theorem was the following:

---

[37] See, *e.g.*, George B. Thomas, *Calculus and Analytic Geometry* (Reading, Mass.: Addison-Wesley, 1960), pp. 790–791.

[38] *Oeuvres* IX, pp. 74–75.

[39] See Chapter 2, above.

[40] "If a prime function of z, such as $f'(z)$, is always positive for all values of z from $z = a$ to $z = b$, b being greater than a, the difference of the primitive functions corresponding to these two values of z, that is $f(b) - f(a)$, will necessarily be a positive quantity." Note that z is a real number; the notation $f'(x)$ is used instead in the second edition. *FA* 1797, p. 45; *Oeuvres* IX, p. 78. It is not clear whether the interval is regarded as open or closed.

[41] See Chapter 4, below.

If $F'(z) = z^m Z(z)$ between $z = a$ and $z = b$, and if M and N are maximum and minimum values of Z there,

$$F(a) + N\left(\frac{b^{m+1} - a^{m+1}}{m+1}\right) \leq F(b) \leq F(a) + M\left(\frac{b^{m+1} - a^{m+1}}{m+1}\right).$$

He needed for this theorem, though he did not say, that a and b are greater than or equal to zero. This is no problem for the application he made, where $a = 0$ and $b = 1$.[42] He proved the theorem by applying the Lemma to the function f defined by the condition

$$f'(z) = z^m(M - Z). \tag{3.3}$$

By the Lemma, since $f'(z)$ is positive between a and b,[43] $f(b) - f(a)$ is positive also. (That is, $\int_a^b z^m(M - Z)\,dz > 0$.) Integrating (3.3)—or, in Lagrange's language, taking the "primitive functions"—

$$f(z) = Mz^{m+1}/m + 1 - F(z)$$

and therefore $f(b) - f(a) = Mb^{m+1}/m + 1 - F(b) - Ma^{m+1}/m + 1 + F(a) > 0$; thus, $F(b) < F(a) + \frac{M(b^{m+1} - a^{m+1})}{m+1}$.

Similarly, applying the Lemma to another function f, defined by

$$f'(z) = z^m(Z - N).$$

$$F(b) > F(a) + \frac{N(b^{m+1} - a^{m+1})}{m+1}.$$ [44]

Thus, the theorem is proved.

Returning now to the recursively computed remainder term, Lagrange recalled that

$$\begin{aligned} f(x) &= f(x - xz) + xP \\ &= f(x - xz) + xz\, f'(x - xz) + x^2 Q \\ &= f(x - xz) + xz\, f'(x - xz) + \frac{1}{2}(xz)^2 f''(x - xz) + x^3 R, \end{aligned}$$

and so on. Lagrange derived bounds for P, Q, and R. We will illustrate this by giving his procedure for Q.

$$f(x) = f(x - xz) + xz\, f'(x - xz) + x^2 Q.$$

Set $Q = F(z)$. Then $Q' = F'(z) = zf''(x - xz)$.[45]

If, in the equation $F'(z) = z^m Z$, m is set equal to 1,

$$Z = f''(x - xz).$$

---

[42] Lagrange used ">" though he meant "≥." I use "≥" for clarity; Lagrange had no notation to express the difference between strict and weak inequality.

[43] Actually, f' so defined could be zero, and *is* zero at the maximum, the minimum, and at $z = 0$. It should be noted that the proof of Lagrange's Lemma assumes f' to be strictly positive. See p. 87, Chapter 4, for further remarks on the confusion between strict and weak inequalities.

[44] *FA* 1797, p. 47; *Oeuvres* IX, p. 81.

[45] Compare p. 70, above. *FA* 1797, p. 44; *Oeuvres* IX, p. 72.

Now Lagrange set $a = 0$, $b = 1$. Since Q is zero when $z = 0$, $F(a) = 0$ and, when $z = 1$, $Q = F(b)$.

Letting M_1 and N_1 be the maximum and minimum values of $f''(x - xz)$ for z between 0 and 1, the above theorem gives

$$N_1/2 \leq F(b) \leq M_1/2$$

and thus $M_1/2$ and $N_1/2$ are the bounds on Q, when z ranges between 0 and 1. Similarly, M and N are bounds on P, and $M_2/3$ and $N_2/3$ are bounds on R, where M and N are maximum and minimum of $f'(x - xz)$, M_2 and N_2 of $1/2\ f'''(x - xz)$ over the same interval.

Lagrange now set $x - xz = u$. Then u ranges between 0 and x as z ranges between 0 and 1. Thus, $f'(u)$, $f''(u)$, $f'''(u)/2$, will range over all values, including those for which $f'(u)$ will be M or N, $f''(u)$ will be M_1 or N_1, $f'''(u)/2$ will be M_2 or N_2. There is some value for u between 0 and x for which $f'(u)$ will be exactly equal to P, some value for which $f''(u)/2!$ will equal Q, some for which $f'''(u)/3! = R$. This follows from the intermediate-value property of continuous functions.[46] Thus,

$$\begin{aligned} f(x) &= f(0) + xf'(u) \\ &= f(0) + xf'(0) + x^2/2!\ f''(u) \\ &= f(0) + xf'(0) + x^2/2!\ f''(0) + x^3/3!\ f'''(u), \text{ etc.,} \end{aligned}$$

where u is some quantity between 0 and x. (The u's are clearly not the same in the different equations.) Or, by the transformation taking x into $x + z$,

$$\begin{aligned} f(z + x) &= f(z) + xf'(z + u) \\ &= f(z) + xf'(z) + x^2/2!\ f''(z + u) \\ &= f(z) + xf'(z) + x^2/2!\ f''(z) + x^3/3!\ f'''(z + u), \text{ etc.} \end{aligned}$$

Again, u is between 0 and x; of course the u's in the different lines are not the same, though Lagrange did not say this. These are what we now call the Taylor series with Lagrange remainder;[47] it is in these forms that Lagrange applied the remainder term to the calculus.

There are today two standard methods of deriving the Lagrange remainder. One is by way of the integral remainder, and resembles the method just discussed.[48] A form of the other method, which views the remainder term as an extension of the Mean Value Theorem, first appeared in Lagrange's *Leçons sur le Calcul des Fonctions.*[49] Thus Lagrange himself discovered the two major methods of finding the remainder term which bears his name.

It should be remarked that Lagrange's derivation of the remainder term in *FA* does not employ the assumption that $f(x + i)$ has an infinite series development in powers of i. The derivation is generally acceptable by modern standards, though it does contain three gaps. We have noted that it might be more convenient to use definite integrals, but this would be a simplification, not a correction.

---

[46] Lagrange stated this, but apparently considered it too obvious to prove. Cauchy and Bolzano did not agree; see Chapter 4, below.

[47] *FA* 1797, P. 49; *Oeuvres* IX, p. 84.

[48] Thomas, *op. cit.*, pp. 794–796.

[49] Discussed at length in Chapter 4, below. Compare Thomas, *op. cit.*, pp. 153–154.

The first of the deficiencies is that his proof that a function f with positive derivative on an interval satisfies $f(b) > f(a)$ is not completely valid; furthermore, the result he *used* is that a function f with a non-negative derivative on an interval $[a, b]$ satisfies $f(b) \geq f(a)$. Second, he assumed that a continuous function has a maximum and minimum on a closed interval, and that it has the intermediate-value property. Finally, he assumed that if $g(z)$ is given, $F(z)$ can always be found uniquely (except for a constant) from $F'(z) = g(z)$.[50] The results needed to fill these lacunae involve the basic properties of the real numbers; thus, acceptable proofs were not given until the work of Bolzano, Cauchy, and Weierstrass.

Results of the Calculus: Application of the Remainder Term

Lagrange viewed his remainders as explicit expressions for P, Q, $R, \ldots$ in the formulas

$$\begin{aligned} f(x+i) &= f(x) + ip \\ &= f(x) + if'(x) + i^2 Q \\ &= f(x) + if'(x) + i^2/2!\, f''(x) + i^3 R, \ldots. \end{aligned}$$

Q, for instance, is $f''(x + j)/2!$, where j is some number between 0 and i. Thus

$$f(x+i) = f(x) + if'(x) + i^2/2!\, f''(x+j). \tag{3.4}$$

In this finite formula, it is clear that $i_0 > 0$ can be chosen sufficiently small so that

$$|if'(x)| > |i^2/2!\, f''(x+j)| \quad \text{for} \quad |i| < i_0, \tag{3.5}$$

i.e., that $|if'(x)| > |i^2 Q|$.[51] By the same procedure,

$$|i^2/2!\, f''(x)| > |i^3/3!\, f'''(x+j)| = |i^3 R| \tag{3.6}$$

for $|i|$ sufficiently small, and similarly for higher order remainders. It is formulas like (3.5), (3.6), that Lagrange used in applying the remainder term to the calculus.[52]

(3.4) and (3.5) together, and the corresponding results for f'', f''', ... are the basis for Lagrange's derivation of many of the results of the calculus. Several of his derivations are, as we shall see, perfectly valid if we substitute for (3.4) and (3.5) the Cauchy or Weierstrass definition of the derivative of $f(x)$ as the function $f'(x)$ which satisfies

$$f(x+i) = f(x) + if'(x) + iV \quad \text{where} \quad \lim_{i \to 0} V = 0. \tag{3.6a}$$

But, of course, $f''(x)$ need not exist.

It should be emphasized that the Lagrange remainder is still the most useful tool in those applications of the calculus in which only derivatives of order higher than two may be neglected.

---

[50] Which of course follows from the fundamental theorem of calculus: for instance, $F(z) = \int_a^z g(t)\,dt$. The Mean Value Theorem is commonly used to prove that, once the constant is specified, the function is unique.

[51] The i_0 notation is mine; he merely said "i." Lagrange sometimes cited as a reason for this step his "theorem" that i can be chosen so that $|if'(x)| > |i^2/2!\, f''(x) + i^3/3!\, f'''(x) + \cdots|$; but it is immediate from (3.5) that taking $|i_0| < \frac{2|\min f'(x)|}{|\max f''(x+j)|}$ will suffice. Fortunately, i_0 can be chosen independently of x as long as $x + i$ ranges over a closed interval—fortunately, because Lagrange never conceived of his i_0 as depending on x.

[52] Formula (3.6), for instance, is the inequality which means that the second order remainder, $i^2/2!\, f''(x+j)$, is continuous at $i = 0$—or uniformly continuous, since i_0 can be chosen independently of x.

Thus, the Lagrange remainder is used both in *FA* and in modern texts to discuss sufficient conditions for maxima and minima, or second and higher orders of contact between curves.[53]

The example of computing maxima and minima of a function $f(x)$ illustrates Lagrange's technique. If the function $f(x)$ has a maximum at x, for small values of i "positive or negative," $f(x) > f(x+i)$; thus $f(x+i) - f(x) < 0$.[54]

But, since $f(x+i) - f(x) = if'(x) + i^2/2!\, f''(x+j)$, an equivalent condition for $f(x)$ to be a maximum at x is

$$if'(x) + i^2/2!\, f''(x+j) < 0. \tag{3.7}$$

But i can be taken small enough so that, if $f'(x) \neq 0$, "the absolute value of the term $if'(x)$ will be greater than that of the term $i^2/2!\, f''(x+j)$."[55] If i is taken that small, the expression $if'(x) + i^2/2!\, f''(x)$ will have the same sign as $if'(x)$, and will change signs when i does. Thus, it will be impossible for (3.7) to be satisfied "unless $f'(x) = 0$." Consequently, for $f(x)$ to be a maximum, $f'(x) = 0$.[56] Similarly, Lagrange showed that $f'(x) = 0$ at a relative minimum.

When $f'(x) = 0$, Lagrange used an analogous procedure to see whether $f(x)$ is a maximum or a minimum at that point. For, if $f'(x) = 0$, Lagrange said that

$$f(x+i) - f(x) = i^2/2!\, f''(x) + i^3/3!\, f'''(x+j)$$

for some j between 0 and i. Now i can be chosen small enough so that $|i^2/2!\, f''(x)| > |i^3/3!\, f'''(x+j)|$.

Then $i^2/2!\, f''(x) + i^3/3!\, f'''(x+j)$ will have the sign of $f''(x)$. Hence, if $f''(x) < 0$, $f(x+i) - f(x) < 0$ also and $f(x)$ is a maximum; if $f''(x) > 0$, $f(x)$ is a minimum.[57]

Lagrange went on to consider the case in which $f''(x) = 0$ also. He showed by the same procedure that $f(x)$ is then neither a maximum nor a minimum unless $f'''(x) = 0$ as well; in this case, $f^{iv}(x) < 0$ for a maximum, $f^{iv}(x) > 0$ for a minimum, and so on.[58]

The application of the theory of analytic functions to geometry depends on these same techniques. For instance, let us consider in some detail Lagrange's treatment of tangents. A straight line is tangent to a curve, according to Lagrange, if it has a common point with that curve, and if no other straight line lies between that curve and the line.[59] He began his discussion by considering under what conditions a third curve may be interposed between two intersecting curves. Suppose that f and F represent the two curves, and that for a fixed x, $F(x) = f(x)$, giving the two a point in common. Near this point, the difference between the ordinates (values

---

[53] For a modern treatment, see D. V. Widder, *Advanced Calculus* (New York: Prentice-Hall, 1947), pp. 76–77; p. 99.

[54] *FA* 1797, p. 152; *Oeuvres* IX, p. 233.

[55] *FA* 1797, p. 152; *Oeuvres* IX, p. 234. This is (3.5), from above. Yes, he says '*valeur absolue*'.

[56] *Ibid.*

[57] *FA* 1797, p. 153; *Oeuvres* IX, p. 235.

[58] *FA* 1797, pp. 153–154; *Oeuvres* IX, pp. 235–236.

[59] *FA* 1797, p. 117; *Oeuvres* IX, p. 183. This, he said, is the way the ancient geometers considered tangents. He did not specify which ancient geometers he meant; Euclid defined a tangent to a circle as a line which met it, but did not cut it. The "no other line may be interposed" property was then proved; see Euclid, *Elements*, III, Def. 2 and Prop. 16. Archimedes sometimes spoke of tangents as "touching," sometimes as "touching without cutting," which is like Euclid's definition; see Thomas L. Heath, *The Works of Archimedes* (New York: Dover, n.d.), p. clxxiii. Lagrange was not the first to use the definition he gave, though; see Colin Maclaurin, *Treatise of Fluxions*. Volume I, p. 179. See also the discussion in Chapter 1, p. 32, above.

of y if the curve is graphed according to $y = f(x)$, $y = F(x)$) is $f(x+i) - F(x+i)$. Lagrange then postulated a third curve ϕ passing through the same point, so that $\phi(x) = f(x) = F(x)$. He set

$$D = f(x+i) - F(x+i)$$
$$\Delta = f(x+i) - \phi(x+i).$$

For ϕ to pass between F and f, he said that for small i we must have "the value of D surpass that of Δ, abstraction being made of the signs."[60] (He presumably meant that $D \geq \Delta \geq 0$ or $D \leq \Delta \leq 0$, and *not* the weaker $|D| \geq |\Delta|$.) By using Taylor series with Lagrange remainders for F, f, and ϕ, and then subtracting, Lagrange obtained

$$D = i[f'(x) - F'(x)] + i^2/2!\,[f''(x+j_f) - F''(x+j_F)]$$

$$\Delta = i[f'(x) - \phi'(x)] + i^2/2!\,[f''(x+j_f) - \phi''(x+j_\phi)].$$[61]

Now he supposed that $f'(x) = F'(x)$. Then

$$D = i^2/2!\,[f''(x+j_f) - F''(x+j_F)].$$

Thus, $|i|$ can be taken small enough so that $|\Delta| > |D|$ if $\phi'(x) \neq f'(x)$. All that is necessary, said Lagrange, is to choose i so that

$$|f'(x) - \phi'(x)| > |(i/2)(\phi''(x+j_\phi) - F''(x+j_F))|.$$[62]

Lagrange's procedure has shown that if $f'(x) = F'(x)$, ϕ cannot lie between f and F near x unless $f'(x) = \phi'(x)$. After a discussion of the analogous results when $f''(x) = F''(x)$, etc., Lagrange considered the case of a straight line tangent to a curve. He let F be the line $F(x) = a + bx$, and $f(x)$ be the curve. Since the tangent was defined by Lagrange by the property that no curve lies between the curve f and its tangent line F, what was done above gives $F'(x) = f'(x)$ at the point of tangency.[63] He easily showed algebraically that there is only one such line, since after b is determined by $F'(x) = f'(x) = b$, a is determined by the value of f at x, the point of tangency.

The last major application I shall consider is that of areas under curves. Suppose the area between x and $x+i$ under the curve $y = f(x)$ is expressed by $F(x+i) - F(x)$, while the total area under the curve up to $(x, f(x))$ is given by $F(x)$.[64] Lagrange explicitly stated that, in this procedure, he considered f as either increasing or decreasing between x and $x+i$. Lagrange characteristically did not draw a diagram, but said that even without one it is evident that the area sought lies between the quantities $i(f(x))$ and $i(f(x+i))$.[65] This area is $F(x+i) - F(x)$. But

---

[60] *FA* 1797, p. 119; *Oeuvres* IX, p. 186.

[61] Where j_f, j_F, j_ϕ, are my notation, used without apology hereinafter. Lagrange used j for all, but specifically said that the j's are not the same.

[62] *FA* 1797, p. 120; *Oeuvres* IX, p. 187. Actually he would need $|i| < |i_0|$ when $|f'(x) - \phi'(x)| > 1/2|(i_0)|| \max_{[x,x+i]} (\phi'' - F'')|$. As long as ϕ'' and F'' are bounded on the interval, such i_0 can be found.

[63] *FA* 1797, p. 122; *Oeuvres* IX, p. 190.

[64] *FA* 1797, p. 155; *Oeuvres* IX, p. 238.

[65] *Ibid.* I have supplied the diagram, illustrating the case where the function f is increasing (for definiteness).

$F(x+i) - F(x)$, by Taylor's theorem with Lagrange remainder, yields

$$F(x+i) - F(x) = iF'(x) + i^2/2!\, F''(x + j_F),$$

and $f(x+i) = f(x) + if'(x + j_f)$.[66]

Thus, $iF'(x) + i^2/2!\, F''(x + j_F)$ is between $if(x)$ and $if(x) + i^2 f'(x + j_f)$. Subtracting $if(x)$ from each of these two expressions,

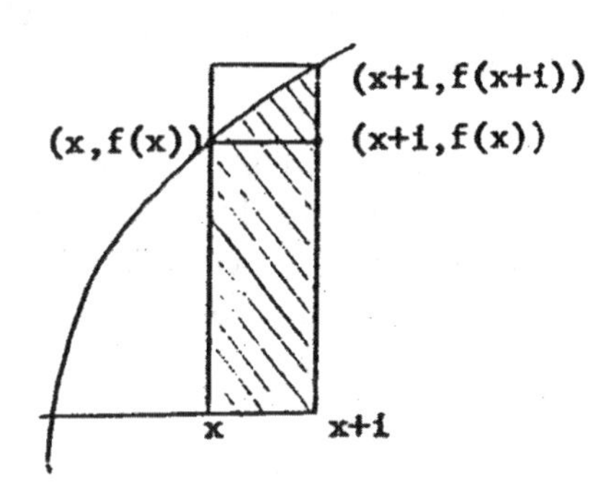

$$|i[F'(x) - f(x)] + (i^2/2!)(F''(x + j_F))| < i^2 |f'(x + j_f)|. \tag{3.8}$$

For small values of i, this is only possible if the term $i[F'(x) - f(x)]$ disappears; otherwise, Lagrange said, if

$$|i| < \left| \frac{F'(x) - f(x)}{f'(x + j_f) - 1/2F''(x + j_F)} \right|,$$

the inequality (3.8) would be false. Therefore he concluded

$$F'(x) = f(x).^{67}$$

Thus, from the definition of $F(x)$ as the area under the curve $f(x)$ from some initial value of x up to x, Lagrange derived the relation that $F(x)$ is "nothing but the primitive function of $f(x)$."[68] The computations of area stemming from this definition of course take the form of finding $F(x)$ such that $F'(x) = f(x)$.

Lagrange's procedure suffers chiefly from having no definition of area, and from assuming that the function $F(x)$ has both a first and a second derivative—if not an infinite Taylor series. His procedure was rejected by Cauchy in favor of defining the integral as the limit of a sum. This alternative was not open to Lagrange, since he rejected the limit-concept.

Lagrange used similar procedures to find arc lengths, surfaces of rotation, volumes, and so on. For instance, he computed the arc length Φ of a curve by treating $\Phi(x+i) - \Phi(x)$ as the length of the arc between x and $x+i$.[69] Again, he obtained the well-known results of the calculus.

---

[66] Again, Lagrange wrote merely j for quantities he recognized as distinct. *FA* 1797, p. 155; *Oeuvres* IX, p. 239.

[67] *FA* 1797, p. 156; *Oeuvres* IX, p. 239. Lagrange did not use absolute value notation, but said "abstraction being made of the sign."

[68] *FA* 1797, p. 156; *Oeuvres* IX, p. 239.

[69] *FA* 1797, p. 157ff; *Oeuvres* IX, p. 241ff.

Finally, Lagrange's consideration of mechanics "as a geometry of four dimensions,"[70] that is, in terms of functions of x, y, z, and t, enabled him to apply the ideas of *FA* to mechanics as well.[71] The rigorous basis for an analytical mechanics was thus the theory of analytic functions, though he had used the differential calculus in the *Mechanique Analitique*.

The wealth of applications to geometry and mechanics was in part responsible for the great vogue of *FA* in Britain and on the Continent.[72] But for our purposes the most striking feature of Lagrange's work is his deep interest in showing that everything which could be done by means of the differential calculus could be deduced from his foundation, and deduced with greater attention to rigorous considerations than ever before. The remainder term was a means to this end, though it later became an important result in its own right. The examples we have surveyed are only a few, but the most essential.[73] Flowing naturally from Lagrange's attachment to algebra as a model of mathematical rigor, the power series foundation for the calculus appears as a fitting climax to the results obtained in the calculus by Newton, Leibniz, the Bernoullis, and Euler; this was in fact the judgment of the eighteenth century.

Refinements of Lagrange's Ideas: 1799–1813

Lagrange continued after 1797 to teach the calculus at the Ecole Polytechnique. In its *Journal* in July, 1799, he explained what his goal would be in the course of lectures planned for the next academic year.[74] He still held that the purpose of the theory of analytic functions was to make the differential calculus a part of algebra. But in the lectures, he planned to give more detail than in his published work, *FA* of 1797.

The course described in the "Discours" was given in 1799 and published as the *Leçons sur le Calcul des Fonctions*. *CF* contains only the basic theory of analytic functions, not their applications to geometry and mechanics. This emphasis, and the greater length of treatment, facilitated a more careful and detailed presentation of several subjects in *CF*. But there is no difference in principle between the two published works. Lagrange himself described *CF* as "a commentary and supplement" to Part I of *FA*.[75]

The re-naming of the subject "calculus of functions," I believe, was intended to emphasize again the central role the function-concept played in Lagrange's view of the calculus. By retaining the old word "calculus" Lagrange was making absolutely certain that nobody would think of his theory of analytic functions as "also a calculus of differentials" or as "also a calculus of fluxions";

---

[70] *FA* 1797, p. 223; *Oeuvres* IX, p. 337.

[71] This is the subject of the second subdivision of Part II of the 1797 (first) edition of the *FA*. Part I is "Exposition of the theory, with its principal uses in analysis," II is "Application of the theory to geometry and mechanics." The second (1813) edition has three parts, the third of which deals with mechanics; thus, Part I is as in the first edition, Part II deals with applications to geometry.

[72] See pp. 79–80 below.

[73] Omitted, but worth mentioning, are the survey of techniques for solving differential equations, the application to the calculus of variations, the analysis of "singular points" at which there is no Taylor series (e.g., $f(x) = \sqrt{a^2 - x^2}$ for $x = a$), the treatment of functions of several variables.

[74] "Discours sur l'objet de la théorie des fonctions analytiques," *Journ. Ecole Polytechnique*, VIe Cah. t. II, Thermidor an VII (July, 1799); reprinted in *Oeuvres* VII, pp. 323–328.

[75] Joseph-Louis Lagrange, "Sur le Calcul des Fonctions," *Séances des Ecoles Normales*, Tome 10e, an IX (1801) pp. 5–6; "Avertissement" in second edition, *Leçons sur le calcul des fonctions* (Paris: Courcier, 1806) reprinted in *Oeuvres* X, p. 13.

the subject is a calculus of *functions*. Its central aim is to find functions, such as $f'(x)$, from other functions, and to apply the relations between these functions to various problems of analysis, mechanics, and geometry.[76]

The second edition of *FA* (1813) shows little change in the essential features from the first edition of 1797. The second edition is more attractively organized; it is divided into chapters, and references are made to relevant portions of *CF* for clarifications of some points. There are new sections on surface areas and volumes of solid figures, and on the many-body problem. Other new sections point out explicitly how some of Lagrange's results are equivalent to those of the differential calculus; for instance, after showing that the area under a curve $y = f(x)$ is given by a function $F(x)$ such that $F'(x) = f(x)$, Lagrange remarked that if the area was called u, $y = u' = du/dx$, so that $du = y\,dx$ and $u = \int y\,dx$, "the known formula for the quadrature of curves."[77]

Lagrange's foundations were widely hailed.[78] Whether they became familiar to a mathematician through attendance at Lagrange's courses, through reading *FA* of 1797, *FA* of 1813, *CF* of 1799–1800, or *CF* of 1806, made little difference. The same doctrine would be found in all.

Impact of the Fonctions Analytiques

Following the publication of *FA*, many works appeared which explained the calculus on the basis of general analytic expressions—usually power series.[79] Perhaps the most influential was the encyclopedic textbook of Silvestre François Lacroix, which based the calculus on the principles of *FA*.[80] The power series foundation was highly praised, by mathematicians[81] and by philosophers.[82] The remainder term of the Taylor series was widely used—for instance, by Lacroix[83] and by Pierre Simon Laplace.[84]

FA's impact on mathematics in Great Britain was of great importance. *FA* was received with favor and enthusiasm.[85] The Cambridge Analytical Society considered Lagrange's method the epitome of Continental analysis. They reformed the teaching of mathematics at Cambridge using the Lagrangian approach, chiefly through their extensively annotated translation of S. F. Lacroix's

[76] The name "calculus of functions" was later adopted by the Cambridge analysts, notably Charles Babbage, for the subject of functional equations (i.e., where equations like $f^2(x) = f(x)$ were solved for f).

[77] *Oeuvres* IX, p. 241.

[78] See immediately below.

[79] See Dickstein, *op. cit.*, for an extensive list; compare Vivanti, *op. cit.* We have already mentioned L. F. A. Arbogast, *Du Calcul des Derivations* (Strasbourg: Levrault, Frères, An. VIII, 1800). See also F. J. Servois, "Essai sur un nouveau mode d'exposition des principes du calcul différentiel," *Annales de Mathématiques*, V (1814–1815), pp. 93–141; and August Leopold Crelle, *Versuch einer allgemeinen Theorie der analytischen Facultäten* (Berlin: G. Reimer, 1823), especially pp. 85–101. A more complete list will be found in the Bibliography, pp. 108–109.

[80] *Traité du calcul différentiel et du calcul intégral.* 3 volumes. (Paris: J. B. M. Duprat, 1797–1800).

[81] E.g., by J. B. Delambre, *Rapport historique sur les progtés des sciences mathématiques depuis 1789, et sur leur état actuel* (Paris: de l'imprimérie impériale, 1810). See also Bibliography.

[82] August Comte, In *Cours de philosophie positive* (Paris: Bachelier, 1830–42), praised Lagrange's method highly in Tom.I, Leçon 6e. This leçon ends by calling Lagrange's method "la plus rationnelle et la plus philosophique de toutes."

[83] *Traité du calcul...*, Volume III, pp. 397–399.

[84] Laplace, *Théorie analytique des probabilités* (3d edition, Paris: Courcier, 1820), reprinted in *Oeuvres complètes de Laplace*, Tome 7e (Paris: Gauthier-Villars, 1886), p. 179.

[85] See the review in *Monthly Review*, London, N. S., XXVIII (1799), Appendix, pp. 481–499, especially pp. 492–499.

Traite élémentaire.[86] Thus, it was the method of *FA* which brought Continental analysis into England. In Scotland, Lagrange's approach was recommended to students by William Wallace, professor of mathematics at the University of Edinburgh from 1819 to 1838.[87]

The praise for *FA* in the works mentioned above stressed the belief that the calculus should be reduced to algebra, and that unclear concepts like "limit" and "infinitesimal" should be avoided. The widespread acceptance of the Lagrangian point of view paved the way for the eventual reduction, by Cauchy and Weierstrass, of the calculus to algebra.

Conclusion

The mathematical objections which can be raised against Lagrange's foundation for the calculus are many: in particular, the series expansion for $f(x+i)$ in powers of i may not exist, and the concept of convergence is virtually absent from his books. But Lagrange's theory is open only to mathematical objections—as the foundations for the calculus given by others were not. Lagrange cannot be accused of being vague, or of substituting words for algebra. Nor did he relegate his definition of $f'(x)$ to an introductory section, never to be seen again, so that he could get on with the business of finding results. His attempt set a standard which would have to be satisfied before any further improvements could be made. By demolishing with philosophical criticisms all prior attempts to found the calculus, and by teaching and convincing so many people, he left Cauchy only one competitive theory to destroy and surpass—Lagrange's theory of analytic functions.

Cauchy's own foundation for the calculus could almost be described in Lagrange's words: the principles of the differential calculus were "freed from any consideration of infinitely small quantities, of evanescent quantities, . . . of fluxions," and although "limits" were used as a basic concept by Cauchy, his principles were indeed "reduced to the algebraic analysis of finite quantities."[88] Lagrange's phrase "analyse algébrique" reappears in the title of Cauchy's great work of 1821, *Cours d'analyse de l'Ecole Royale Polytechnique. Ire Partie: Analyse algébrique*. The use of this phrase symbolizes Lagrange's role as a forerunner of nineteenth century analysis. It is now time to consider Lagrange in that role, and to see by what steps his proof-methods contributed to the establishment of Weierstrassian rigor.

---

[86] Silvestre François Lacroix, *Traité élémentaire de calcul différentiel et de calcul intégral*, second edition, (Paris: Courcier, 1806); the translation was *An Elementary Treatise on the Differential and Integral Calculus* (Cambridge: J. Smith for J. Deighton *et al*, 1816); translation, appendix, and notes, by Charles Babbage, George Peacock, and J. F. W. Herschel.

[87] Wallace recommended *FA* to Mary Somerville, according to Martha Somerville, *Personal Recollections, from Early Life to Old Age, of Mary Somerville* (Boston: Roberts Brothers, 1874), p. 79.

[88] From the full title of *FA*: "Théorie des fonctions analytiques, contenant les principes du calcul différentiel, dégagés de toute considération d'infiniment petits ou d'évanouissans, de limites ou de fluxions et réduits à l'analyse algébrique des quantités finies."

4

From Proof-Technique to Definition: The Pre-History of Delta-Epsilon Methods

Introduction

The nineteenth century is often called "the Age of Rigor in Analysis." This name can be understood as "the age of delta and epsilon." In defining limits, in proving theorems, in exemplifying clear thinking, the rigor known as "Weierstrassian" represents to us the greatest achievement of the nineteenth-century analysts.

The development of these principles from the less adequate foundations for the calculus of the eighteenth century illustrates a trend common to many branches of nineteenth century mathematics, a trend toward increasing generality and abstraction. In the area of the foundations of analysis, this trend can be viewed as a retreating of the frontier of the "obvious." For instance, Lagrange believed that the principles of the algebra of finite polynomials could be extended to the algebra of infinite series; André-Marie Ampère rejected this, but, like Lagrange, accepted the intermediate-value property as a self-evident attribute of continuous functions; Augustin-Louis Cauchy and Bernhard Bolzano realized that the intermediate-value property had to be proved, but implicitly assumed in their proofs the convergence of bounded monotone sequences.[1] Again, after Richard Dedekind had worked out a rigorous theory of the irrational, while Karl Weierstrass had completed the task of reducing the concepts of the calculus to those of arithmetic, the search for the foundations of arithmetic itself was begun.

The working out of the principles and techniques of delta-epsilon methods lies at the heart of the rigorization of analysis. The essential principles of these rigorous considerations, as understood from the time of Cauchy, are the following:

(1) Every statement about limits or other concepts of the calculus is either immediately expressed in the algebraic language of finite quantities (as in the work of Weierstrass), or at least translatable into algebraic language for use in proofs (as in the work of Cauchy). In fact, every equation involving limits is reducible to a set of inequalities.

---

[1] See Note 48 for Bolzano, 60 for Cauchy. A monotonic sequence converges to a limit if and only if it is bounded. On all points regarding the properties of convergence discussed in this chapter, the reader may consult Tom M. Apostol, *Mathematical Analysis* (Reading, Mass.: Addison-Wesley, 1957).

(2) The definition of "limit" must be broad enough to allow the basic concepts of the calculus—derivative, integral, sum of a series—to be defined as limits, and broad enough to allow the proof of as many of the received results of analysis as possible. In particular, a function should be allowed to oscillate about its limit.

The concepts of "limit" employed in the definitions of the concepts of the calculus in the seventeenth and eighteenth centuries were always given verbally. Even if translated into algebra, they would not have corresponded to all that the modern definitions of "limit," designed to provide a firm basis for the calculus, require.

These drawbacks did not appear to disturb most of the men of the eighteenth century. Little use was really made of the "foundations." They were explanations, useful analogies, plausibility-arguments—not an essential part of the calculus. The differential calculus, the integral calculus, the fluxional calculus, were used as algorithms; the foundations might make the mathematician more comfortable with ideas like 0/0, but he could use the calculus to solve most problems that interested him without any need for firm foundations.

Jean-le-Rond D'Alembert's definition of limit is typical of eighteenth century treatments of this concept.[2] It is worth citing here, because the treatment was meant to be systematic:

> One magnitude is said to be the *limit* of another magnitude, when the second can approach nearer to the first than a given magnitude, as small as that [given] magnitude can be supposed; nevertheless, without the magnitude which is approaching ever being able to surpass the magnitude which it approaches; so that the difference between a quantity and its limit is absolutely inassignable.... Properly speaking, the limit never coincides, or never becomes equal, to the quantity of which it is a limit; but it can always approach closer and closer, and can differ from it by as little as desired.[3]

Now what does it mean to say that one magnitude "approaches" another? The kinematic image may have intuitive advantages, but it is not mathematics, since "approach" is not given a precise definition. The requirement that the quantity never coincide with its limit reminds one of the obvious fact that the area of the inscribed polygons of n sides, as n gets large, never coincides with the area of the circle. Thus, to avoid considering 0/0, D'Alembert correctly did not allow h to be zero in computing the limit, as h goes to 0, of $\frac{f(x+h)-f(x)}{h}$. Unfortunately, he also seems not to have allowed $\frac{f(x+h)-f(x)}{h}$ ever to equal the final limit, as it does in the case of a linear function. More serious, the related stipulation that the magnitude can never surpass its limit means that we cannot say that the limit of the successive terms $-1, 1/2, -1/4, 1/8, -1/16, \ldots$ is zero; it means that in the calculus, we cannot consider the derivative of functions like $x^2 \sin \frac{1}{x}$ near $x = 0$. In this connection, Lagrange pointed out that it could not be said that the subtangent was the limit of the subsecant, since "nothing prevents the subsecant from increasing still when it has become the subtangent."[4] This is not analogous to the circle-polygon example; the subsecant is not so

---

[2] Florian Cajori, *A History of the Conceptions of Limits and Fluxions in Great Britain from Newton to Woodhouse* (Chicago and London: Open Court, 1919) and Carl B. Boyer, *The History of the Calculus and its Conceptual Development* (New York: Dover, 1959) can be consulted for the limit-concept in the eighteenth century.

[3] *Dictionnaire encyclopédique des mathématiques*, by d'Alembert, L'abbé Bossut, de la Lande, le Marquis de Condorcet, &c. (Paris: Hôtel de Thou, 1789), article "Limite," written by D'Alembert.

[4] Joseph-Louis Lagrange, "Discours sur l'object de la théorie des fonctions analytiques," in *Oeuvres de Lagrange*, ed. J.-A. Serret, Volume VII, p. 325. The article first appeared in the *Journal de l'Ecole Polytechnique*, VIe Cah., t. II, an VII

defined that it cannot go right past the subtangent, in a way that the inscribed polygons can never get outside of the circle.

Principle (1) excludes limit-considerations like D'Alembert's. His definition is never translated into algebraic language, since it is not used in proofs, except once (Article "Différentiel") by way of illustration. His definition also violates (2). In the attempt to avoid the question—much debated in the eighteenth century—of whether or not a quantity ever reaches its limit, D'Alembert made his definition too restrictive. He was too firmly wedded to the example of the circle as the limit of the inscribed polygons to see that other examples demanded a broader definition.

Principle (1) requires that any system which uses infinitesimals[5] or quantities which are "really zero"[6] will be excluded. This is the sense in which the methods of Leibniz, the Bernoullis, and Euler do not meet modern standards.

The Greek "method of exhaustion" in a sense satisfies Principle (1), although the inequalities used by the Greeks are not given an algebraic form. However, only one-sided approaches to the limiting area are employed: first from one side, then from the other; no oscillation is considered. The subject matter is geometric—areas, rather than integrals—and "area" is never explicitly defined. While a rigorous geometrical basis for the calculus based on the "method of exhaustion" is theoretically possible, the definitions would be unwieldy, and the formulation and proof of the received results of analysis would be a superhuman task.

Attempts to rigorize the calculus in the eighteenth century took many approaches.[7] Finite geometrical magnitudes, evanescent time-intervals, and quantities absolutely zero dance before one's eyes. Each had its virtues and defenders, but none gained universal acceptance; in the presence of competing systems, this proved impossible. None of the systems of the eighteenth century was able to lead to new results—say, a theory of convergence—or to expose past fallacies.

The first successful basis for the calculus is in the work of Augustin-Louis Cauchy, and is presented in the lectures he delivered at the Ecole Polytechnique in Paris in the 1820's. His treatment incorporates both Principles (1) and (2). These did not first appear full-blown in Cauchy's lectures. They grew out of special techniques used in the proofs of particular theorems, in the work of Lagrange and Ampère. The proof-techniques in question will be followed in the proofs of a basic inequality in the differential calculus, by Lagrange, by Ampère, and by Cauchy.

The evolution of delta-epsilon proof techniques can be conceived as having passed through several stages. When inequalities are first introduced into analysis, they are generally restricted to positive quantities, as in Lagrange's early work.[8] At the second stage, reached by Lagrange,

---

[1799]. The *subtangent* is the portion of the x-axis cut off by the perpendicular to the axis from the point of tangency, and the intersection of the axis with the tangent. The *subsecant* is analogously defined. See D'Alembert, *op. cit.*, article "Soustangente."

[5] Such as that given in G. F. A. de L'Hospital, *Analyse des infiniment petits pour l'intelligence des lignes courbes* (Paris: Imprimérie Royale, 1696). This is based on the work of Johann Bernoulli; see J. O. Fleckenstein, "L'école mathématique baloise des Bernoulli a l'aube du XVIIIe siècle," Les Conférences du Palais de la Découverte, Série D. No 62, Paris, 1958, page 17.

[6] Leonhard Euler, *Institutiones calculi differentialis* (Impensis Academiae Imperialis Scientiarum Petropolitanae, 1755) tried to use such quantities. This work is Volume X of Euler's *Opera Omnia*, ed. Ferdinand Rudio, Series I (Leipzig and Berlin: B. G. Teubner, 1913).

[7] See Cajori, *op. cit.*, and Boyer, *op. cit.*

[8] See, for example, Joseph-Louis Lagrange, *Théorie des fonctions analytiques*, 1797. An example is given in the Appendix, pp. 103–104.

this stricture is abandoned, either by the consideration of general upper and lower bounds, or, what is equivalent, by the treatment of absolute values—as, for instance, in the proof we shall be discussing. This improvement assures that the purely algebraic parts of the argument will be free from obvious types of error. However, the distinctions between the concepts "greater than" and "greater than or equal to," and between "greater than" and "bounded away from" were not always clearly made before Cauchy's work.

The final stage in the development which we shall consider was the realization of which concepts were essential to the calculus. This meant that the properties of the inequalities had to be seen as fundamental to the concepts of the calculus, and had to be used in deriving and proving results: the magisterial achievement of Cauchy.

After these advances had been made, there were further important refinements in the foundations of the calculus: The complete arithmetization of the basic concepts due to Weierstrass and his school, the distinction between point-wise and uniform convergence in the 1840's and the treatment of iterated limits, the elucidation of the nature of the real numbers, and the topology of Euclidean space. But these—which the definitions and proof-techniques used in Cauchy's lectures made possible—are beyond the scope of this book.

"Errours, tho' never so small, are not to be neglected in Mathematicks":[9] Lagrange and the Lagrange Remainder

Joseph-Louis Lagrange made the first important use of algebraic inequalities to obtain precision in analysis. But his use of them was not meant to give a rigorous foundation to the basic concepts of the calculus. Lagrange wanted to show how the error term in Taylor's formula could be computed. In so doing, he unwittingly forged the tool which enabled Ampère and Cauchy to abandon the old "foundations" of the calculus and to develop the rigorous basis for nineteenth century analysis.

Lagrange spoke of finding the "limits" of infinite series. The word "limit" in this use by Lagrange has nothing to do with D'Alembert's foundation for the calculus. As discussed in Chapter 2, in eighteenth-century algebra, the finding of the "limits" of the roots of an equation meant finding bounds in between which the values of the roots must lie. These bounds were used as first approximations to the roots in the solution of equations by various approximation techniques. In our notation, to say as Lagrange did that the remainder term R_n of the Taylor series has as its limits R and S is to say that $R \leq R_n \leq S$. This inequality-characterization of "limit" appears in many eighteenth-century works on the solution of equations; Lagrange brought it into the calculus; Cauchy established its kinship when used in the calculus with the more vague, kinematic notion of "limit" propounded by D'Alembert.

The calculation of the bounds of the remainder term of the Taylor series will not concern us here.[10] Here I shall concentrate on the Lemma which Lagrange used to find his remainder term. The method of proof used in this Lemma is one of the most important positive contributions of the *Leçons sur le Calcul des Fonctions* to the development of the calculus.

---

[9] Isaac Newton, "Quadrature of Curves," from *Lexicon Technicum. Or, an Universal Dictionary of Arts and Sciences* by John Harris, Volume 2 (London, 1710), reprinted in D. T. Whiteside, ed., *The Mathematical Works of Isaac Newton*, Volume 1 (New York and London: Johnson Reprint Corporation, 1964), p. 141.

[10] See Chapter 3, above.

The Lemma states:

> A function which is zero when the variable is zero will have necessarily, while the variable increases positively, finite values of the same sign as its derived function, or of opposed sign if the variable increases negatively, as long as the values of the derived function keep the same sign and do not become infinite.[11]

That is, suppose $g(0) = 0$. If, when x is between 0 and a, $0 < g'(x) < \infty$, then $g(x) > 0$, for a positive, $g(x) < 0$ for a negative. If $g'(x)$ were negative, $g(x) < 0$ for a positive, $g(x) > 0$ for a negative. (Notation mine)

This is the sort of theorem which is obvious when one looks at a diagram. It is remarkable that Lagrange felt called upon to give a proof, especially since he—and everybody else in the eighteenth century—usually accepted geometric intuitions about "continuity."[12] I shall give the argument in some detail; it is the kind of proof later used by Ampère and Cauchy, and it is as close to a delta-epsilon proof as one can find in the eighteenth century.

The proof begins with the assertion, for any function f, that

$$f(x+i) = f(x) + i(f'(x) + V)$$

where V is a function of x and i such that, when i becomes zero, so does V. This assertion is equivalent to saying that, as i approaches zero, the ratio $\frac{f(x+i)-f(x)}{i}$ converges to $f'(x)$, since V is a function going to zero with i. The equation

$$f(x+i) = f(x) + if'(x) + iV$$

can only be justified by an appeal to the definition of $f'(x)$; for us, or for Cauchy, it just restates the definition; for Lagrange, the justification comes through an appeal to the position of $f'(x)$ in the Taylor series, arguing from the fact that V is an infinite series in i.[13]

Given this equation, some i can be found such that the corresponding value of V, "abstraction being made of the sign,"[14] will be less than any given quantity. This verbal statement is followed by an algebraic treatment:

> Let D be a given quantity which we can take as small as we please; one can then always give to i a value small enough for the value of V to be included between the limits D and $-D$, thus, since
>
> $$f(x+i) - f(x) = i(f'(x) + V)$$
>
> it follows that the quantity $f(x+i) - f(x)$ will be included between these two: $i(f'(x) \pm D)$.[15]

---

[11] Joseph-Louis Lagrange, *Leçons sur le calcul des fonctions* (Paris: Courcier, 1806) is in *Oeuvres*, Volume X. Leçons 1–19 were given at the Ecole Polytechnique in 1799, and first published with an additional Leçon, in *Séances des Ecoles Normales*, 1801. I shall give references to the edition published in the *Oeuvres*, and shall cite it as *CF*. Compare *Fonctions analytiques, Oeuvres* IX, pp. 78–80, where the proof is more primitive. The *Calcul des fonctions* contains some refinements and explanations not in *FA*, but does not differ essentially from it in the exposition of the basic theory. See *CF*, p. 86, for the statement of the Lemma. (The *FA* proof is given in the Appendix, below.)

[12] Compare the geometric characterization of continuity given in *FA*, discussed pp. 67–68, above.

[13] A "proof" is to be found in *FA, Oeuvres* IX, pp. 28–29. Compare Chapter 3, above.

[14] *Calcul des fonctions, Oeuvres* X, p. 87. He had the idea of absolute value, but he did not call it that here.

[15] *Idem.*

In modern terms, for a fixed value of x, $V(x, i)$ is continuous in i as i goes to zero. In the version of this Lemma given in the *Fonctions analytiques*, Lagrange did not make clear whether or not he meant to restrict V to positive values.[16] No trace of this restriction remains in the *Calcul des fonctions*. The appreciation of the significance of absolute value by Lagrange, here and elsewhere, is an important step forward. Only through the use of absolute values could the requirement that a function or sequence never surpass its limit be discarded.

Lagrange then said that $f(x+2i) - f(x+i)$ lies between $i(f'[x+i] \pm D)$, $f(x+3i) - f(x+2i)$ lies between $i(f'[x+2i] \pm D)$, etc. Here, D is given and Lagrange assumed (though not explicitly) that the same i will work always, for any x in the interval under consideration. Further, he assumed explicitly that none of the $f'(x+ki)$ is infinite (k-notation mine).

Now, if all the $f'(x+ki)$ have the same sign, so will their sum. The sum $\sum_{k=1}^{n} f(x+ki) - f[x+(k-1)i]$, which equals $f(x+ni) - f(x)$, "will have for limit the sum of the limits, that is, the quantities $if'(x) + if'(x+i) + if'(x+2i) + \cdots + if'(x+[n-1]i) \pm niD$."[17] (By "limit," Lagrange means bound.)

Since D is arbitrary, Lagrange said, it can be taken as less than the absolute value of $\frac{f'(x)+f'(x+i)+\cdots+f'(x+[n-1]i)}{n}$. He gave no detailed reason for the possibility of so choosing D. Possibly he considered D to depend on i and n, in which case it can be calculated. In this case, the rest of the proof is invalid, since i, and therefore n, must be chosen after D. Alternatively, he may have had in mind the fact that, if $|f'(x)|$ is bounded away from zero on the interval, it has a non-zero minimum. If D is taken as less than that minimum, it will be less than $\left|\frac{f'(x)+f'(x+i)+\cdots+f'(x+[n-1]i)}{n}\right|$.[18]

If D is so chosen, Lagrange continued, $f(x+ni) - f(x)$ must lie between zero and $2i[f'(x) + f'(x+i) + \cdots f'(x+[n-l]i)]$. Now, if P is the greatest positive or negative value of the quantities of form $f'(x+ki)$, he stated that $f(x+ni) - f(x)$ lies between 0 and $2niP$. In taking i as small as desired, he finished, n can be taken as large as desired and he can write $ni = z$, where z has the same sign as i. Thus,

$$f(x+z) - f(x) \quad \text{will lie between 0 and } 2zP.$$

The Lemma is thus proved, if $f(x+z) - f(x)$ be considered as a function of z.

In this proof, small positive quantities have been treated by means of the algebra of inequalities, and a delta-epsilon calculation undertaken. We see here the major feature distinguishing Lagrangian proofs from previous proofs concerning the properties of the derivative. Lagrange did not finish his proof by saying "the sum of a finite number of positive infinitesimals is again positive," nor by an appeal to geometry. He worked with finite, though small quantities, and was able to apply all the formal power of algebra to their mutual relationships.

The proof does assume implicitly that $f'(x)$ is both bounded and bounded away from zero; Lagrange seems to have thought it sufficient for $f'(x)$ to be finite and never equal to zero. There are two more serious objections. First, the proof is based on the equation

$$f(x+i) = f(x) + if'(x) + iV, \tag{4.1}$$

---

[16] *FA*, 1797, p. 45. Compare Appendix.

[17] *CF, Oeuvres* X, p. 88. In his terminology, the statement that the limit of the sum is the sum of the limits merely means that, if $-a < b < a$ and $-c < d < c$, then $-a - c < b + d < a + c$.

[18] If Lagrange had this in mind, this is an example of the confusion between "greater than 0" and "bounded away from 0" which was first resolved by Cauchy.

which Lagrange could only prove (and, even then, incorrectly) by using the full Taylor series expansion of $f(x+i)$ in powers of i, in which all the derivatives must be bounded.[19] Cauchy overcame this objection by defining $f'(x)$ so that it would satisfy (4.1). Second, Lagrange, in assuming that one choice of i would make V small for all values of x, confused convergence and uniform convergence. We shall see later that Cauchy fell into the same confusion.

Lagrange applied the Lemma to find the limits of the remainder term of the Taylor series. The procedure is essentially this: let the maximum of $f'(x)$ on the interval in question be $f'(q)$, the minimum $f'(p)$. Now define

$$g'(i) = f'(x+i) - f'(p)$$
$$h'(i) = f'(q) - f'(x+i).$$

Integrating with respect to the initial condition $g(0) = h(0) = 0$,

$$g(i) = f(x+i) - f(x) - if'(p)$$
$$h(i) = if'(q) - f(x+i) + f(x).$$

As long as f' is never infinite on the interval in question, he said that he could apply the Lemma just proved to conclude that $g(i)$ and $h(i)$ are positive. Actually, the application of the Lemma to find the remainder term requires that the Lemma hold for $f(x+z) - f(x) \geq 0$, since $g'(i) = f'(x+i) - f'(p) \geq 0$ is all that can be claimed.[20] Lagrange had no special notation to distinguish between strict inequalities and those allowing equality.

If the Lemma is applied in this way, $g'(i)$ and $h'(i)$ are greater than or equal to zero. Thus,

$$f(x+i) - f(x) - if'(p) \geq 0$$
$$f(x) - f(x+i) + if'(q) \geq 0$$

which together yield

$$f(x) + if'(p) \leq f(x+i) \leq f(x) + if'(q),$$

setting "limits" or bounds on the value of $f(x+i)$.[21]

Lagrange showed in general that $f(x+i)$ will always lie between

$$f(x) + if'(x) + i^2/2!f''(x) + \cdots + i^u/u!f^{(u)}(p)$$

and

$$f(x) + if'(x) + i^2/2!f''(x) + \cdots + i^u/u!f^{(u)}(q),$$

---

[19] See Chapter 3, above.

[20] He could have avoided this difficulty by considering functions like

$$g'(i) = f'(x+i) - f'(p) + \varepsilon \text{ and}$$
$$h'(i) = f'(q) - f'(x+i) + \varepsilon$$

which would yield

$$-\varepsilon i + f(x) + if'(p) < f(x+i) < f(x) + if'(q) + \varepsilon i.$$

Realizing ε to be arbitrary then yields Lagrange's result. We shall see, below, that Cauchy used this procedure.

[21] *CF, Oeuvres* X, p. 91. Compare *FA, Oeuvres* IX, pp. 80–81, and Chapter 3, above.

where p and q are, respectively, the minimum and maximum points of the uth derivative on the interval.[22]

$f^{(u)}(x)$ will, Lagrange claimed, take on all the intermediate values between $f^{(u)}(p)$ and $f^{(u)}(q)$. Thus, there is a quantity X in the interval $[x, x+i]$ such that

$$f(x+i) = f(x) + if'(x) + i^2/2!f''(x) + \cdots + i^u/u!f^{(u)}(X),$$

exactly.[23] The intermediate-value property Lagrange stated as something obvious; by means of this property and his Lemma, Lagrange obtained again the Lagrange form of the remainder term.[24]

Lagrange's "true metaphysic" for the calculus—reducing calculus concepts to power series-satisfied his yearning for system, but could not lead him to the remainder term. So, instead of neglecting this vital point, Lagrange devised a new method which would yield this term. Although he made errors in its application, he deserves credit for seeing the method of proof which was to insure rigor in analysis.

What Lagrange did for the Taylor series, Cauchy was to do for the derivative: to reduce the question of its value to a sequence of inequalities which include it. What for Lagrange was a stepping-stone to a first-order error-estimate in the Taylor series became a defining property in the hands of Cauchy—and, even before and in a different way, in the hands of Ampère.

Ampère's Forgotten Contribution: A New Definition

André Marie Ampère (1775–1836) is better known for his work in electricity than for his researches in analysis. Yet early in his career his interests were primarily mathematical, culminating in publications on probability theory, differential equations, and the concept of the derivative.[25] It was on the strength of this work, rather than on that of the electrical researches which he began in 1820, that he became a member of the Institut de France in 1814.[26] Born in Lyon, Ampère received his early mathematical training through reading the works of the great: D'Alembert, Euler, Lagrange. It was in the method of proof that Lagrange had used to derive the remainder term of the Taylor series that Ampère saw the most elegant and economical definition of the derivative.

If the remainder term of the Taylor series could be derived by this method, Ampère considered, why not the Taylor series itself? Could he not dispense with the Taylor series as well as the other ineffectual definitions of $f'(x)$? Why not define the derived function in terms of the ratio $\frac{f(x+i)-f(x)}{i}$ and the inequalities which that ratio must satisfy when i is very small? Ampère, in his paper of 1806: *Recherches sur quelques points de la théorie des fonctions derivées qui conduisent à une nouvelle démonstration de la série de Taylor, et à l'expression finie des termes qu'on néglige lorsqu'on arrête cette série a un terme quelconque*, has the honor of being the first person to make such a definition:

---

[22] *CF, Oeuvres* X, p. 94. See also *FA, Oeuvres* IX, p. 84, and Chapter 3.

[23] *CF, Oeuvres* X, pp. 91–95, *FA, Oeuvres* IX, pp. 80–85.

[24] Compare Chapter 3, above. Lagrange regarded this derivation as "simpler" than that in *FA*. See *FA, Oeuvres* IX, p. 85.

[25] Ampère's life is sympathetically viewed in Louis de Launay, *Le Grand Ampère*, (Paris: Perrin et Cie, 1925). See also "Ampère," in *Oeuvres complètes de François Arago* (Paris: Gide et J. Baudry, Leipzig: T. O. Weigel, 1854), Tom. II, pp. 1–116.

[26] De Launay, *op. cit.*, p. 176.

> The derived function of $f(x)$ is a function of x such that $\frac{f(x+i)-f(x)}{i}$ is always included between two of the values that this derived function takes between x and $x+i$, whatever x and i may be.[27]

That is, $f'(x)$ is the function such that, given x and i, there exist p and q between x and $x+i$ such that

$$f'(p) \leq \frac{f(x+i)-f(x)}{i} \leq f'(q). \quad \text{(My notation)} \tag{4.2}$$

The paper begins by treating the derivative as if it were $\frac{f(x+i)-f(x)}{i}$ "when $i = 0$."[28] In discussing some of the properties of $f'(x)$ so defined, the property given by (4.2) arises, after some ten pages. Having shown that the function satisfying (4.2) was unique, he could state his new definition, for now it was clear that it was, to him, completely equivalent to the old one.

Ampère did not give the modern definition of the derivative used by Cauchy. Ampère chose to follow Lagrange's signposts in another direction. It was the properties of the maximum $f'(q)$ and of the minimum $f'(p)$, and not of arbitrarily close bounds about $f'(x)$, on which Ampère focussed his attention. And it was in the properties of its maximum and minimum on any interval that Ampère saw the essential property of the derivative. It should be remarked that a continuous function satisfying (4.2) is $f'(x)$ as Cauchy defined it.

Before he could even begin to use any non-Lagrangian definition of the derived function, Ampère had to assure himself that Lagrange's procedures would work. He did this by attempting to prove that $f'(x)$ would not be infinite or zero except at a few isolated points (except, of course, for a constant function). But he stated that he would demonstrate the *existence* of $f'(x)$.[29] As a result, this paper has been called Ampère's attempt to show that any continuous function has a derivative.[30] But really Ampère only wanted to prove that Lagrange's procedure was sound. That $f'(x)$ existed, except perhaps at a few points, meant that it was non-zero and finite. Given his use of the term, one might say that the limit of e^{1/x^2} as x approaches zero does not exist, or that the limit of x^2 does not exist as x approaches 0. The sense in which $f'(x)$ does not exist for $f(x) = |x|$ at $x = 0$ did not occur to him—perhaps because of the long-standing habit of considering only one-sided limits. This paper of 1806 was written at a time when many of the concepts used—existence, continuity, derivative—were still to have their exact senses fixed; Professors Pringsheim and Molk simply took the modern meaning of Ampère's word.

What is the derivative? Ampère's procedure was not the systematic "definition, axiom, theorem," to which we have become accustomed and to which Cauchy adhered in his *Cours d'analyse*.[31] At first Ampère described, or characterized, the derivative as the value of the quotient of differences $\frac{f(x+i)-f(x)}{i}$ "when $i = 0$."[32] This enabled any reader to recognize the

---

[27] *Journal de l'école polytechnique*, 13e Cahier, Tom. VI, 1806, pp. 148–181. The definition is on page 156.

[28] Ampère, *op. cit.*, p. 149.

[29] *Ibid.*

[30] This description comes from "Principes fondamentaux de la théorie des fonctions," by A. Pringsheim, to be found in a French exposition of the German original by Jules Molk in *Encyclopédie des Sciences Mathématiques Pures et Appliquées* Tom. 2, Vol. 1, Fascicule 1 (Paris: Gauthier-Villars, Leipzig: B. G. Teubner, 1909), ed. Jules Molk, p. 44.

[31] Augustin-Louis Cauchy, *Cours d'analyse de l'Ecole Royale Polytechnique. Ire Partie: Analyse algébrique* (Paris: Imprimérie Royale, 1821).

[32] Ampère, *op. cit.*, p. 149.

object $f'(x)$ to which he was referring. He did not use the language of limits—$f'(x)$ is not the limit of the ratio, but its value when $i = 0$. Apparently he inherited Lagrange's distaste for the concept "limit" to the point of preferring the 0/0 characterization of $f'(x)$. When he had proved that the function so characterized had the property he wanted, he turned about and used his new property to define $f'(x)$.

Ampère's argument was firmly rooted in the terminology of his predecessors. Clearly for any fixed $i \neq 0$, $\frac{f(x+i)-f(x)}{i}$ is a function of x; hence, he argued, this must be true for $i = 0$.[33] Thus $f'(x)$ is a function of x, which has the form 0/0 to be sure, but the sort of 0/0 which has a finite value. The treatment of 0/0 as an indeterminate expression whose value depends on how it is generated is basic to the views on the calculus expressed by Leonhard Euler and Lazare Carnot.[34]

To prove that the ratio $\frac{f(x+i)-f(x)}{i}$ could not be infinite or zero except at a finite number of isolated points, the only technique available to Ampère was the indirect one. He stated the denial of the theorem: if the ratio is infinite or zero at more than a finite number of points, it must, according to him, be infinite or zero on a whole interval (unless, of course, it is a constant). He chose to show that denying the theorem led to a contradiction. In his proof, he used the techniques developed by Lagrange, in the proof discussed above.

Ampère began by assuming that there was some interval $[a, k]$ such that, for all x in that interval, $\frac{f(x+i)-f(x)}{i}$ is in fact either zero or infinity, when $i = 0$, although $f(x)$ is finite in the interval. Let $f(a) = A$, $f(k) = K$, $a \neq k$, $A \neq K$. These last two conditions mean that the interval is not a point and the function not a constant.

Now if $\frac{f(x+i)-f(x)}{i}$ goes to infinity as i approaches zero, i can always be taken small enough so that, for all x in the interval, $\frac{f(x+i)-f(x)}{i}$ exceeds any given quantity—in particular, exceeds $\frac{K-A}{k-a}$.[35] This interpretation of "goes to infinity" as meaning "exceeds any given quantity" is very old—perhaps the first systematic use being by Johann and Jakob Bernoulli in investigating convergence.[36]

Suppose that the points $b, c, d, e, \ldots$ are in the interval and are such that $a < b < \cdots < e < k$ and $f(b) = B, \ldots f(e) = E$. Ampère then proved algebraically that, since $a < e < k$, $\frac{K-A}{k-a}$ lies between $\frac{K-E}{k-e}$ and $\frac{E-A}{e-a}$.[37]

---

[33] Ampère, *op. cit.*, pp. 148–149.

[34] Euler's work is mentioned in Note 6. See Lazare Carnot, *Réflexions sur la Métaphysique du calcul infinitésimal* (Paris: Duprat, 1797). See also A. P. Juschkewitsch, "Euler und Lagrange Uber die Grundlagen der Analysis," in *Sammelband der zu ehren des 250 Geburtstages Leonhard Eulers.* Deutschen Akademie der Wissenschaften zu Berlin (Berlin: Akademie-Verlag, 1954), for a masterly discussion of Euler's zeros, a few remarks on Carnot's work, and the role of infinite series for Euler and Lagrange.

[35] Ampère, *op. cit.*, p. 140.

[36] Jakob Bernoulli's publications on this subject, incorporating some of his brother Johann's work, span the period 1689–1704. They are collected and translated into German by G. Kowalewski as Jakob Bernoulli, *Über unendliche Reihen* 1689–1704 (Leipzig: Wilhelm Engelmann, 1909).

[37] Ampère, *op. cit.*, pp. 151–154. This is a simple result about inequalities though it may have been suggested to him by geometry. Since

$$\frac{K-E}{k-e} - \frac{K-A}{k-a} = \frac{aE - Ae + Ak - ak + eK - Ek}{(k-e)(k-a)} \quad \text{and}$$

$$\frac{K-A}{k-a} - \frac{E-A}{e-a} = \frac{aE - Ae + Ak - ak + eK - Ek}{(k-a)(e-a)},$$

Similarly, there are more of these fractions with the same properties; for instance, if $e < g < k$, then $\frac{K-E}{k-e}$ is between $\frac{K-G}{k-g}$ and $\frac{G-E}{g-e}$. Finally, he had $\frac{B-A}{b-a}, \frac{C-B}{c-b}, \ldots \frac{K-H}{k-h}$, as many as desired. (A modern mathematician would obtain this result by induction on the number of fractions; to Ampère it is obvious.) From these one can always select a pair, of which one will be greater than $\frac{K-A}{k-a}$ while the other will be less. Or, in clearer notation than Ampère's, there exist P, Q in $[a, k]$ such that

$$\frac{f(P+i)-f(P)}{i} \leq \frac{K-A}{k-a} \leq \frac{f(Q+i)-f(Q)}{i} \tag{4.3}$$

where now Ampère let $b-a=c-b=\cdots=k-h$, and set $b-a=\cdots=k-h=i$.

Ampère then applied to the ratio $\frac{f(x+i)-f(x)}{i}$ a property often used by his contemporaries: the intermediate value property of continuous functions. He asserted that there must be some x (which I shall write X) between P and Q such that

$$\frac{f(X+i)-f(X)}{i} = \frac{K-A}{k-a}. \tag{4.4}$$

This is of course true since, for fixed i, $\frac{f(x+i)-f(x)}{i}$ is a continuous function of x on $[a, k]$.

But, Ampère continued, $\frac{f(x+i)-f(x)}{i}$ is supposed to become infinite on the interval as i goes to zero. Therefore we can choose an i (I shall write it j to avoid the confusion of notation that Ampère's i causes) such that $\frac{f(x+i)-f(x)}{i} > \frac{K-A}{k-a}$ for all $i < j$ and for all x on the interval—including X. This is a contradiction: if i is such that $b-a=c-b=\cdots=k-h=i<j$, he had already shown that he can find X such that $\frac{f(X+i)-f(X)}{i} = \frac{K-A}{k-a}$.

Similarly, he argued that the ratio could not be zero all through the interval (if $A \neq K$) by assuming the ratio to be positive and less than any assignable quantity, for small i and all x.

His proof is, as it happens, completely valid in the case that the convergence of the ratio $\frac{f(x+i)-f(x)}{i}$ to infinity (or zero) is *uniform*. Ampère had no more reason to assume that his j would work for all x than had Lagrange—nor than Cauchy would in the future.[38] In fact, Ampère did not show that $\frac{f(x+i)-f(x)}{i}$ must have a finite nonzero limit, even if it is uniformly convergent. It may have no limit at all. Ampère did not even consider this possibility—proof enough that he was not proving what we would call an existence theorem.

Ampère went on to consider the relationship between the value of $f'(x)$ at a particular x and the behavior of $\frac{f(x+i)-f(x)}{i}$ for small values of i. This was based, to be sure, on Lagrange's approach to the Taylor series remainder, but went beyond Lagrange in seeing these properties as fundamental to the concept of $f'(x)$.

Ampère said that he could write

$$f'(x) + I = \frac{f(x+i)-f(x)}{i}$$

---

notice that the numerators of the two fractions on the right are the same and their denominators are positive. Therefore, both fractions must have the same sign. Thus, either the two differences on the left are positive together, or they are negative together; either $\frac{K-E}{k-e} \geq \frac{K-A}{k-a} \geq \frac{E-A}{e-a}$ or else $\frac{K-E}{k-e} \leq \frac{K-A}{k-a} \leq \frac{E-A}{e-a}$. (Ampère, like Lagrange, had no special notation for "less than or equal.")

[38] The distinction between point-wise and uniform convergence was not made in the eighteenth century—nor even in the early work of Cauchy—but was central to the work of Weierstrass. The concepts were first distinguished in the 1840's. The uniform convergence of the ratio $\frac{f(x+i)-f(x)}{i}$ to a finite limit is equivalent to the continuity of $f'(x)$.

where I is a function of x and i, which vanishes when i does (I is the function Lagrange had called V). This follows from thinking of $f'(x)$ as the value of the ratio when $i = 0$. He then proceeded to show that $f'(x)$ satisfies an inequality similar to (4.3):

There exist x_0, x_1 in $[a, k]$ such that

$$f'(x_0) \leq \frac{K - A}{k - a} \leq f'(x_1). \tag{4.5}$$

This is clearly the same as (4.2). He proved (4.5) by assuming that $f'(x) < \frac{K-A}{k-a}$ for all x, and arguing to a contradiction; similarly for $f'(x) > \frac{K-A}{k-a}$. The validity of his proof requires the continuity of $f'(x)$.[39]

This extensive discussion should have made clear Ampère's debt to Lagrange. In the notation $f'(x)$ and $f(x+i) - f(x)$, in the use of the term "derived function," in the equation $\frac{f(x+i)-f(x)}{i} = f'(x) + I$ where I goes to zero with i, and in the concern with showing that $f'(x)$ must be finite and non-zero for non-constant f, the roots of Ampère's paper in Lagrange's Lemma are unmistakable. Ampère also cited some of Lagrange's other work in this paper. He did not rigorously prove here what he set out to prove, but did perform an exercise in the Lagrangian proof-techniques.

Ampère had, like Lagrange before him, all the materials for Cauchy's definition of $f'(x)$ in his hands. But his fancy was caught by the inequality

$$f'(x_0) \leq \frac{K - A}{k - a} \leq f'(x_1). \tag{4.5}$$

Ampère argued that $f'(x)$ was the only function having the property that, in any interval $[a, k]$ on which it is defined, there exist x_0 and x_1 such that (4.5) holds.[40] His argument for this fact assumes explicitly that, if there were another function with the property of (4.5) it could be written on some interval (in my notation, the interval will be $[b, c]$) as $f'(x) + \phi(x)$ or as $f'(x) - \phi(x)$, where $\phi(x)$ is positive on the interval. Necessary also for his proof to be valid would be that $\phi(x)$ be bounded away from zero—for which it suffices for $\phi(x)$ to be continuous on the interval $[b, c]$ and that $f'(x)$ be continuous at some one point in $[b, c]$. By extremely sophisticated arguments, it can be shown that a derivative has at least one point of continuity in any interval. Thus, two functions, each satisfying (4.5), whose difference is continuous, must in fact be equal.

Up to this point, he had not defined $f'(x)$. He used two properties: that $f'(x)$ is $\frac{f(x+i)-f(x)}{i}$ "when $i = 0$," and that it satisfied $f'(x) + I = \frac{f(x+i)-f(x)}{i}$ where I vanishes with i. But these usages led him to the new definition, that $f'(x)$ was the unique function such that, for all intervals

---

[39] The proof is essentially this. Suppose that, for all x in $[a, k]$ $f'(x) < \frac{K-A}{k-a}$. Choose i such that, for all x in the interval, $I < \frac{K-A}{k-a} - f'(x)$. He thought i could be so chosen because $I(x, i)$ vanishes as i goes to zero. Thus, he concluded that $f'(x) + I < \frac{K-A}{k-a}$, the result he wanted for later use. The conditions on I to obtain this result are even stronger than uniform convergence to 0. $I(x, i)$ is made, independently of x, not only less than a given quantity, but less than a quantity which itself depends on x. If $f'(x)$ were continuous, the assumption would be justified; rather than appeal to such criteria, Ampère in effect assumed that, given an $\varepsilon(x)$, there was a δ such that, for all x, $i < \delta$ implies that $I < \varepsilon(x)$.

In fact, uniform convergence of $I(x, i)$ to zero would suffice if $f'(x)$ were bounded away from $\frac{K-A}{k-a}$.

Granting all this, for i sufficiently small, since $f'(x) + I = \frac{f(x+i)-f(x)}{i}$ he obtained $\frac{f(x+i)-f(x)}{i} < \frac{K-A}{k-a}$.

But this is impossible, since for any i, P and Q can be found satisfying (4.3), above. Similarly, assuming $f'(x)$ to be greater than $\frac{K-A}{k-a}$ for all x leads to a contradiction.

[40] Ampère, *op. cit.*, pp. 155–156.

$[x, x+i]$ on which $f(x)$ was defined, there exist x_0, x_1 on the interval such that

$$f'(x_0) \leq \frac{f(x+i) - f(x)}{i} \leq f'(x_1).$$[41]

We can show by the Law of the Mean that the derivative does have this property. Moreover, if the derivative is continuous, Ampère's uniqueness proof is valid: no other continuous function has this property.

To use eighteenth-century algebraic language (which he did not) the maximum and minimum derivatives *limit* the ratio of differences. It is but a short step from this to Cauchy's welding of the two limit concepts: limit-as-bound and the kinematic one used by D'Alembert, into a rigorous definition of the derivative as the limit, in the modern sense, of the ratio of differences. Ampère did not take this step. But he did define the derivative in terms of an explicit algebraic inequality—and he tried to build on his work. Nothing had been done with the older definitions of limit; no deeper understanding could be built on them, since they did not make the concepts of the calculus instances of more general *mathematical* ideas.

Consider, Ampère said, a function $f(x)$. It satisfies

$$f(x+i) = f(x) + if'(x) + iI.$$

He had already shown that $f'(x)$ satisfying that equation also satisfies the inequality-condition he chose to define the derivative. Thus, if we want to compute $f'(x)$ for a given $f(x)$, we need only use the terms $f(x) + if'(x)$ in the development of $f(x+i)$, and neglect the terms in which i remains—not because they are zero, for Ampère knew they are not, but because they do not belong in the equation for $f'(x)$, but in that for I.[42] In postulating the existence of I, and its vanishing with i whatever x may be, he was following Lagrange; if he had made this property the definition of $f'(x)$, he would have anticipated Cauchy.

"This procedure is evidently that of the differential calculus," continued Ampère, if one sets $dx = i$, $dy = if'(x)$, where $y = f(x)$. Thus the differential calculus "can in this way be freed, not only from the consideration of infinitesimals, but also from that of the formula of Taylor."[43]

In other words, not only could he do what Lagrange did—banish all previous foundations for the calculus—but he could also dispense with Lagrange's basic principle. In fact Ampère thought that he had proved Taylor's theorem, by a procedure similar to that used by Lagrange to get the remainder term.[44] Ampère did not use his definition in his discussion of the Taylor series or the differential calculus. He preferred the characterization of $f'(x)$ given by $f'(x) = \frac{f(x+i)-f(x)}{i} - I$, which is easier to work with.

Ampère's originality lay in his treating $f'(x)$ as a function which satisfied a certain inequality on a given interval. Where Lagrange had seen the inequalities satisfied by $f'(x)$ as incidental properties, Ampère saw them as essential defining properties. Ampère preferred to prove that $f'(x)$ as commonly characterized was unique than to assume that there is a Taylor series expansion

---

[41] *Ibid.*, p. 156, x_0, x_1 are my notation.

[42] *Ibid.*, p. 162.

[43] *Idem.*

[44] This derivation, which is not part of the line of development I am tracing here, is given later in Ampère's paper: pp. 163–169; its discussion would take us too far afield.

with as many derivatives as desired. This was not because he had found functions without Taylor series. Rather, it was because the standards of elegance in mathematics demand that as few assumptions be made as possible. Lagrange's "proof" that every function had a Taylor series was, apparently, not convincing to Ampère, who felt that the derivative should be prior to the Taylor series—which, after all, includes it.

The derivative then should be defined uniquely and unexceptionably in the simplest possible way—not by the imprecise concepts rejected by Lagrange, not in terms of the Taylor series—but by an inequality which $f'(x)$ and $f'(x)$ alone could satisfy. Establishing this was Ampère's achievement, while the choice of more fruitful inequalities was left for Cauchy.

Cauchy: Prolegomena to Any Future Mathematics

The achievement of Augustin-Louis Cauchy with respect to the foundations of analysis can be viewed from many perspectives. It would be unfair to view his work merely as the culmination of a tradition stemming from Lagrange and Ampère. This tradition is, of course, the subject of this chapter, but its bringing to fruition represents only a small part of Cauchy's contributions. His landmark papers on many subjects, coupled with wonderfully clear presentations of the rigorous bases for his work, make him interesting and readable even to mathematicians today.

Relevant for our present purposes is his definition of $f'(x)$, first published in 1823, which is close to the characterizations of $f'(x)$ by the equation $f(x+i) = f(x) + if'(x) + iI$ used by Lagrange and Ampère. His definition of "limit" enabled him to fuse the old, non-rigorous notions of the derivative with these characterizations.

> When the successively attributed values of the same variable indefinitely approach a fixed value, finishing by differing from it by as little as desired, the latter is called the limit of all the others.[45]

At first glance, Cauchy's definition of limit, first published in 1821, does not seem to be much better than D'Alembert's. It is also verbal, also speaks of "approach." One advantage, however, appears immediately: the rejection of the "never surpassing" requirement. He did not explicitly reject it, but he did not mention it. Not the difference, but the *absolute value* of the difference, was to become very small. Also, he did not say that the values given to the variable could "never become equal" to the limit; Cauchy's definition thus allows the derivative of a linear function to be the limit of the ratio of its finite differences. In addition, instead of having one "magnitude" approach another, Cauchy had successively attributed values of a variable approach a fixed value. This focussing on the set of values, rather than on a magnitude which moves or changes, brings us away from the intuitive realm of motion and into the realm of arithmetic or algebra. Unfortunately, however, Cauchy did not distinguish here between dependent and independent variable; it is not clear exactly under what conditions a "variable" approaches a limit—that is, what determines the successively attributed values. This confusion may have contributed to his confusion of point-wise and uniform continuity, page 98, below.

---

[45] *Résumé des leçons données a l'école royale polytechnique sur le calcul infinitésimal.* (Paris: Imprimérie Royale, 1823). This is in Cauchy's *Oeuvres Complètes* (Paris: Gauthier-Villars, II Série, Tom. IV, 1899) and it is to the latter edition that all page references are made. The definition of limit quoted here is from page 13; it was also given in the *Cours d'analyse* published in 1821.

Finally, the most important difference is that, although Cauchy's "limit" may look like the same kind of vague verbal notion used by his predecessors, it is in practice much more: it was translated by him into algebra, and applied to proofs concerning the concepts of the calculus.

Cauchy used his definition of a limit to define the derivative, and systematically used his definition of the derivative in proofs. The derivative is the limit of a ratio; the integral is the limit of a sum; and, if Cauchy used the term "infinitely small quantity" he had rigorously defined it as a variable whose limit is zero.[46]

Cauchy gave a rigorous account of "continuity":

> When, the function $f(x)$ admitting a unique and finite value for all values of x included between two given limits, the difference $f(x+i) - f(x)$ is always between these limits an infinitely small quantity, we say that $f(x)$ is a *continuous function* of the variable x between the limits in question.[47]

No longer are there intuitive appeals to curves, or to quantities generated by motion. And his infinitely small quantity is a variable (he should specify, a dependent variable) with zero for its limit, not a special metaphysical entity.

It should be noted that Bernhard Bolzano, in 1817, had given almost the same definition of continuity in his proof that a continuous function possessed the intermediate-value property.[48] Bolzano later adopted Cauchy's definition of the derivative in his *Functionenlehre*[49] and gave the first example of a continuous non-differentiable function.[50] Unfortunately, Bolzano, a resident of Prague and a religious mystic, was out of contact with the main centers of European mathematics and little notice was taken of him until the late nineteenth century.[51] His *Functionenlehre* remained unpublished until the twentieth. His work was highly rigorous: he gave a critique of Lagrange's proof of the Lemma we have discussed, and, as already mentioned, echoed Lagrange's call for the foundation of the calculus purely on algebra, without recourse to intuition or ideas of motion.[52]

Let us consider in detail Cauchy's treatment of the derivative of a continuous function. Suppose $y = f(x)$ is continuous between two given values of the variable x. Then "an infinitely small increment attributed to the variable produces an infinitely small increment of the [value of the] function itself."[53] So "the two terms of the ratio of differences $\frac{f(x+i)-f(x)}{i}$ will be infinitely small quantities." Thus far, the language could be that of L'Hospital. "But, while the two terms approach the limit zero indefinitely and simultaneously, the ratio itself can converge to another

---

[46] *Calcul infinitésimal*, p. 16.

[47] *Calcul infinitésimal*, pp. 19–20.

[48] Bernhard Bolzano, *Rein analytischer Beweis des Lehrsatzes, dass zwischen je zwey Werthen, die ein entgegensetztes Resultat gewähren, wenigstens eine reele Wurzel der Gleichung liege* (Leipzig: Wilhelm Engelmann, 1905) (originally Prague, 1817).

[49] Published by Karel Rychlík (Prag: Königliche Böhmische Ges ellschaft der Wissenschaften, 1936) as Band I of *Bernhard Bolzano's Schriften*. A reference to Cauchy's *"Cours d'algèbre"* (evidently the *Cours d'analyse... Analyse algèbrique*) is to be found on p. 94. Compare Bolzano's *Paradoxes of the Infinite*, first published in 1851 (London: Routledge & Kegan Paul, 1950), where he praised Cauchy before defining $f'(x)$, p. 37.

[50] Boyer, *op. cit.*, p. 269.

[51] See Otto Stolz, "B. Bolzano's Bedeutung in der Geschichte der Infinitesimalrechnung," *Mathematische annalen XVIII* (1881) pp. 255–279. However, Niels Hendrik Abel may have known some of his work; see Oystein Ore, *Neils Hendrik Abel: Mathematician Extraordinary* (Minneapolis: U. of Minn., 1957), p. 96.

[52] *Rein analytischer Beweis...*, pages 6 and 7.

[53] Cauchy, *op. cit.*, p. 22.

limit, either positive or negative." This sentence could have been written by Newton, D'Alembert, Carnot. But the next sentence is unmistakably Cauchy's, and its message is of unprecedented and fundamental importance: "This limit, when it exists, has a determined value for each particular value of x."[54] It is a function of x, which depends on $f(x)$, and is called the derived function $f'(x)$.[55]

Only in the work of Cauchy do we find the clause "when the limit exists" in defining $f'(x)$. Lagrange and Ampère saw the possibility that the derivative might not be well-behaved at isolated points; Cauchy saw that the limit of $\frac{f(x+i)-f(x)}{i}$ might simply not exist. This is not the same as saying that it is infinite or zero. Perhaps Cauchy only had in mind that the limit might not exist at "isolated points," but his language is more general. This made it possible for others to ask under which conditions the limit existed and under which it did not. Cauchy followed Ampère in making plausible that two quantities might go to zero and yet have, taken in ratio, a finite limit—in the same notation and in similar language. But he did not attempt a proof that this was almost always so.

That Cauchy knew the work of Lagrange and Ampère is clear, and not just because of the similarities in notation. We have several anecdotes related by Cauchy's biographer Valson on how Lagrange supervised Cauchy's early mathematical education.[56] Cauchy explicitly rejected Lagrange's foundation for the calculus in the *Avertissement* (Foreword) to his lectures on the Infinitesimal Calculus,[57] the only previous account of the foundations that he bothered to discuss.

More relevant to this chapter is Cauchy's debt to Ampère. This was personal as well as through Ampère's writings. In the Introduction to his *Cours d'analyse*, published in 1821,[58] Cauchy acknowledged that he had "profited several times from the observations of M. Ampère, as well as from the methods that he has developed in his lectures on analysis." And in discussing the relations between the derivative $f'(x)$ and the ratio of the finite differences $\frac{f(x+i)-f(x)}{i}$, Cauchy cited Ampère's paper of 1806.[59] One should not underestimate Cauchy's originality, or suggest that he owes everything to Ampère. But the influence of the proof-technique of that 1806 paper is very striking.

Cauchy, like Lagrange and Ampère, felt that everything provable in the calculus should be proved, even if it seemed obvious. And less seemed obvious to Cauchy. He went so far as to give a proof for the intermediate-value theorem for continuous functions.[60] This result he then used to prove the Mean Value Theorem for $f'(x)$, which is a corollary to the major theorem of Leçon 7 of Cauchy's Lectures on the Infinitesimal Calculus, published in 1823.

The theorem of Leçon 7 is of interest for several reasons. First, from our present point of view: it is, almost exactly, the result used by Ampère to define the function $f'(x)$. Moreover,

---

[54] *Idem.*

[55] *Ibid.*, p. 23.

[56] C.-A. Valson, *La Vie et les Travaux du Baron Cauchy* (Paris: Gauthier-Villars, 1868), *passim.*

[57] See Note 45. The "Avertissement" has no page numbers.

[58] *Cours d'analyse*, pp. vii–viii.

[59] *Calcul infinitesimal,* p. 45.

[60] *Cours d'analyse*, Note III, p. 460. The theorem says that, if $f(x)$ is continuous on $[x_0, X]$ and if b lies between $f(x_0)$ and $f(X)$, we can always satisfy the equation $f(x) = b$. A geometrical argument was given in the text, but the Note contains an analytic proof. The proof implicitly assumes—as indeed it must—a form of the completeness axiom of the real numbers (monotone-sequence property). Thus, Cauchy took much less for granted than Lagrange and Ampère in dealing with continuity—the intermediate-value property is no longer self-evident.

its corollary—the Mean Value Theorem—is an important result, and depends also upon the intermediate-value property that Cauchy proved continuous functions to have. These considerations make it a good choice to illustrate Cauchy's delta-epsilon proof-procedure.

This proof-procedure employs the translation of his verbal definitions into the algebraic language of inequalities. The actual technique of the proof of this theorem is just like Ampère's; the essential difference is that it is not just an ad hoc technique. Cauchy's proof-method is dictated by his definitions—because his definitions are in fact a systematization of the proof-procedure worked out by Lagrange and Ampère.

Perhaps it was the teaching of a course at the Ecole Polytechnique that stimulated Cauchy to give systematic definitions of the concepts of the calculus, to justify already accepted and established proof-procedures. Teaching requires making explicit for one's students concepts which the teacher may be able to deal with "in use." In the eighteenth and nineteenth centuries, the elegant, systematic treatments often appear in courses of lectures: witness Weierstrass's lectures at Berlin, or Lagrange's *FA*. In any case, the appearance of new standards of rigor in widely read texts and in lectures by leading figures leads to the rapid acceptance of these standards and of the proof-techniques that go with them.

These remarks suggest one more reason for focussing attention on the theorem of Leçon 7. Although it is not the first proof to use delta-epsilon techniques, it is the first such proof to use the delta and epsilon. This is more than a curiosity, for the spread of this notation testifies to the hold that the details of Cauchy's work quickly gained on the methods of proof used by nineteenth century analysts.

The theorem in question is the following:

> If, $f'(x)$ being continuous between the limits $x = x_0$, $x = X$, we designate by A the smallest, and by B the largest value which the derived function $f'(x)$ receives in the interval, the ratio of the finite differences $\frac{f(X)-f(x_0)}{X-x_0}$ will necessarily be between A and B.[61]

This is what Ampère took as the defining property of $f'(x)$. In his proof, Ampère made use of the fact that $f'(x)$ was very close in value to $\frac{f(x+i)-f(x)}{i}$; in fact, that the difference, I, could be made arbitrarily small for i sufficiently small. His justification was modelled on Lagrange's work, and implicitly assumed that $f'(x)$ was the limit of the ratio of differences.

For Cauchy, this procedure is justified by his own definition of $f'(x)$. However, the proof is substantially the same—and a great deal easier to follow. I shall preserve his notation and terminology throughout:

Choose δ, ε as "two very small numbers." δ is chosen such that, for all values of i less than δ, and for x between x_0 and X, the ratio $\frac{f(x+i)-f(x)}{i}$ is greater than $f'(x) - \varepsilon$ and less than $f'(x) + \varepsilon$. Here Cauchy assumed that the derivative $f'(x)$ exists and is finite on the interval $[x_0, X]$ and applied his definition of $f'(x)$. In saying that he can, given $\varepsilon > 0$, find δ such that

$$f'(x) - \varepsilon < \frac{f(x+i) - f(x)}{i} < f'(x) + \varepsilon \quad \text{if} \quad i < \delta,$$

he merely translated his definition of the derivative into an algebraic inequality.

---

[61] *Calcul infinitésimal*, p. 44. Cauchy said that the ratio "sera... comprise entre A et B," meaning $A \leq \frac{f(X)-f(x_0)}{X-x_0} \leq B$. For strict inequality, he used the term "renfermée."

It must be pointed out that Cauchy took his definition of $f'(x)$ for a particular x and applied it to the whole interval. To say what is implicit here, that given an ε we can find a δ that works for *every* x in the interval, is to assume that $f'(x)$ is the uniform limit of the quotients $\frac{f(x+i)-f(x)}{i}$ in the interval (or, what can be shown to be equivalent by methods not available to Cauchy, that the derivative is continuous there). The confusion arises from not precisely specifying on what the variable δ depends.

Having made this algebraic statement, Cauchy proceeded to interpose $n-1$ new values of the variable x, namely, $x_1, x_2, \ldots, x_{n-1}$, between x_0 and X. Now if the interval $X - x_0$ is divided into elements $x_1 - x_0, x_2 - x_1, \ldots, X - x_{n-1}$, all of the same sign and each with numerical value less than δ (not necessarily equal as they were for Ampère),

$$f'(x_0) - \varepsilon < \frac{f(x_1) - f(x_0)}{x_1 - x_0} < f'(x_0) + \varepsilon$$

$$f'(x_1) - \varepsilon < \frac{f(x_2) - f(x_1)}{x_2 - x_1} < f'(x_1) + \varepsilon$$

$$\cdots\cdot$$

$$f'(x_{n-1}) - \varepsilon < \frac{f(X) - f(x_{n-1})}{X - x_{n-1}} < f'(x_{n-1}) + \varepsilon$$

and therefore all the fractions of the form (in my notation) $\frac{f(x_k)-f(x_{k-1})}{x_k-x_{k-1}}$ are greater than $A - \varepsilon$ and less than $B + \varepsilon$, since A and B are the minimum and maximum values of $f'(x)$ on the interval.

Now he invoked a lemma on fractions. It is general for n fractions, and he gave the case $n = 2$. If x/a and y/b have positive denominators, $\frac{x+y}{a+b}$ lies between them. This is easily generalized to n fractions: the result was derived in the *Cours d'analyse* of 1821, and is the same as the fraction-inequalities worked out by Ampère.[62]

Applying this to the fractions $\frac{f(x_k)-f(x_{k-1})}{x_k-x_{k-1}}$, all of which have positive denominators, we have that $\frac{f(X)-f(x_0)}{X-x_0}$ will lie between the quantities $A - \varepsilon$, $B + \varepsilon$. Since this is true no matter how small ε is, we can conclude

$$A \leq \frac{f(X) - f(x_0)}{X - x_0} \leq B, \text{ and the theorem is proved.}$$[63]

There are important differences between Cauchy's proof and that of Ampère. First, in notation: using the delta instead of saying "a value of i" avoids the use of i for two separate concepts, and makes it much easier to follow the proof. Similarly, the index notation for the values of the variable $x_0, x_1, \ldots$ is clearer than the $a, b, \ldots$ used by Ampère. Cauchy had the knack of devising clear notation.

He made his hypotheses explicit; the proof is crystal clear, and when he erred (as in assuming what was later defined as uniform convergence) the reader can see exactly where. The actual *use* in a proof of the definition of $f'(x)$ as a limit is typical of Cauchy—and everyone who follows him. Finally, the fact that, if for any ε, $A - \varepsilon < x < B + \varepsilon$, we can conclude $A \leq x \leq B$, is very useful. Cauchy's use of this shows that he does not confuse "$<$" with "$\leq$" or with "bounded away from."

---

[62] *Cours d'analyse*, p. 368 (Note II, Theorem XII). Compare Ampère, *op. cit.*, pp. 151–154, and my Footnote 37.

[63] *Calcul infinitésimal*, p. 45. Cauchy did not use the symbols "$<$" and "$\leq$," but expressed the distinction verbally. Compare Note 61.

The corollary follows: if $f'(x)$ is itself continuous between x_0 and X, then, in passing from one limit (Cauchy here used "limit" to mean bound) to the other, this function varies in such a way as to remain always between the two values A and B, and to take successively all the intermediate values. This corollary requires also the intermediate-value theorem for continuous functions.[64]

Cauchy realized that by using inequalities he could give the first rigorous definition of limit, and thus base the calculus on the concept of limit. Though his definition was verbal, the algebraic translation to deltas and epsilons was obvious to him. The use of delta and epsilon was therefore justified in proofs about the concepts of the calculus. Seeing this and exploiting the method of proof constitute a great contribution of Cauchy to the foundations of the calculus.

Cauchy did the job of synthesizing previous work so well that his successors never felt the need to go back to the pioneering efforts which contributed to the synthesis. In seeing the roots of his achievement in the work of Lagrange and Ampère, we can better understand how such a monumental work of synthesis is made possible.

A large part of Cauchy's achievement, even in the restricted view taken by the present chapter, is certainly in deriving and proving new results. But another important part is the result of his ability to cull precise concepts from work written with ill-defined and hazy ideas, and to present proofs which are models of clarity in the notation and in the ideas.

Lagrange had forged the powerful tool of delta-epsilon techniques; Ampère had related it to a definition of $f'(x)$. Cauchy had made its use the starting point for rigorous proofs concerning the concepts of the calculus. Weierstrass exploited the delta-epsilon methods still further. Beginning with a purely arithmetic characterization of limit, he and his school were able to prove theorems on limits *per se*. In addition, he clearly distinguished between point-wise and uniform continuity and convergence. Thus many general results concerning the foundations of the calculus and convergence of infinite series could be rigorously proved and a more powerful theory perfected. As we look at the advances made, we see again and again the interplay between the questioning of the obvious and its proof—or rejection. Rigor in analysis has, since Cauchy, been identified as the science of delta and epsilon.

---

[64] See Note 60. The corollary is to be found in the *Calcul infinitésimal*, p. 45.

Conclusion

Lagrange's commitment to the necessity of an algebraic foundation for the calculus led him to the major accomplishments of the *FA* and *CF*: the sharply argued critique of the prevailing eighteenth century foundations for the calculus, the study of functions by means of their power series expansions, the derivation and use of the remainder term of the Taylor series, and the development of what are essentially delta-epsilon proofs. His influence on the development of analysis in the nineteenth century rests on these accomplishments.[1]

The development of the foundations of the calculus from Newton and Leibniz to Weierstrass, and the length of time this development took, may illustrate how hard to understand the concepts involved are. The difficulties are clear to anyone who has taught the calculus. Mathematics did not develop in the way in which we teach it in our elementary courses; nevertheless, ideas which are difficult for beginning students today were often sources of great difficulty in the past. The connection between equality of derivatives and a set of complicated delta-epsilon inequalities, and the whole motivation for the rigorous defintions, still troubles many a student.

Delta-epsilon proofs, once one has seen a few examples, are easy to construct.[2] The verbal limit-concept is intuitively plausible as soon as the student begins thinking about tangents, or about rates of change. But the difficulty lies in bringing these two ideas together. That the basic equations between limits of ratios are actually sets of possible inequalities seems contrary to all expectation. Not until the calculus was over a century and a half old was this conclusion finally accepted.

There was a tension between the expectations of mathematicians in the seventeenth and eighteenth century and the apparent properties of the concepts of the calculus. Since Plato, mathematics had been regarded as the science of that which is unchanging and certain; in eighteenth-century analysis, the "unchanging and certain" could only mean fixed and finite magnitudes. But the calculus, after a brief flirtation with the idea of "fixed infinitesimal," appeared as a mathematics of change. A differential quotient or a fluxion could be interpreted as a velocity. In apparent contrast, the differential quotient or fluxion could be interpreted as a limit; but taking a limit was considered a process, and dy/dx the culmination of a limiting process "as numerator and denominator simultaneously become zero." These interpretations are unsatisfactory; a mathematical object was expected to remain fixed, not undergo change. Thus the mathematical status of an equality between limits was uncertain. Writing equality between quantities which appear to be always approaching, but never reaching, their limits, makes little mathematical sense from the

---

[1] See pp. 79–80, above; also, see Chapter 4.

[2] Eighteenth-century algebraists had no trouble manipulating inequalities more complex than those used in the foundations of analysis today.

point of view of eighteenth century algebra. Yet equalities between differential quotients are the subject of the calculus.

Lagrange wanted to base the calculus on the algebra of fixed and finite quantities. Here he was on the right track, and his insistence on this point sufficed to discredit the old verbal notion of limits, the velocity-based fluxion, Euler's ratios of zeros, and first and last ratios. But Lagrange did not realize that *inequalities* between fixed and finite quantities, not equalities, would eventually become the foundation for the calculus.

Cauchy used inequalities to recast, and thus resurrect, the old terminology of limits, infinitesimals, and differentials. His *Leçons sur le Calcul Infinitésimal*, as the title shows, is conservative in language. He retained the word "infinitesimal," though defining it in terms of limits, and preserved a verbal definition of limit, though refining it. But although his verbal statements involve equalities between limits, his proofs about limits involve inequalities. Thus he kept continuous the limit-tradition of Newton and D'Alembert, and made it possible for nineteenth-century analysts to interpret eighteenth-century work in terms of the rigorous limit-concept. His use of algebraic inequalities makes his proofs acceptable by modern standards, and gives mathematical meaning to his verbal definition.[3]

Cauchy's *Cours d'Analyse* and *Calcul Infinitésimal* still left unsolved the problems of changing the order of limiting operations, and the fundamental topological properties of the real number system and the related properties of continuous functions. These problems were taken up by men like Bolzano, Abel, and Cauchy himself. They were completely resolved by Weierstrass and his school.[4] Moreover, Weierstrass' precise algebraic definitions of the concepts of the calculus eliminated any vagueness which remained. The reduction of the calculus to algebra thus had a historical importance greater than a mere contribution to the economy of thought. It led to the solution of problems which could not have been solved in the eighteenth century, and which often cannot even be formulated in eighteenth-century language. The reduction of the calculus to algebra was a long and complex process, but the first major steps were taken in the *Théorie des fonctions analytiques*; it is to Lagrange that we owe the conception of the calculus as algebra.

---

[3] Though of course there are gaps in his proofs due to other causes—for instance, his failure to distinguish between point-wise and uniform continuity.

[4] For Weierstrass' school, see the Bibliography.

Appendix

This appendix will reproduce Lagrange's proof that, if $f'(x)$ is positive from $z = a$ to $z = b$, for $b > a$, then $f(b) - f(a)$ is positive.[1]

Lagrange supposed that $f', f'', \ldots$ are all finite. Then there is an i such that the value of $if'(z)$ will exceed that of the rest of the series $i^2/2!\, f''(z) + i^3/3!\, f'''(z) + \cdots$.[2] If $f'(z)$ is positive, i (also positive) can be found small enough so that the entire series.

$$if'(z) + i^2/2!\, f''(z) + i^3/3!\, f'''(z) + \cdots$$

will be positive also. But this series is just $f(z+i) - f(z)$. Thus, he had proved that, if $f'(z)$ is positive, $i > 0$ can be taken sufficiently small so that $f(z+i) - f(z)$ is also positive.

Lagrange did not say whether he conceived this i to depend on the choice of z, but assumed implicitly in what follows that it is independent of z.[3]

He divided the interval (which we would write $[a, b]$) into $n+1$ parts, each of length i, so that $i = \frac{b-a}{n+1}$. By what was done above, $f(a+1) - f(a)$ is positive, $f(a+2i) - f(a+i)$ is positive,... $f(a+(n+1)i) - f(a+ni)$ is positive, by setting successively $z = a,\ a+i, \ldots, a+ni$, as long as the derivatives $f'(a + \frac{k(b-a)}{n+1})$ are positive for $k = 0, 1, \ldots n$. (k-notation is mine) If this is so, then *a fortiori* he had that the sum $[f(a+(n+1)i) - f(a+ni)] + \cdots + [f(a+2i) - f(a+i)] + [f(a+i) - f(a)]$ will be greater than zero also. But that sum is $f(b) - f(a)$; so he had proved that $f(b) - f(a) > 0$.[4]

This proof has assumed the following:

1. $f(z), f'(z), f''(z) \ldots$ are defined on $[a, b]$ and finite for these values of z.
2. $f'(z)$ is strictly greater than 0 on $[a, b]$.
3. The series $f(z) + if'(z) + i^2/2!\, f''(z) + \cdots$ converges to the limit $f(z+i)$ for all z, $z+i$ on $[a, b]$. Therefore, there exists some δ such that, for all z between a and b, if $0 < i < \delta$,

$$if'(z) > \sum_{k=2}^{\infty} i^k/k!\, |f^{(k)}(z)|.$$

These assumptions are of course stronger than we now know to be necessary. For instance, it suffices for $f'(z) \geq 0$ to be defined on (a, b) while $f(z)$ is continuous at the end points a and b.

---

[1] *FA* 1797, p. 45; *Oeuvres* IX, p. 78, is somewhat modified.

[2] *FA* 1797, p. 45. Presumably, absolute values are intended, but he did not say so.

[3] The theorem could be proved without the assumption of independence, but would require the Heine-Borel property or something equivalent. If $f(z)$ is analytic, in the modern sense, on $[a, b]$, then i can be shown to be independent of z.

[4] *FA* 1797, p. 46.

A correct proof from these assumptions of the result $f(b) \geq f(a)$ would somehow involve the completeness of the real numbers.

Of course, the generalization to $f'(z) \leq 0$ implies $f(b) - f(a) \leq 0$ is obvious.

The version of this theorem given in the second edition of *FA* (1813) makes one important change. He no longer appealed to the series $f(z+i) = f(z) + if'(z) + i^2/2!\, f''(z) + \cdots$ to argue: $f'(z) > 0$ means that there is some i sufficiently small for which $f(z+i) - f(z) > 0$ also. Instead, he worked from the formula

$$f(x+i) = f(x) + iP. \text{ (Using, in 1813, } x \text{ instead of } z.)$$

He said that, since when $i = 0$, $P = f'(x)$, $f'(x) > 0$ means that $P(x, i)$ "must be positive from $i = 0$ up to a certain value of i which can be taken as small as desired."[5] He gave no justification for this, but what he evidently had in mind is the continuity of the function P with respect to i at $i = 0$. (In fact, uniform continuity is required.) As discussed in Chapter 4, the version of this Lemma in the *Calcul des Fonctions* makes the delta-epsilon arguments for this continuity-consideration explicit.

---

[5] *Oeuvres* IX, p. 78.

Bibliography

Analytical Bibliography: 1966

I. Lagrange

A. *Oeuvres de Lagrange.* Publiées par les soins de M. J.-A. Serret. 14 vols. (Paris: Gauthier-Villars, 1867–1892).

The arrangement is as follows:

Vol. I: Delambre's *Eloge*; material printed in the publications of the Turin Academy.
Vol. II: Material printed in the publications of the Turin and Berlin Academies.
Vol. III: Material printed in the publications of the Berlin Academy.
Vol. IV: Material printed in the publications of the Berlin Academy.
Vol. V: Material printed in the publications of the Berlin Academy.
Vol. VI: Material printed in the publications of the Académie des Sciences, Paris, and Institut de France.
Vol. VII: Diverse pieces not included in Academy publications.
A table of contents at the end of Vol. VII covers the first seven volumes.
Vols. VIII–XII: *Ouvrages didactiques*
Vol. VIII: *De la résolution des equations numériques* (1808).
Vol. IX: *Théorie des fonctions analytiques* (reprinted from second edition, 1813).
Vol. X: *Leçons sur le calcul des fonctions* (reprinted from second edition, 1806).
Vols. XI–XII: *Mécanique analytique* (reprinted from second edition, 1811–1815).
Vols. XIII–XIV: Published correspondence. These two volumes, edited by Ludovic Lalanne, are the only volumes in the *Oeuvres* which are indexed.

B. Published works of Lagrange used for this book

1. *De la résolution des équations numériques de tous les degrés* (Paris: chez Duprat, An VI).
2. "Discours sur l'objet de la théorie des fonctions analytiques," *Journal de l'Ecole Polytechnique* VIe cah. t. II An VII (1799); *Oeuvres* VII, pp. 323–328.
3. "Leçons élémentaires sur les mathématiques données à l'école normale en 1795," *Séances des Ecoles Normales* An III (1794–5); *Oeuvres* VII, pp. 181–288.
4. *Leçons sur le calcul des fonctions*, nouvelle édition (Paris: Courcier, 1806); *Oeuvres* X [see No. 15].
5. *Lettera di Luigi de la Grange Tournier, Torinese, all' illustrissimo Signor Conte Giulio Carlo da Fagnano, contenente una nuova serie per i differenziali ed integrali di qualsivoglia grado, corrispondente alla Newtoniana per le potestà e la radici* (Torino: Nella Stamperia reale, 1754); *Oeuvres* VII, 583–588.
6. *Lettres Inédites de Joseph-Louis Lagrange à Léonard Euler.* Tirées des Archives de la salle des conférences de l'Académie Impériale des Sciences de Saint-Pétersbourg. Published by B. Boncompagni (St. Pétersbourg: Expédition pour la confection des papiers de l'état, 1877).
7. *Mechanique analitique* (Paris: Desaint, 1788).
8. *Mécanique analytique* (Paris: Courcier, 1811–15); Oeuvres XI–XII.

9. "Note sur la métaphysique du calcul infinitésimal," *Miscellanea Taurinensia, II*, pp. 17–18, (1760–1761); *Oeuvres* VII, 597–599.
10. "Nouvelle méthode pour résoudre les équations littérales par le moyen des aéries," *Mémoires de l'Acad. royale des Sciences et Belles-Lettres de Berlin, 24*, pp. 251–326 (1768); *Oeuvres* III, pp. 3–73.
11. "Recherches d'Arithmétique," *Nouv. Mém. Berl.*, Part I (1773), Part II (1775); *Oeuvres* III, pp. 695–795.
12. "Réflexions sur la résolution algébrique des équations," *Nouv. Mém. Berl.*, pp. 134–215 (1770); pp. 138–253 (1771); *Oeuvres* III, pp. 205–424.
13. "Sur la forme des racines imaginaires des équations," *Nouv. Mém. Berl*, Classe de mathématique pp. 222–258 (1772); *Oeuvres* III, pp. 477–516.
14. "Sur la résolution des équations numériques," et "Additions au mémoire sur la résolution des équations numériques," *Mém. Acad. Berl., 23*, pp. 311–352 (1767); *24*, pp. 111–180 (1768); *Oeuvres* II, pp. 539–580, pp. 581–654.
15. "Sur le Calcul des Fonctions," *Séances des Ecoles Normales, 10* An IX (1801).
16. "Sur une nouvelle espèce de calcul relatif à la différentiation et à l'intégration des quantités variables," *Nouv. Mém. Berl.*, Classe de mathématiques, pp. 185–221 (1772); *Oeuvres* III pp. 439–476.
17. *Théorie des fonctions analytiques.* Contenant les principes du calcul différentiel, dégagés de toute considération d'infiniment petits ou d'évanouissans, de limites ou de fluxions, et réduits a l'analyse algébrique des quantités finies (Paris: Imprimérie de la république, An V [1797]).
18. *Théorie des fonctions analytiques*, nouvelle édition, révue et augmentée par l'Auteur (Second edition, Paris: Courcier, 1813); *Oeuvres* IX.
19. *Traité de la résolution des équations numériques de tous les degrés* (second edition, Paris: chez Courcier, 1808); *Oeuvres* VIII.
20. Lagrange *et al*, "Prix proposés par l'Académie Royale des Sciences et Belles-Lettres pour l'année 1786," *Nouveaux Mémoires de l'Academie Royale des Sciences et Belles-Lettres* Année 1784, Avec l'Histoire pour la même année, *Histoire* 1784, pp. 12–14.
21. Lagrange *et al*, "Prix proposés par l'Académie Royale des Sciences et Belles-Lettres pour l'Année 1788," *Histoire* 1786, pp. 8–9.

C. Lagrange's manuscripts from his Paris period are at the Bibliothèque de l'Institut, MSS 886–916.

Those relating specifically to analysis:

Infinitesimal calculus	904
Series	906
Theory of functions	907
Algebraic analysis	908
Arithmetic and geometry	909

Most closely resemble published work. An exception is "Sur la théorie des parallèles," MS 909, ff 17–43.

Professor J. E. Hofmann has informed me that a correspondent in East Berlin has told him that there are some Lagrange manuscripts at the Akademie der Wissenschaften in East Berlin, but these are drafts for published work.

D. Other Published Lagrange Letters.

Only Fagnano's letters (1754–1759) have interesting mathematical material.

1. Beltrami, Eugenio, "Comunicazione di una lettera di Lagrange a F. M. Zanotti," *Rendiconto della Sessioni dell' Accademia delle Scienze dell' Instituto di Bologna*, 1872–1873, pp. 97–100.
2. Boatner, Charlotte H., "Certain unpublished letters from French scientists of the revolutionary period taken from the files of Joseph Lakanal," *Osiris, 1*, pp. 173–183, (1936).

3. Bopp, K., "Eine Schrift von Ensheim "Recherches sur les calculs differentiel et intégral" mit einem sich darauf beziehenden, nicht in die "Oeuvres" übergegangenen Brief von Lagrange," *Sitzungsber. der Heidelberger Akademie der Wissenschaften*, 1913, pp. 1–49.
4. Fagnano, Giulio Carlo, Conte di, *Opere matematiche del Marchese Giulio Carlo de' Toschi di Fagnano*, ed. V. Volterra, G. Loria, D. Gambioli, *III* (Milano: Albrighi, Segati,... 1912), pp. 179–213.
5. Favaro, Antonio, "Sette lettere di Lagrange al P. Paolo Frisi, tratte dagli autografi nella Biblioteca Ambrosiana di Milano," *Atti R. Acc. Torino, 31*, pp. 182–194, (1895).
6. Henry, Charles, "Sur quelques billets inédits de Lagrange," *Bullettino di bibliografia e di storia delle Scienze matematiche e fisiche, XIX*, pp. 129–135, (1886).
7. Riccardi, Pietro, "Alcune lettere di Lagrange, di Laplace e di Lacroix dirette al matematico Pietro Paoli," *M. R. Acc. Modena*, Ser. 3, *1*, Sci., pp. 105–129, (1897).
8. Pittarelli, G., "Duo lettere inedite di Lagrange all' abate di Caluso esistente nell' Archivio storico municipale di Asti," *Atti del IV Congresso internazionale dei Matematici*, Roma, 1908, *III* (Roma: 1909) pp. 554–556.
9. Sarton, George; Taton, René; Beaujouan, Guy, "Documents nouveaux concernant Lagrange," *Rev. d'Hist. des. Sci. 3*, pp. 110–132, (1950).
10. Sclopis, Federigo, "Lettera di Luigi Lagrange al marchese D. Caracciolo," *Atti della R. Accad. di Torino VII*, pp. 431–437, (1871).

II. Articles about Lagrange.

The best article on his mathematical career is Loria (No. 21). The best on his personality is Sarton, who has used the published correspondence as well as contemporary accounts. Both Loria and Sarton have extensive bibliographies. Hankel and Vuillemin deal with variour aspects of his mathematics. Juskewitch's article, relating *FA* to Euler's *ICD*, is the best thing ever written on Lagrange's mathematics. Dickstein summarizes the immediate reaction to *FA*. Biot, Maurice's two articles, Delambre, Denina, Virey and Potel are contemporary accounts. The first three contain many interesting anecdotes. The other biographical articles on this list have little that is new.

1. [Anon.] ["Review of *FA*"], *Monthly Review*, London, N.S., *XXVIII* Appendix, pp. 481–499 (1799).
2. Biermann, Kurt-R., "Lagrange im Urteil und in der Erinnerung A. v. Humboldts," *Monatsberichte der deutschen Akademie der Wissenschaften zu Berlin*, B. 5, Heft 7, pp. 445–450, (1963).
3. Biot, Jean-Baptiste, "Notice historique sur M. Lagrange," *Mélanges scientifiques et littératires III* (Paris: Michel Lévy Frères, 1858) pp. 117–124.
4. Bortolotti, Ettore, "Lagrange (O Lagrangia), Giuseppe Luigi" *Enciclopedia Italiana* (Roma: Treccani, 1933) *XX*, pp. 380–381.
5. Briano, Giorgio, *Giuseppe Luigi Lagrangia*, (Torino: Unione Tipografico-Editrice Via Carlo Alberto, No. 33, Casa Pomba, 1861).
6. Burzio, Filippo, *Lagrange* (Torino: Unione Tipografico-Editrice Torinese, 1942).
7. Chiò, Felice, "Luigi Lagrange," *Gazzetta di Torino* N. 165 (Torino: Tipografia Letteraria, 1867).
8. Delambre, Jean Baptiste Joseph, "Notice sur la vie et les ouvrages de M. le Comte J.-L. Lagrange," *Oeuvres de Lagrange, I* (Paris: Gauthier-Villars, 1867) pp. viii–li.
9. Denina, Carlo, "Louis de la Grange," *La Prusse littéraire sous Fréderic II, II* (Berlin: H. A. Rottmann, 1790) pp. 140–147.
10. Dickstein, S., "Zur Geschichte der Prinzipien der Infinitesimal-rechnung. Die kritiker der 'théorie des fonctions analytiques' von Lagrange," *Abhandlungen zur Geschichte der Mathematik, 9*, pp. 65–79 (1899), or in *Zeitschrift für Mathematik und Physik* Supplement to *44*, pp. 65–79 (1899).
11. Forti, Achille, *Intorno alla vita e alle opere di L. Lagrange Discorso* (Pisa: Tip. Nistri, 1868).
12. Genocchi, Angelo, "Luigi Lagrange," *Il primo secolo della Reale Accademia della Scienze di Torino* (Turin: Stamperia Reale di G. B. Paravia e C., 1883) pp. 86–95.
13. Guareschi, Icilio, "Notizie storiche Intorno a Luigi Lagrange," *Mem. R. Acc. Torino 64* (1) pp. 1–13 (1913).

14. Hankel, Hermann, "Lagrange's Lehrsatz," Ersch und Gruber, *Allgemeine Encyklopädie der Wissenschaften und Künste, 79* (A-G) (Leipzig: Brockhaus, 1865) pp. 353–367.
15. Herrmann, Dieter, "Joseph-Louis Lagrange," *Sterne, 39*, pp. 58–63, (1963).
16. Jourdain, P. E. B., "The Ideas of the 'fonctions analytiques' in Lagrange's Early Work," *Proceedings, International Congress of Mathematicians. II* (Cambridge: University Press, 1912) pp. 540–541.
17. Juschkewitsch, A. P., "Euler und Lagrange über die Grundlagen der Analysis," *Sammelband der zu ehren des 250 Geburtstages Leonhard Eulers*. Deutschen Akademie der Wissenschaften zu Berlin (Berlin: Akademie-Verlag, 1954), pp. 224–244.
18. Korn, A., "Joseph-Louis Lagrange," *Sitzungsberichte der Berliner Mathematischen Gesellschaft, 12*, pp. 90–94 (1913).
19. Külb, Ph. H., "Joseph Louis la Grange," Ersch und Gruber, *Allgemeine Encyklopädie der Wissenschaften und Künste, 79*, (Leipzig: Brockhaus, 1865), pp. 339–353.
20. Krylov, A. N., ed., *Josef Louis Lagrange, 1736–1936*. Svornik Statei K 200-Letiu So Dnya Rojdenia (Moskva: Izd. Akademia Nauk SSR., 1937), in Russian. Table of contents in *Isis 28*, p. 199, (1938).
21. Loria, Gino, "G. L. Lagrange nella vita e nelle opere," *Annali di matematice XX*, pp. ix–lii. Reprinted in Loria, Gino, *Scritti, Conferenze, discorsi*, (Padova: Antonio Milani, 1937), pp. 293–333.
22. Loria, G., "Lagrange e la storia delle matematiche," *Bibl. Math.* III Ser., T. 13, pp. 333–338, (1912–13).
23. Loria, G., "Lagrange nelle Accademie di Berlino e Parigi," A. Lorgna, *Memorie pubblicate nel secondo centenario della nascita, a cura dell' Accademia Scienze Lettere di Verona* (Verona: La Tipografica Veronese, 1937), pp. 71–80.
24. Loria, G., "Nel secondo centenario della nascita di G. L. Lagrange," *Isis, 28*, pp. 366–375 (1938).
25. Lorey, Wilhelm, "Josef Louis Lagrange," *Journ. f. reine und angew. Math.*, T. 175, pp. 224–239 (1936).
26. [Maurice, Frédéric], "Lettre à M. le Redacteur du *Moniteur Universel* sur l' Eloge de Lagrange, par M. Delambre, . . . ," *Le Moniteur Universel*, Samedi, 26 Février 1814, No. *57*, pp. 226–228.
27. Maurice, Frédéric, "Lagrange," *Biographie universelle, 22* (Paris: Ch. Delagrave et Cie, n.d.), pp. 523–534.
28. Menabrea, Luigi Federigo, "Luigi Lagrange," *Atti delle reale Accademia delle Scienze di Torino*, II, Disp. 7[a], pp. 540–556 (1867).
29. Nielsen, Niels, "Lagrange," *Géomètres français sous la révolution* (Copenhague: Levin and Munksgaard, 1929), pp. 136–152.
30. Riccardi, Pietro, "Lagrange, Luigi da Torino, 1736–1813," *Biblioteca Matematica Italiana* (Modena: Coi Tipi della Società Tipografica, 1893), Section 2, pp. 1–11.
31. Sarton, George, "Lagrange's Personality," *Proc Amer. Phil. Soc., 88*, pp. 457–96, (1944).
32. Ugoni, Camillo, "Giuseppe Luigi Lagrange," *Commentari dell' Ateneo di Brescia*, (1845–46) pp. 352–360.
33. Vacca, Giovanni, "Sui Primi anni di Giuseppe Luigi Lagrange," *Boll. di bibl. e storie della matematiche*, IV, pp. 1–4, (1901).
34. Virey, Julien Joseph et Potel, *Précis historique sur la vie et la morte de Lagrange*, (Paris: Mme Ve Courcier, 1813).
35. Vuillemin, Jules, *La philosophie de l'algèbre de Lagrange (Réflexions sur la mémoire de 1770–1771)*. (Paris: Palais de la Découverte, 1960).

III. Works on the Calculus of a Lagrangian Tendency.

1. Crelle, August Leopold, *Versuch einer allgemeinen Theorie der analytischen Facultäten, nach einer neuen Entwickelungs*-Methode (Berlin: G. Reimer, 1823).

2. Froberg, J. P., *et al, De analytica calculi differentialis et integralis theoria, inventa a cel. La Grange* (Upsaliae: 1807–1810).
3. Gruson, Jean Philippe, "Le Calcul d'Exposition," *Mémoires de l'Académie Royale des Sciences et Belles-Lettres de Berlin* Classe de mathématique, (1798), pp. 151–216, (1799–1800), pp. 157–188.
4. Hoene-Wronski, Jozef Maria, *Philosophie de l'infini, contenant des contre-réflexions et des réflexions sur la métaphysique du calcul infinitesimal* (Paris: P. Didot l'aîné, 1814).
5. Lacroix, Silvestre François, *Traité du Calcul Différentiel et du Calcul Intégral*, 3 vols. (Paris: J. B. M. Duprat, 1797–1800).

On Lacroix, see René Taton, "Sylvestre-François Lacroix (1765–1834) mathématicien, professeur et historien des sciences," *Actes du VIIe Congrès international d'Histoire des Sciences* (Paris: Hermann, 1953), pp. 588–593.

René Taton, "Condorcet et Sylvestre-François Lacroix," *Rev. Hist. Sci., 12*, pp. 127–158, pp. 243–262 (1959).

6. Lacroix, Silvestre François, *An elementary treatise on the differential and integral calculus*, tr. from the French with an Appendix and notes by Charles Babbage, George Peacock, J. F. W. Herschel, (Cambridge: J. Smith for J. Deighton *et al*, 1816).

On the Cambridge Analytical Society, see W. W. Rouse Ball, *A History of the Study of Mathematics at Cambridge* (Cambridge: University Press, 1889).

7. Laplace, Pierre Simon, *Théorie analytique des probabilités*, (3d edition, Paris: Courcier, 1820) reprinted in *Oeuvres complètes de Laplace*, Tome 7e (Paris: Gautheir-Villars, 1886) (used the Taylor series with Lagrange remainder, p. 179).
8. Pasquich, Johann, "Anfangsgründe einer neuen Exponentialrechnung," *Archiv der reinen und angewandten Mathematik, II*, (8), pp. 386–424 (1798).
9. Servois, F. J., "Essai sur un nouveau mode d'exposition des principes du calcul différentiel," *Annales de Mathématiques, V*, pp. 93–141, (1814–15).
10. Valperga-Caluso, Tommaso, "Sul paragone del calcolo delle funzioni derivate coi metodi anteriori," *Memorie di Matematica e di fisica delle Società Italiana delle Scienze, XIV* (Part 1), pp. 201–224, (1809).

IV. History of Mathematics

A. Bibliographies.

The most useful bibliographical work is Sarton, which includes not only lists of the major research tools but also a list of biographies of eminent nineteenth century mathematicians; Archibald is an outline history containing extensive bibliographical footnotes; Loria may be used in the same way as Sarton; *Mathematical Reviews* and *Zentralblatt* are indexing journals in mathematics, which include sections on history. Hofmann, listed under Histories, contains an extensive index of names with lists of works by, and works about, mathematicians up to about 1800. Standard bibliographical tools in the History of Science, such as the Critical Bibliographies in *Isis*, and the *Bulletin Signaletique*, are also of use in the history of mathematics.

1. Archibald, R. C., "Outline of the History of Mathematics," *Amer. Math. Monthly 561* (Jan. 1949).
2. Loria, Gino, *Guida allo studio della storia delle matematiche* (Milano: Ulrico Hoepli, 1946).
3. *Mathematical Reviews*, published by the American Mathematical Society (1940–).
4. Sarton, George, *The Study of the History of Mathematics* (Cambridge: Harvard, 1936), reprinted (New York: Dover, 1957).
5. *Zentralblatt für Mathematik und ihre Grenzgeblete* (Berlin, 1931–).

B. Histories

Cantor is the most comprehensive history of mathematics available, up to 1800; technical summaries of most important publications make it especially useful for the eighteenth century. Struik is the most reliable English work, and has good chapter-by-chapter bibliographies. The German edition has more up-to-date bibliographical information. Bell is indispensable for his treatment of nineteenth and early twentieth century topics. Bourbaki is a collection of historical articles from the Bourbaki *Eléments de mathématiques*; the articles have a very modern point of view, and many were written after taking a fresh look at the primary sources. Klein is an excellent, school-oriented, view of nineteenth century mathematics; table of contents is given in Sarton (IV. A. 4.). Boutroux gives a suggestive philosophical interpretation of the history of mathematics; Hofmann, though very concise, is reliable and has an extremely useful bibliography; Moritz's collection of quotations gives sources and page numbers for many "familiar quotations" by and about mathematicians.

1. Bell, Eric Temple, *The Development of Mathematics*, Second edition (New York and London: McGraw-Hill, 1945).
2. Bourbaki, Nicolas, *Eléments d'histoire des mathématiques* (Paris: Hermann, 1960).
3. Bonola, Roberto. *Non-Euclidean Geometry* (New York: Dover, 1955).
4. Boutroux, Pierre, *L'Idéal scientifique des mathématiciens dans l'antiquité et dans les temps modernes* (Paris: F. Alcan, 1920).
5. Cajori, Florian, *A History of Mathematical Notations*, 2 vols. (Chicago: Open Court, 1928–29).
6. Cantor, Moritz, *Vorlesungen über Geschichte der Mathematik*, 4 vols. (Leipzig: B. G. Teubner, 1900–08).
7. Carruccio, Ettore, *Mathematics and Logic in History and in Contemporary Thought* (London: Faber, 1964).
8. Hofmann, Joseph Ehrenfried, *Geschichte der Mathematik*, 3 vols. (Berlin: Walter de Gruyter, 1953–1957).
9. Klein, Felix, *Vorlesungen über lie Entwicklung der Mathematik im 19. Jahrhundert*, 2 vols. (Berlin: J. Springer, 1926–27).
10. Moritz, Robert Edouard, *Memorabilia Mathematica, or the Philomath's Quotation-Book* (New York: Macmillan, 1914).
11. Struik, Dirk J., *A Concise History of Mathematics* (New York: Dover, 1948).
12. Struik, Dirk J., *Abriss der Geschichte der Mathematik* (Berlin: VEB Deutscher Verlag der Wissenschaften, 1963).
13. *Encyklopdie der mathematischen Wissenschaften mit Einschluss ihrer Anwendungen*, (Leipzig: Teubner, 1898–1935). French revision (but only in part): *Encyclopédie des sciences mathématiques pures et appliquées* (Paris: Gauthier-Villars, and Leipzig: B. G. Teubner, 1904–1915). Description and table of contents in Sarton, *The Study of the History of Mathematics*, pp. 56–57.

V. General Background Works

A. Primary (Philosophy and General Mathematics)

1. Apostol, Tom M., *Mathematical Analysis* (Reading, Mass.: Addison-Wesley, 1957).
2. Berkeley, George, *The Works of George Berkeley.* Ed. A. C. Fraser (Oxford: Clarendon Press, 1871).
3. Comte, Auguste, *Cours de philosphie positive* (Paris: Bachelier, 1830–42).
4. Descartes, René, *Discours de la méthode pour bien conduire sa raison, et chercher la verité dans les sciences...* (Leyde: I. Maire, 1637).
5. Diderot, Denis, *Encyclopédie, ou Dictionnaire raisonné des sciences, des arts et des métiers, par une société de gens de lettres.* Mis en ordre et publié par M. Diderot... et quant à la partie mathématique, par M. D'Alembert (Paris: chez Briasson, David l'ainé, le Breton, Durand, 1751–65).
6. Euler, Leonhard, *Lettres à une princesse d'Allemagne sur divers sujets de physique et de philosophie* (St. Petersburg: Mietau, etc., 1770).
7. Hume, David, *An Enquiry Concerning the Human Understanding* (Chicago: Open Court, 1900)

8. Laplace, Pierre Simon, *Essai philosophique sur les probabilités.* Reprinted (Paris: Gauthier-Villars, 1921).
9. Leibniz, G. W., *Nouveaux essais sur l'entendement humain.* Ed. Emile Boutroux (Paris: C. Delagrave, 1899).
10. Locke, John, *An Essay Concerning Humane Understanding* (London: Printed by Eliz. Holt, for Thos. Bassett [1689]).
11. Newton, Isaac, *Philosophiae Naturalis Principia Mathematica* (Londini: J. Streater, 1687).
12. Saccheri, Girolamo, *Euclides ab omni naevo vindicatus.* Ed. and tr. George B. Halsted (Chicago: Open Court, 1920).
13. Somerville, Martha, *Personal Recollections, from Early Life to Old Age, of Mary Somerville* (Boston: Roberts Brothers, 1874).
14. Thomas, George B., *Calculus and Analytic Geometry* (Reading, Mass.: Addison-Wesley, 1960).
15. Widder, David V., *Advanced Calculus* (New York: Prentice-Hall, 1947).

B. Secondary (History of Philosophy and Related Subjects)

Boyer gives brief accounts of L. N. M. Carnot, Condorcet, Monge, Lagrange, Laplace, and Legendre; Merz vols. 1–2 gives a lot of material on nineteenth century science that can be found nowhere else; vols. 3–4 are also interesting, on philosophy; Randall is a good intellectual history; Enriques discusses the history of problems bearing on science.

1. Boyer, Carl B., "Mathematicians of the French Revolution" *Scripta Math.*, 25, pp. 11–31, (1960).
2. Enriques, Federigo, *The Historic Development of Logic* (New York: Henry Holt and Co., 1929)
3. Merz, John T., *A History of European Thought in the Nineteenth Century*, 4 vols., reprinted (New York: Dover, 1965).
4. Randall, John Herman, Jr., *The Making of the Modern Mind* (Cambridge, Mass.: Houghton Mifflin, 1954).

C. Academies and Schools

Harnack is an immensely valuable history of the Berlin academy. Vol. 1 of the Ecole Polytechnique Livre, "L'Ecole et la Science," is especially useful.

1. Amburger, Erik, *Die Mitglieder der Deutschen Akademie der Wissenschaften zu Berlin:* 1700–1950 (Berlin: Akademie-Verlag, 1950).
2. Biermann, Kurt-R., and Dunken, Gerhard, eds., *Deutsche Akademie der Wissenschaften zu Berlin: Biographischer Index der Mitglieder* (Berlin: Akademie-Verlag, 1960).
3. Harnack, Adolf, *Geschichte der Königlich Preussischen Akademie der Wissenschaften zu Berlin*, 3 vols. (Berlin: Reichsdruckerei, 1900).
4. Ecole Polytechnique, *Livre du centenaire* 1794–1894 (Paris: Gauthier-Villars et Cie, 1894–7).

D. Mechanics

Dugas is the standard history of mechanics; Jouguet has good short comments together with selections from important works in the history of mechanics; Mach's positivist history of mechanics is stimulating and valuable; Truesdell is illuminating on the importance of Euler.

1. Dugas, René, *A History of Mechanics*, tr. J. R. Maddox (Neuchâtel: Editions du Griffon, 1955).
2. Jouguet, Emile, *Lectures de Mécanique: La Mécanique enseignée par les auteurs originaux* (Paris: Gauthier-Villars et Cie, 1924).
3. Mach, Ernst, *The Science of Mechanics.* Sixth American edition (Chicago: Open Court, 1960).
4. Truesdell, C., *The Rational Mechanics of Flexible or Elastic Bodies*, 1638–1788. Introduction to Leonhardi Euleri *Opera Omnia*, X and XI, seriei secundae (Turici: Orell Füssli, 1960).

VI. Algebra

A. Primary Sources on Eighteenth Century Algebra

Newton's *Universal Arithmetic*, Maclaurin, and Euler's *Introductio* set most of the major problems. See also Lagrange, Nos. 1, 3, 10, 11, 12, 13, 14, 19.

1. Bernoulli, Daniel, "Observationes de seriebus quae form-antur ex additione vel subtractione quacunque terminorum se mutuo consequentium, ubi praesertim earundem insignis usus pro inveniendis radicum omnium Aequationum Algebraicarum ostenditur," *Commentarii Academiae Scientiarum Imperialis Petropolitanae*, III, pp. 85–100 (1728).
2. Bezout, E., "Mémoire sur plusieurs classes d'équations de tous les degrés qui admettent une solution algébrique" *Mémoires de l'académie royale des sciences*, Paris, 1762, pp. 17–52.
3. Bezout, E., "Recherches sur le degré des Equations résultantes de l'évanouissement des inconnues et sur les moyens qu'il convient d'employer pour trouver ces Equations," *Mem. Ac. Roy. Sci. Par.*, 1764, pp. 288–338.
4. Bezout, E., "Mémoire sur la résolution génerale des équations de tous les degrés," *Mem. Ac. Roy. Sci. Par*, 1765, pp. 533–552.
5. Courtivron, Gaspard le Compasseur, le Marquis de, "Sur une manière de résoudre par approximation les équations de tous les degrés," *Mem. Ac. Roy. Sci. Par.*, 1744, pp. 405–414.
6. Clairaut, Alexis Claude, *Elémens d'Algèbre*, 6e édition (Paris; chez Courcier, An X (1801)).
7. De Gua de Malves, Jean Paul, "Recherche du nombre des racines réeles ou imaginaires, réeles positives ou réeles negatives, qui peuvent se trouver dans les Equations de tous les degrés," *Mem. Ac. Roy. Sci. Par.* 1741, pp. 435–494.
8. Euler, Leonhard, "Analysis facilis et plana ad eas series maxime abstrusas perducens, quibus omnium aequationum algebraicarum non solum radices ipsae, sed etiam quaevis earum potestates exprimi possunt," *Nov. Acta Ac. Sci. Imp. Petrop., IV*, pp. 55–73 (1786).
9. Euler, Leonhard, "De Integratione Aequationum Differentalium altiorum gradum," *Miscellanes Berolinensia, VII* pp. 193–242 (1743).
10. Euler, Leonhard, "De innumeris generibus seriekum maxime memorabilium, quibus omnium aequationum algebraicarum non solum radices ipsae sed etiam quaecunque earum potestates exprimi possunt," *Nov. Act. Petrop, IV*, pp. 74–95 (1786).
11. Euler, Leonhard, *Elements of Algebra*, tr. John Hewlett (London: Longman, Orme, and Co., 1840).
12. Euler, Leonhard, *Introductio in analysin infintorum* (Lausanne: Bousquet, 1748); *Opera Omnia* I, vols. 8–9.
13. Euler, Leonhard, "Methodus generalis investigandi radices omnium aequationum per approximationem," *Nov. Act. Acad. Sci. Imp. Petrop., VI*, pp. 16–24 (1788).
14. Euler, Leonhard, "Nova ratio quantitates irrationalis proxime exprimendi," *Nov. Comm. Ac. Imp. Sci. Petrop., XVIII*, pp. 136–170 (1773).
15. Euler, Leonhard, "Obserationes circa radices aequationum," *Nov. Comm. Acad. Sci. Imp. Petrop., XV*, pp. 51–74 (1770).
16. Euler, Leonhard, "Recherches sur les racines imaginattes des équations," *Histoire de l'Academie Royale des Sciences et Belles-Lettres, Berlin*, 1749, pp. 222–288.
17. Euler, Leonhard, *Vollständige Anleitung zur Algebra* [First edition: St. Petersburg, 1770] (Stuttgart: Reclam-Verlag Stuttgart, 1959).
18. Fontaine, Alexis des Bertins, "Sur la résolution des équations," *Mém. Ac. Roy. Sci, Par.* 1747, pp. 665–677.
19. Gauss, Carl Friedrich, *Die vier Gauss' schen Beweise für die Zerlegung ganger algebraischer Functionen in reele Factoren ersten oder zweiten Grades* (1799–1849) Herausgegeben von E. Netto (Leipzig and Berlin: Wilhelm Engelmann, 1913).
20. Girard, Albert, *Invention houvelle en l'algèbre* (Amsterdam; chez Guillaume Iansson Blaeuw, 1629). Reprinted by D. Bierens de Haan (Leiden: chez Muré Frères, 1884).
21. Harriot, Thomas, *Artis analyticae praxis ad aequationem algebraicanova, expedita, et generali methodo, resolvendas* (Londini: Apud Robertum Barker . . . et Haered. Io. Billii, 1631).

22. Lacroix, S. F., *Complément des élémens d'algèbre à l'usage de l'école centrale des quatre-nations.* Third edition (Paris: chez Courcier, 1804).
23. Lambert, J. H., "Observations analytiques," *Nouv. Mém. Ac. Roy. Sci. et B.-L., Berlin*, 1770, pp. 225–244.
24. Lambert, J. H., "Observations sur les équations d'un degré quelconque," *Nouv. Mém. Ac. Roy. Sci. B.-L., Berlin*, 1763 pp. 278–291.
25. Lambert, J. H., "Observationes Variae in mathesin puram," *Acta Helvetica, III*, pp. 128–168 (1758).
26. Maclaurin, Colin, *A Treatise of Algebra, in Three Parts* (Second edition; London: Printed for A. Millar and J. Nourse, 1756).
27. Newton, Isaac, *Arithmetica Universalis, sive de compositione et resolutione arithmetics liber.* (Cantabrigiae: Typis Academicis, 1707; Londini: Benjamin Tooke, 1707).
28. Rolle, Michel, *Traité d'algèbre, ou principes généraux pour résoudre les questions de mathématique* (Paris; chez Estienne Michallet, 1690).
29. Segner, Joh. Andr. de., "Methodus Simplex et universalis, omnes omnium aequationum radices detegendi," *Nov. Coma. Acad. Scient. Imp. Petrop., VII*, pp. 211–226 (1758–1759).

B. Secondary Works

Bachmacova is good, but deals mostly with the nineteenth century; Matthiessen gives a technical summary of the contents of many major papers–an excellent place to begin; Netto and Le Vavasseur give an introduction to rational functions with many useful historical notes, more modern in treatment than Matthiessen; Reiff has some details not contained in Cantor; Tropfke's treatment of elementary algebra is useful.

1. Bachmacova, Isabella, "Le Théorème fondamentale de l'algèbre et la construction des corps algèbriques," *Arch. int. Hist. Sci, 13*, pp. 211–222, (1960).
2. Matthiessen, L., *Grundzüge der antiken und modernen Algebra der litteralen Gleichungen* (Leipzig: B. G. Teubner, 1878)
3. Netto, E., and Le Vavasseur, R., "Les Fonctions Rationnelles," *Encyclopédie des sciences Mathématiques pures et appliquées*, Tome I, Vol. 2, Fascicule 1, pp. 1–232 (Paris: GauthierVillars, 1907; Leipzig: B. G. Teubner, 1907).
4. Reiff, Richard, *Geschichte der unendlichen Reihen* (Tübingen: Verlag der H. Laupp'schen Buchhandlung, 1889).
5. Tropfke, Johannes, *Geschichte der Elementar-Mathematik*, 4 vols. (Leipzig: Walter de Gruyter, 1921–24).

VII. The Calculus

A. General Works on the History of Analysis

Boyer is the standard history, and contains an extensive, though unannotated, bibliography especially useful for the nineteenth century German literature. Boutroux, Tom. II, Ch. II, "L'Analyse infinitésimale" pp. 268–353 is a valuable exposition with many historical notes; Tom. II has a historical appendix with capsule biographies and lists of works by important "analysts" from Ahmes to Hermite (pp. 461–482). Brill and Noether, and Markuschewitsch, are excellent and thorough histories of their subjects; both stress the nineteenth century. Cajori (A. 8) is valuable for England; Cajori (A. 7) traces the origin of D'Alembert's limits, and discusses eighteenth century texts; Hankel has an important historical sketch of the function-concept; Pringsheim is an excellent treatment of the Taylor series, starting, however, with Taylor rather than Newton and Leibniz. The articles by Pringsheim and Molk, and Voss and Molk, are typical *Encyclopédie* treatments of topics: good expositions with abundant historical comments and notes; all are immensely valuable. Toeplitz is illuminating for the period up to Newton; Vivanti contains a thorough technical analysis of many papers, with some good interpretation, and is the only source besides Boyer which covers the whole subject systematically and in depth.

1. Bohlmann, G., "Übersicht über die wichtigsten Lehrbücher der Infinitesimal-rechnung von Euler bis auf die heutige Zeit," *Jahresbericht der Deutsch. Math. Verein.*, *6*, pp. 91–110 (1897).
2. Boutroux, Pierre, *Les principes de l'analyse mathématique: exposé historique et critique*, 3 vois. (Paris: A. Hermann et Fils, 1914–1919).
3. Boyer, Carl B., *The History of the Calculus and its Conceptual Development* (New York: Dover, 1959).
4. Boyer, Carl B., "Analysis: Notes on the evolution of a subject and a name," *Math. Teacher, 47*, pp. 450–462 (1954).
5. Brill, A., and Noether, M., "Die Entwickelung der Theorie der algebraischen Funktionen in älterer und neuerer Zeit," *Jahresbericht der Deutsch. Math.-Verein.*, *III*, pp. 107–566 (1892–3).
6. Cajori, Florian, "Discussion of Fluxions: From Berkeley to Woodhouse," *Am. Math. Monthly, 24*, pp. 145–54 (1917).
7. Cajori, Florian, "Grafting of the theory of limits on the calculus of Leibniz," *Am. Math. Monthly, 30*, pp. 223–34 (1923).
8. Cajori, Florian, *A History of the Conceptions of Limits and Fluxions in Great Britain from Newton to Woodhouse* (Chicago: Open Court, 1931).
9. Cajori, Fiorian, "Indivisibles and 'Ghosts of Departed Quantities' in the History of Mathematics," *Scientia, 37*, pp. 301–306 (1925).
10. Delambre, J.-B., *Rapport historique sur lea progrès des sciences mathématiques depuis 1789, et sur leur état actuel.* (Paris: de l'imprimérie impériale, 1810).
11. Hankel, Hermann, *Untersuchungen über die unendlich oft oszillierenden und unstetigen Funktionen* (Leipzig: Wilhelm Engelmann, 1905).
12. Markuschewitsch, A. I., *Skizzen zur Geschichte des analytischen Funktionen* (Berlin: Deutscher Verlag der Wissenschaften, 1955).
13. Pringsheim, A., "Zur Geschichte des Taylorschen Lehrsatzes," *Bibliotheca Mathematics*, III Ser., *1*, pp. 433–479 (1900).
14. Pringsheim, A., revised J. Molk, "Algorithmes Illimités," *Encyclopédie des Sciences Mathématiques Pures et Appliquées*, Tome 1, Vol. 1, Fascicule 4 (Paris: Gauthier-Villars, 1904; Leipzig: B. G. Teubner, 1904), pp. 209–328.
15. Pringsheim, A., revised J. Molk, "Nombres irrationnels et notion de limite," *Enc. des Sci. Math.*, Vol. I, Part 1, Fascicule 1, pp. 133–160.
16. Pringsheim, A., revised J. Molk, "Principes fondamentaux de la théorie des fonctions," *Enc. des Sci. Math.*, Tome 2 Vol. 1, Fascicule 1.
17. Toeplitz, Otto, *Die Entwichlung der Infinitesimalrechnung I* (Berlin: Springer-Verlag, 1949).
18. Vivanti, G., "Infinitesimalrechnung," pp. 639–869 in Moritz Cantor, *Vorlesungen über Geschichte der Mathematik*, IV.
19. Voss, A., and Molk, J., "Calcul différentiel," *Enc. des Sci. Math.*, Vol. II, Part 1, Fascicule 2, Part 3, pp. 242–336.

B. A List of Primary Sources Illustrating the History of the Foundations of the Calculus in the Eighteenth Century.

1. Collections of the Work of Newton and Leibniz

Much scholarship on Newton and Leibniz has dealt with the controversy over who invented the calculus; Boyer's bibliography may be consulted. Useful for understanding their work are the articles by Cajori. Eneström compares Leibniz's 1684 paper with Gerhardt's publication; Hofmann examines its topic with the help of MSS; Hochstetter contains several interesting articles, notably Hofmann's "Leibniz' mathematische Studien in Paris,"; Whiteside's superb study gives the seventeenth century background to Newton's achievement.

1. Leibniz, G. W., *The Early Mathematical Manuscripts of Leibniz*: tr. J. M. Child (Chicago and London: Open Court, 1920).

2. Leibniz, G. W., *Leibniz über die Analysis des Unandlichen* [1684–1703] ed. Gerhard Kowalewski (Leipzig: Wilhelm Engelmann, 1908).
3. Leibniz, G. W., "Analysis Infinitorum" [Papers publ. 1684–1713] *Mathematische Schriften*, ed. G. I. Gerhardt, Zweite Abtheilung, Band 1; *Gesammelte Werke*, ed. Georg Heinrich Pertz, Dritte Folge, Mathematik, Fünfter Band, pp. 220–392.
4. Leibniz, G. W., "Symbolismus memorabilis calculi algebraici et infinitesimales in comparatione potentiarum et differentiarum . . . " *Miscellanea Berolinensia, 1*, 1710.
5. Newton, Isaac, *Mathematical Principles of Natural Philosophy*, ed. Florian Cajori (Berkeley and Los Angeles: University of California Press, 1934). Compare (V. A. 11).
6. Whiteside, D. T., ed., *The Mathematical Works of Isaac Newton*, Vol. 1 (New York and London: Johnson Reprint Corporation, 1964). Contains:

   Quadrature of Curves from "Lexicon Technicum Or, an Universal Dictionary of Arts and Sciences" by John Harris, Vol. 2 (London, 1710).

   A Treatise of the Method of Fluxions and Infinite Series (London: T. Woodman and J. Millan, 1737).

   Of Analysis by Equations of an infinite Number of Terms, Published by John Stewart (London, 1745).

7. Cajori, F., "Leibniz, the Master Builder of Mathematical Notations," *Isis, 7*, pp. 412–429 (1925).
8. Cajori, F., "The Spread of Newtonian and Leibnizian Notations of the Calculus," *Bull. Am. Math. Soc., 27* pp. 453–458 (1921).
9. Eneström, Gustav, "Über die erste Aufnahme der Leibnizschen Differentialrechnung" *Bibliotheca Mathematica (3) IX*, pp. 309–20 (1908–9).
10. Hofmann, J. E., *Die Entwicklungsgeschichte der Leibnizschen Mathematik* (München: Leibniz-Verlag, 1949).
11. Hochstetter, E., ed., *Leibniz zu seinem 300. Geburtstag*, 1646–1946 (Berlin: W. de Gruyter, 1946).
12. Whiteside, Derek Thomas, "Patterns of Mathematical Thought in the Later Seventeenth Century," *Archive for History of Exact Sciences 1*, pp. 179–388 (1961).

2. Primary sources Illustrating the History of the Foundations of the Calculus from the Bernoullis to Lacroix.

   The most useful secondary sources are cited for each author. More work on Johann Bernoulli, on Arbogast, on Carnot, and on L'Huilier would be welcome; no complete study of Nieuwentijt exists, nor of Taylor. Cantor may be consulted on all these figures.

   1. D'Alembert, Jean-le-Rond, l'abbe Bossut, de la Lande, le Marquis de Condorcet, *Dictionnaire encyclopédique des mathématiques* (Paris: Hotel de Thou, 1789). Mathematical articles from the *Encyclopédie*.

      See Taton, René, "Les mathématiques selon l'encyclopédie," *Revue d'histoire des sciences, 4*, pp. 255–266 (1951).

   2. D'Alembert, Jean-le-Rond, *Recherches sur differens points importans du système du monde* (Paris: chez David, 1754).
   3. Arbogast, L. F. A., *Du Calcul des Dérivations* (Strasbourg: Levrault, Frères, An VIII (1800)).
   4. Arbogast, L. F. A., "Essai sur de nouveaux principes de calcul différentiel et intégral, indépendans de la théorie des infiniment-petits et celle des limites," Biblioteca Medicea-Laurenziana, Florence, MS Codex Ashburnham Appendix Sign. 1840.

      See Fréchet, Maurice, "Biographie du mathématicien alsacien Arbogast," *Thalès, 4*, pp. 43–55 (1940).

Zimmermann, Karl, *Arbogast als Mathematiker und Historiker der Mathematik*, Inaugural-Dissertation zur Erlangung der Doktorwürde der Hohen Naturwissenschaftlich-Mathematischen Fakultät der Ruprecht-Karls-Universität zu Heidelberg, 1934 [Abridged].

5. Berkeley, George, "The Analyst," *The Works of George Berkeley*, Vol. 3, ed. A. C. Fraser (Oxford: Clarendon Press, 1901).

   Gibson, G. A., "Berkeley's *Analyst* and its Critics: an Episode in the Development of the Doctrine of Limits," *Bibl. Matem.* N.S. *XIII* pp. 65–70 (1899).

6. Bernoulli, Jakob, *Über unendliche Reihen 1689–1704* (Leipzig: Wilhelm Engelmann, 1909).

   Hofmann, Josef E., *Über Jakob Bernoullis Beiträge zur Infinitesimalmathematik* (Genève: Institut de Mathématiques, Université, 1958).

7. Bernoulli, Johann, "Additamentum affectionis omnium quadraturarum et rectificationum curvarum per seriem quandam generalissimam," *Acta Eruditorum*, 1694, pp. 437–441.
8. Bernoulli, Johann, *Opera Omnia*, 2 vols. (Lausannae et Genevae: Bousquet, 1742).

   Fleckenstein, J. O., "L'école mathématique baloise des Bernoulli à l'aube du XVIIe Siècle," (Paris: Palais de la Découverte, 1958).

9. Carnot, L. N. M., *Reflexions sur la métaphysique du calcul infinitésimal* (Paris: chez Duprat, 1797).
10. Euler, Leonhard, *Institutiones Calculi Differentialis cum eius usu in analysi finitorum ac doctrina serierum* (Petropolitanae: Impensis Academiae imperialis scientiarum). *Opera Omnia* Series Prima, Vol. 10 (Leipzig and Berlin: B. G. Teubner, 1913).

    Sammelband der zu ehren des 250. Geburtstages Leon-hard Eulers. Redaktion von Kurt Schröder (Berlin: Akademie-Verlag, 1959). See especially the articles by J. E. Hofmann, and A. P. Juschkewitsch (II.17).

    Pasquier, L. G. du, *Léonard Euler et ses Amis* (Paris : J. Hermann, 1927).

    Spiess, Otto, *Leonhard Euler* (Frauenfeld und Leipzig: Verlag von Huber and Co., 1929).

11. Foncenex, Daviet de, "Sur les principes fondamentaux de la mechanique," *Miscellanea Taurinensia*, Tomus Alter, *Mélanges de philosophie et de mathématique de la Societé Royale de Turin* 1760–1761, Section "Mécanique," pp. 299–322.
12. Landen, John, *A Discourse concerning the Residual Analysis: A new branch of the algebraic art, of very extensive use, both in Pure Mathematics and Natural Philosophy* (London: J. Nourse, 1758).
13. Landen, John, *Observations on converging series, occasioned by Mr. Clarke's translation of Mr. [Antonio Maria] Lorgna's treatise on the same subject* (London: Printed for the author, 1781).
14. Landen, John, *The Residual Analysis: A new branch of the algebraic art, of very extensive use, both in pure mathematics, and natural philosophy*, Book 1 [all published] (London: Printed for the author, 1764).

    Green, H. G., and Winter, H. J. J., "John Landen, F.R.S. (1719–1790), Mathematician" *Isis*, *35* (6–10) 1944.

15. L'Hospital, G. F. A. de, *Analyse des infiniment petits pour l'intelligence des lignes courbes* (Paris: Imprimérie Royale, 1696).
16. L'Huilier, *Exposition élémentaire des principes des calculs supérieurs* (Berlin: chez George Jacques Decker, [1787]).
17. L'Huilier, Simon, *Principiorum calculi differentialis et integralis expositio elementaris* (Tubingae: apud J. G. Cottam, 1795).

Wolf, Rudolf, "Simon Lhuilier von Genf," *Biographien zur Kulturgeschichte der Schweiz*, 4 vols. (Zürich: Orell, Füssli, and Comp., 1858–1862), *I*, pp. 401–422.

18. Maclaurin, Colin, *A Treatise of Fluxions in Two Books* (Edinburgh, T. W. and T. Ruddimans, 1742).

    Turnbull, Herbert Westren, *Bi-centenary of the Death of Colin Maclaurin* (1698–1746). Aberdeen University Studies No. 127 (Aberdeen: The University Press, 1951).

    Tweedie, Charles, "The 'Geometria Organica' of Colin Maclaurin: A Historical and Critical Survey," *Proceedings of the Royal Society of Edinburgh, 36*, pp. 87–150 (1915–1916).

    Tweedie, Charles, "A Study of the Life and Writings of Colin Maclaurin," *The Mathematical Gazette, 8*, pp. 133–151 (1915–1916).

    Tweedie, Charles, "Notes on the Life and Works of Colin Maclaurin," *Idem.*, *9*, pp. 303–305, (1917–1919).

19. Nieuwentijt, Bernhard, *Analysis infinitorum, seu curvilineorum proprietates et polygonorum natura deductae* (Amstelaedami: Apud Joannem Wolters, 1695).
20. Nieuwentijt, Bernhard, *Considerationes circa analyseas ad quantitates infinitè parvas applicatae principia, et calcul differentialisusum* (Amstelaedami: Apud Joannem Wolters, 1694).

    Beth, E. W., "Nieuwentijt's Significance for the Philosophy of Science," *Synthèse, 9*, pp. 447–453 (1955).

    Freudenthal, Hans, "Nieuwentijt und der teleologische Gottebeweis," *Synthèse, 9*, pp. 454 464 (1955).

21. Taylor, Brook, Methodus incrementorum directa et inversa (Londini: Impensis Gulielmi Innys, 1717).

    Auchter, Heinrich, *Brook Taylor, der Mathematiker und Philosoph* (Marburg: Druckerei und Verlag Wissenschaftlichen Werke Konrad Triltsch, Würzburg, 1937).

C. Ampère, Bolzano, and Cauchy

There is no study of Ampère's mathematics. Much good work is now coming out of Eastern Europe on Bolzano. Stolz's account of Bolzano's published mathematical works is classic. Surprisingly, no study of Cauchy's mathematics exists, though there are articles on special subjects (e.g., analytic functions) which deal at length with his work.

1. Ampère, André Marie, "Recherches sur quelques points de la théorie des fonctions dérivées qui conduisant à une nouvelle démonstration de la série de *Taylor*, et à l'expression finie des termes qu'on néglige lorsqu'on arrète cette série à un terme quelconque," *Journal de l'école polytechnique* 13e Cahier, Tome VI, pp. 148–181 (1806).

    Célébration à Lyon du Centenaire de la Mort d'André-Marie Ampère 1836–1936 Tome I. Compte rendu des Journées des 5, 6, 7, et 8, Mars 1936 publié sous la direction de M. C. Chalumeau.

    Arago, F., "Ampère," *Oeuvres Complètes de François Arago* ed. J.-A. Barral (Paris: Gide et J. Baudry, 1854; Leipzig: T. O. Weigel, 1854), II, pp. 1–116.

    Launay, Louis de, *Le Grand Ampère* (Paris: Perrin et Cie, 1925).

    Launay, Louis de., *Correspondance du Grand Ampère* (Paris: Gauthier-Villars, 1936–1943)

2. Bolzano, Bernard, *Functionenlehre, Bernhard Bolzano's Schriften*, Publ. by Karel Rychlík (Prag: Königliche Böhmische Gesellschaft du Wissenschaften, 1930).
3. Bolzano, Bernhard, *Paradoxes of the Infinite* (London: Routledge and Kegan Paul, 1950).
4. Bolzano, Bernhard, *Rein analytischer Beweis des Lehrsatzes, dass zwischen je zwey Werthen, die ein entgegengesetztes Resultat gewähren, wenigstens eine reele Wurzel der Gleichung liege* (Leipzig: Wilhelm Engelmann, 1905).

Berg, Jan, *Bolzano's Logic* (Acta Universitatis Stock-holmiensis Stockholm: Studies in Philosophy 2.) (Stockholm: Almquist and Wiksell, 1962).

Kowalewski, Gerhard, "Über Bolzanos nichtdifferenzierbare stetige Funktion," *Acta Mathematica XLIV*, pp. 315–319 (1923).

Rychlik, Karel, "La théorie des nombres réels dans un ouvrage posthume manuscrit de Bernard Bolzano," *Rev. Hist. Sci., 14*, pp. 313–327 (1961).

Stolz, Otto, "B. Bolzano's Bedeutung in der Geschichte der Infinitesimalrechnung" *Mathematische Annalen, XVIII*, pp. 255–279 (1881).

Rootselaar, B. van, "Bolzano's Theory of real numbers," *Arch. Hist, exact. Sci., 2*, pp. 168–180 (1963).

5. Cauchy, Augustin-Louis, *Cours d'Analyse de l'Ecole Royale Polytechnique, Ire Partie: Analyse Algébrique* (Paris: Imprimérie Royale, 1821). *Oeuvres Complètes d'Augustin Cauchy*, IIe Série, Tom. III (Paris: Gauthier-Villars, 1897).
6. Cauchy, Augustin-Louis, *Résumé des leçone données à l'école royale polytechnique sur la calcul infinitesimal* (Paris: De l'imprimérie royale, 1823). *Oeuvres Complètes d'Augustin Cauchy*, IIe Série, Tome IV (Paris: Gauthier-Villars, 1899).
7. Cauchy, Augustin-Louis, *Leçons sur les applications du calcul infinitésimal à la géometrie*, 2 vols. (Paris: De l'imprimérie royale, 1826). *Oeuvres*, IIe Série, Tome V (Paris: Gauthier-Villars, 1903).

Boncompagni, B., "La Vie et les travaux du Baron Cauchy . . . par C.-A. Valson," [Review] *Bullettino di Bibliografia e di Storia delle Scienze Matematiche e Fisiche, II*, pp. 1–102 (1869).

Jourdain, P. E. B., "On the Origin of Cauchy's Conception of a Definite Integral and of the Continuity of a Function," *Isis, 1*, pp. 661–703 (1913).

Jourdain, P. E. B., "The Theory of Functions with Cauchy and Gauss," *Bibliotheca Mathematica* (3) *VI*, pp. 190–207.

Juschkewitsch, A. P., "On the origin of the concept of Cauchy's Definite Integral," [in Russian] *Akademiia Nauk SSSR, Institut istorii estestvognaniia, Trudy, 1* pp. 373–411 (1947).

D. Weierstrass and his School

Weierstrass' work on foundations was expounded in his lectures, which have never been published. Many samples of Weierstrassian rigor can be found in his *Mathematische Werke*, 7 vols. (Berlin: Mayer & Miller, 1894–1927). But the Weierstrassian foundation for the calculus can best be studied through the work of his students.

1. Borel, Emile, *Leçons sur la théorie des fonctions* (Paris: Gauthier-Villars et fils, 1898).
2. Hankel, Hermann, "Grenze," *Allgemeine Encyklopädie der Wissenschaften und Künste* herausgegeben von Johann Samuel Ersch und Johann Gottfried Gruber (Leipzig: J. A. Brockhaus, 1871), Erste Section A-G, *90* pp. 185–211.
3. Heine, E., "Die Elemente der Funktionenlehre," *Journ. f. die Reine u. Angewandte Mathematik, LXXIV*, pp. 172–88 (1872).
4. Kossak, Ernst, *Die Elemente der Arithmetik* (Berlin: Programmabhandlung des werderschen Gymnasiums in Berlin, 1872)

5. Pincherle, Salvatore, "Saggio di una introduzione alla teoria delle funzioni analitiche secondo i principii del Prof. C. Weierstrass," *Giornale di Matematiche, XVIII*, pp. 178–254, pp. 317–57 (1880).
6. Stolz, Otto and Gmeiner, J. Anton, *Einleitung in die funktionentheorie.* Second edition (Leipzig: B. G. Teubner, 1904–05).
7. Stolz, Otto, *Grundzüge der Differential- und Integral-rechnung*, 3 vols. (Leipzig: B. G. Teubner, 1893–99).

Lampe gives the best account of his life; Poincaré's magnificent article places his mathematical work in its proper perspective. The works listed under VII. A. are all useful for Weierstrass and his school. See also IV.B.1, 9, and 13.

8. Lampe, Emil, "Karl Weierstrass," *Jahresbericht der Deutschen Mathematiker-Vereinigung, 6*, pp. 27–44 (1897).
9. Poincaré, Henri, "L'Oeuvre mathématique de Weierstrass," *Acta Mathematica, 22*, pp. 1–18 (1898).

Bibliography: 1966–Present

Note on the Updated Bibliography: In addition to items important to the study of Lagrange's work in general and his calculus in particular, I have included the works I consider most useful among more general histories, bibliographic aids, and sourcebooks. In addition, the reader's attention is called to the two principal journals of the history of mathematics, *Archive for History of Exact Sciences* and *Historia Mathematica*, the latter of which includes abstracts of books and articles published in the history of mathematics as well as book reviews.

Barroso-Filho, Wilton, and Comte, Claude. 1988. "La formalisation de la dynamique par Lagrange." In [Rashed 1988a], pp. 329–348.

Belhoste, Bruno. 1985. *Cauchy: Un mathématicien légitimiste au XIXe siècle.* Paris: Belin.

Belhoste, Bruno. 1991. *Augustin-Louis Cauchy: A Biography*. Berlin: Springer.

Belhoste, Bruno, Dahan-Dalmedico, Amy, Picon, Antoine. 1994. *La formation polytechnicienne: Deux siècles d'histoire*. Paris, Dunod.

Birkhoff, Garrett (Ed.). 1973. *A Source Book in Classical Analysis.* Cambridge, Mass.: Harvard.

Bolzano, B. 1969. *Gesamtausgabe*. Ed. Eduard Winter u. a. Stuttgart-Bad Cannstatt: Fromann-Holzboog. Vols. 1–

Bolzano, Bernhard. 2004a. *Functionenlehre* [1830]. In B. Bolzano, *Schriften*, Prague: Königlichen Böhmischen Gesellschaft der Wissenschaften, 1930. English translation in Russ 2004, *Mathematical Works of Bolzano*, pp. 429–590.

Bolzano, Bernhard. 2004b. *Rein analytischer Beweis des Lehrsatzes dass zwischen je zwey [sic] Werthen, die ein entgegengesetztes Resultat gewaehren, wenigstens eine reele Wurzel der Gleichung liege*, Prague, 1817. English translation in Russ, S. B., "A translation of Bolzano's paper on the intermediate value theorem," *Historia Mathematica* 7 (1980), pp. 156–185. Also in Russ 2004, *Mathematical Works of Bolzano*, pp. 251–278.

Borgato, Maria Teresa; Pepe, Luigi. 1987. "Lagrange a Torino (1750–1759) e le sue lezioni inedite nelle R. Scuole di Artiglieria. *Boll. Stor. Sci. Mat. 7*, pp. 3–43.

Borgato, M. T. and Pepe, L. 1988. "Una memoria inedita di Lagrange sulla teoria delle parallele," *Boll Storia Sci Mat* 9, 307–335.

Borgato, M. T., and Pepe, L. 1990. *Lagrange: Appunti per una biografia scientifica.* Torino: La Rosa.

Bos, H. J. M. 1977. "Calculus in the Eighteenth Century: The Role of Applications," *Bull Inst Math Appl* 13, 221–227.

Bottazzini, Umberto. 1986. *The Higher Calculus: A History of Real and Complex Analysis from Euler to Weierstrass.* Tr. Warren van Egmond. New York et al: Springer.

Bottazzini, Umberto. 1989. "Lagrange et le problème de Kepler," *Rev. Hist. Sci. 42*, pp. 27–42.

Boyer, Carl. 1967. *A History of Mathematics.* New York: Wiley.

Boyer, Carl, and Merzbach, Uta C. 1989. *A History of Mathematics: Second Edition.* New York et al: Wiley.

Bradley, Robert E., and Sandifer, C. Edward. 2007. Eds, *Leonhard Euler: Life, Work, and Legacy*. Amsterdam: Elsevier.

Bradley, Robert E. and Sandifer, C. Edward. 2009. *Cauchy's Cours d'analyse: An Annotated Translation.* New York et al: Springer.

Bradley, Robert E., D'Antonio, L. A., Sandifer, C. Edward. 2007. Eds., *Euler at 300: An Appreciation.* Washington, D. C.: Mathematical Association of America.

Calinger, Ronald. 1982. Ed. *Classics of Mathematics.* Oak Park, IL: Moore.

Dauben, Joseph. 1982. "Progress of Mathematics in the Early 19th Century: Context, Contents, and Consequences." *Acta Historiae Rerum Naturalium nec non Technicarum.* Special Issue. Prague.

Dauben, Joseph. 1985. *The History of Mathematics from Antiquity to the Present: A Selective Bibliography.* New York and London: Garland.

Dhombres, Nicole, and Dhombres, Jean. 1989. *Naissance d'un nouveau pouvoir: Sciences et savants en France (1793–1824).* Paris: Editions Payot.

Domingues, João Caramalho. 2005. "1797–1800 S. F. Lacroix, *Traité du calcul différentiel et du calcul intégral*," in Grattan-Guinness 2005b, *Landmark Writings in Western Mathematics*, 277–291.

Dugac, Pierre. 1979. *Histoire du théorème des accroissements finis.* Paris: Université Pierre et Marie Curie.

Dugac, Pierre. 1980. *Limite, point d'accumulation, compact.* Paris: Université Pierre et Marie Curie.

Dugac, Pierre. 1990. "La théorie des fonctions analytiques de Lagrange et la notion d'infini," in G. König, ed., *Konzepte des mathematisch Unendlichen im 19. Jahrhundert*, pp 34–46. Göttingen: Vandenhoeck & Ruprecht.

Dunham, William. 1999. *Euler: The Master of Us All.* Washington, D. C.: Mathematical Association of America.

Dunham, William. 2005. *The Calculus Gallery: Masterpieces from Newton to Lebesgue*. Princeton: Princeton University Press.

Dunham, William. 2007. Ed., *The Genius of Euler: Reflections*. Washington, D. C.: Mathematical Association of America.

Englesman, Steven. 1980. "Lagrange's Early Contributions to the Theory of First-Order Partial Differential Equations," *Historia Mathematica 7*, pp. 7–23.

Engelsman, Stephen B. 1984. *Families of Curves and the Origins of Partial Differentiation.* Amsterdam: Elsevier.

Euler, L. 1980. *Leonhardi Euleri Commercium Epistolicum.* With Clairaut, d'Alembert, and Lagrange. Ed. A. P. Juskevič & R. Taton. Basel: Birkhauser. In *Opera omnia*, Ser. 4, Vol. 5.

Euler, Leonhard. 1988–1990. *Introductio in analysin infinitorum* [1748]. Translated by John Blanton as Leonhard Euler, *Introduction to the Analysis of the Infinite*, in 2 volumes. New York: Springer.

Euler, Leonhard. 2000. *Institutiones calculi differentialis* [1755]. Translated by John Blanton as Leonard Euler, *Foundations of Differential Calculus*. New York: Springer.

Fauvel, John, and Gray, Jeremy. 1987. *The History of Mathematics: A Reader.* London: MacMillan.

Ferraro, Giovanni and Panza, Marco. 2003. "Developing into series and returning from series: a note on the foundations of eighteenth-century analysis," *Historia Mathematica* 30, 17–46.

Flett, P. M. 1974. "Some Historical Notes and Speculations concerning the Mean-Value Theorem of the Differential Calculus," *Bulletin of the Institute of Mathematics and Its Applications 10*, pp. 66–72.

Fraser, Craig. 1985. "J.-L. Lagrange's Changing Approach to the Foundations of the Calculus of Variations," *Archive for History of Exact Sciences 23*, pp. 151–191.

Fraser, Craig. 1987. "J.-L. Lagrange's Algebraic Vision of the Calculus," *Historia Mathematica 14*, pp. 38–53.

Fraser, Craig. 1989. "The Calculus as Algebraic Analysis: Some Observations on Mathematical Analysis in the 18th Century," *Archive for History of Exact Sciences 39*, pp. 317–331.

Fraser, Craig. 1990. "Lagrange's analytical mechanics: Its Cartesian origins and reception in Comte's positive philosophy," *Stud Hist Phil Sci* 21, 243–256.

Fraser, Craig. 2005. "1797 Joseph Louis Lagrange, *Théorie des fonctions analytiques*." In Grattan-Guinness 2005b, *Landmark Writings in Western Mathematics*, 258–276.

Gilain, Christian. 1989. "Cauchy et le Cours d'analyse de l'école polytechnique," *Bulletin de la Société des amis de la Bibliothèque de l'école polytechnique* 5, pp. 3–145.

Gillispie, Charles. 1971. *Lazare Carnot Savant*. Princeton: Princeton University Press.

Gillispie, Charles C. 1997. *Pierre-Simon Laplace, 1749–1827: A Life in Exact Science,* in collaboration with R. Fox and I. Grattan-Guinness. Princeton: Princeton University Press.

Goldstine, H. 1977a. *History of Numerical Analysis from the 16th through the 19th Century.* New York et al: Springer.

Goldstinc, H. 1977b. *History of the Calculus of Variations from the 17th through the 19th Century.* New York et al: Springer.

Grabiner, Judith V. 1974. "Is Mathematical Truth Time-Dependent?" *Amer. Math. Monthly 81*, pp. 354–365.

Grabiner, Judith V. 1978. "The Origins of Cauchy's Theory of the Derivative," *Historia Mathematica 5*, pp. 379–409.

Grabiner, Judith V. 1981a. *The Origins of Cauchy's Rigorous Calculus.* Cambridge, Mass., and London, England: M. I. T. Press.

Grabiner, Judith V. 1981b. "Changing Attitudes Toward Mathematical Rigor: Lagrange and Analysis in the Eighteenth and Nineteenth Centuries," in H. N. Jahnke and M. Otte, eds., *Epistemological and Social Problems of the Sciences in the Early Nineteenth Century.* Dordrecht: D. Reidel, pp. 311–330.

Grabiner, Judith V. 1983. "Who Gave You the Epsilon? Cauchy and the Origins of Rigorous Calculus," *Amer. Math. Monthly 90,* pp. 185–194.

Grabiner, Judith V. 1984. "Cauchy and Bolzano: Tradition and Transformation in the History of Mathematics," in Everett Mendelsohn, ed., *Transformation and Tradition in the Sciences*, pp. 105–124. Cambridge, England: Cambridge University Press.

Grabiner, Judith V. 1997. "Was Newton's calculus a dead end? The Continental influence of Maclaurin's *Treatise of Fluxions*," *American Mathematical Monthly* 104, pp. 393–410.

Grabiner, Judith V. 2004. "Maclaurin and Newton: The Newtonian style and the authority of mathematics," *American Mathematical Monthly* 111, pp. 841–852.

Grabiner, Judith V. 2009. "Why Did Lagrange 'Prove' the Parallel Postulate?" *American Mathematical Monthly,* 116, pp. 3–18.

Grattan-Guinness, Ivor. 1970a. "Bolzano, Cauchy and the 'New Analysis' of the Early Nineteenth Century," *Archive for History of Exact Sciences 6*, pp. 372–400.

Grattan-Guinness, Ivor. 1970b. *The Development of the Foundations of Mathematical Analysis from Euler to Riemann.* Cambridge, Mass.: M. I. T. Press.

Grattan-Guinness, Ivor. 1980. (Ed.) *From the Calculus to Set Theory, 1630–1910.* London: Duckworth.

Grattan-Guinness, Ivor. 1985. "A Paris Curiosity, 1814: Delambre's Obituary of Lagrange, and its 'Supplement,'" in M. Folkerts and U. Lindgren, eds., *Mathemata: Festschrift für Helmuth Gericke.* Stuttgart: Steiner, pp. 493–510. See also in C. Mangione, ed., *Scienza e filosofia: Saggi in onore di Ludovico Geymonat.* Milano: Garzanti, 1985.

Grattan-Guinness, Ivor. 1990. *Convolutions in French Mathematics, 1800–1840*, 3 vols. Basel: Birkhauser.

Grattan-Guinness, Ivor. 2005a. "1821, 1823 A.-L. Cauchy, Cours d'analyse and Résumé of the calculus." In Grattan-Guinness 2005b, *Landmark Writings in Western Mathematics,* pp. 341–353.

Grattan-Guinness, Ivor. 2005b. Ed., *Landmark Writings in Western Mathematics, 1640–1940.* Amsterdam: Elsevier.

Gray, Jeremy. 2008. *Plato's Ghost: The Modernist Transformation of Mathematics.* Princeton, Princeton University Press.

Guicciardini, Niccolò. 1989. *The Development of Newtonian Calculus in Britain, 1700–1800.* Cambridge: Cambridge University Press.

Guitard, Thierry. 1987. "On an Episode in the History of the Integral Calculus," *Historia Matheamtica 14*, pp. 215–219.

Hahn, Roger. 2005. *Pierre Simon Laplace, 1749–1827: A Determined Scientist.* Cambridge MA: Harvard University Press.

Hamburg, Robin. 1976. "The Theory of Equations in the 18th Century: The Work of Joseph Lagrange," *Archive for History of Exact Sciences 16*, pp. 17–36.

Itard, Jean. 1973. "Lagrange." *Dictionary of Scientific Biography*, ed. C. C. Gillispie. New York: Scribner. Vol. 8, pp. 559–573.

Jahnke, H. N. 1992. "A structuralist view of Lagrange's algebraic analysis and the German combinatorial school," in Echeverria, J., Ibarra, A., and Mormann, T., eds., *The space of mathematics*, pp. 280–295. Berlin: Walter de Gruyter.

Jahnke, Hans Niels, ed. 2003. *A History of Analysis*. Translated from the German by the authors. Providence, RI: American Mathematical Society.

Katz, Victor. 2009. *A History of Mathematics: An Introduction,* 3d edition. Boston et al: Addison-Wesley.

Kiernan, B. M. 1972. "The Development of Galois Theory from Lagrange to Artin," *Archive for History of Exact Sciences 8*, pp. 40–154.

Kitcher, Philip. 1983. *The Nature of Mathematical Knowledge.* New York and Oxford: Oxford University Press.

Kline, Morris. 1972. *Mathematical Thought from Ancient To Modern Times.* New York: Oxford.

Lagrange, Joseph-Louis. 1973. *Oeuvres de Lagrange.* Ed. M. J.- A. Serret, 14 vols., Paris: Gauthier-Villars, orig. 1867–1892. Reprinted, Hildesheim and New York: Georg Olms Verlag.

Lagrange, Joseph-Louis. 1987. *Principii di analisi sublime.* Ed. Maria Teresa Borgato. *Boll. Stor. Sci. Mat. 7*, pp. 45–200.

Lagrange, J.-L. 1997. *Analytical Mechanics.* English translation of the *Mécanique analytique* [2d ed., 1811–1815] by A. Boissonnade and V. N. Vagliente, trans. and eds. Dordrecht: Kluwer.

Lusternik, L. A., and Petrova, S. S. 1972. "Les premières étapes du calcul symbolique." *Revue Hist. Sci. Applic. 25*, pp. 201–206.

Manning, Kenneth. 1975. "The Emergence of the Weierstrassian Approach to Complex Analysis," *Archive for History of Exact Sciences 14*, pp. 297–383.

May, K. O. 1973. *Bibliography and Research Manual of the History of Mathematics.* Toronto and Buffalo: University of Toronto Press.

Mazzotti, Massimo. 2007. *The World of Maria Gaetana Agnesi: Mathematician of God.* Baltimore, MD: Johns Hopkins University Press.

Newton, Isaac. 1995. *The Principia: Mathematical Principles of Natural Philosophy, A New Translation by I. B. Cohen and A. Whitman.* Berkeley: University of California Press.

Novy, Lubos. 1973. *The Origins of Modern Algebra.* Tr. J. Tauer. Leyden: Noordhoff.

Oravas, G. and Maclean, L. 1966. "Historical Development of Energetical Principles in Elastomechanics. I: From Heraclitos to Maxwell," *Applied Mechanics Reviews 19*, pp. 647–658.

Ovaert, J.-L. 1976. "La thèse de Lagrange et la transformation de l'analyse," in C. Houzel, ed., *Philosophie et calcul de l'infini.* Paris: Maspero, pp. 157–200.

Pambuccian, Victor. 2009. "On the equivalence of Lagrange's axiom to the Lotschnittaxiom," *Journal of Geometry* 95, pp. 165–171.

Pepe, Luigi. 1986. "Tre 'prime edizioni' ed un' introduzione inedita della *Théorie des fonctions analytiques* di Lagrange," *Boll. Stor. Sci. Mat 6*, pp. 17–44.

Pulte, Helmut. 2005. "1788 Joseph Louis Lagrange, *Mechanique analitique*," in Grattan-Guinness 2005b, *Landmark Writings in Western Mathematics*, 208–225.

Rashed, Roshdi. 1988a. (Ed.) *Sciences à l'époque de la Révolution Française: Recherches historiques.* Paris: Blanchard.

Rashed, Roshdi. 1988b. "Lagrange, lecteur de Diophante." In [Rashed 1988a], pp. 39–83.

Richards, Joan L. 1989. "Rigor and Revolution: The Demise of Natural Mathematics," *Proceedings*, Canadian Society for the History and Philosophy of Mathematics, Université de Laval, pp. 135–148.

Richards, Joan L. 2006. "Historical Mathematics in the French Eighteenth Century," *Isis 97* (4), pp. 700–713.

Russ, S. B. 1980. "A Translation of Bolzano's Paper on the Intermedate Value Theorem," *Historia Mathematica 7*, pp. 156–185.

Russ, Steve. 2004. *The Mathematical Works of Bernard Bolzano.* Oxford: Oxford University Press.

Sageng, Erik. 2005. "1742 Colin MacLaurin, A Treatise of Fluxions," in Grattan-Guinness 2005b, *Landmark Writings in Western Mathematics*, 143–158.

Sandifer, C. Edward. 2007. *How Euler Did It.* Washington, D. C.: Mathematical Association of America.

Schubring, Gert. 2005. *Conflicts Between Generalization, Rigor, and Intuition: Number Concepts Underlying the Development of Analysis in 17–19th Century France and Germany.* New York: Springer.

Sinaceur, Hourya. 1973. "Cauchy et Bolzano." *Revue d'histoire des sciences 26*, pp. 97–112.

Smithies, F. 1986. "Cauchy's Conception of Rigour in Analysis," *Archive for History of Exact Sciences 36*, pp. 41–61.

Struik, Dirk J. 1969. *A Source Book in Mathematics, 1200–1800.* Cambridge, Mass.: Harvard.

Taton, René. 1974. "Inventaire chronologique de l'oeuvre de Lagrange," *Revue Hist. Sci. Applic. 27*, pp. 3–36.

Taton, René. 1986. "Lagrange et Leibniz: De la théorie des fonctions au principe de raison suffisante." In A. Heinekamp, ed., *Beiträge zur Wirkunqs- und Rezeptionsgeschichte von Gottfried Wilhelm Leibniz.* Stuttgart: Steiner, pp. 139–147.

Taton, René. 1988a. "Le départ de Lagrange de Berlin et son installation à Paris en 1787. *Rev. Hist. Sci. 41*, pp. 39–74.

Taton, René. 1988b. "Sur quelques pièces de la correspondance de Lagrange pour les années 1756–1758." *Boll. Stor. Sci. Mat. 8*, pp. 3–19.

Terrall, Mary. 2002. *The Man Who Flattened the Earth: Maupertuis and the Sciences in the Enlightenment*. Chicago: University of Chicago Press.

Tiulina, I. A. 1977. *Zhozef Lui Lagranzh, 1736–1813.* [In Russian] Moscow: Nauka.

Truesdell, C. 1960 "The Rational Mechanics of Flexible or Elastic Bodies, 1638–1788." In L. Euler, *Opera Omnia*, Ser. 2, vol. 11, part 2.

Truesdell, C. 1968. *Essays in the History of Mechanics.* New York et al: Springer.

Truesdell, Clifford. 1989. "Maria Gaetana Agnesi," *Arch Hist Ex Sci* 40, 113–147.

Tymoczko, Thomas. 1986. (Ed.) *New Directions in the Philosophy of Mathematics.* Boston et al: Birkhäuser.

Weil, André. 1984. *Number Theory: An Approach through History.* Boston: Birkhäuser.

Wussing, Hans. 1969. *Die Genesis des abstrakten Gruppenbegriffes.* Berlin: VEB Deutscher Verlag.

Youschkevitch, A. P. 1971. "Lazare Carnot and the Competition of the Berlin Academy in 1786 on the Mathematical Theory of the Infinite." In [Gillispie 1971].

Youschkevitch, A. P. 1977. "The Concept of Function up to the Middle of the Nineteenth Century," *Archive for History of Exact Sciences 16*, pp. 37–85.

II

Selected Writings

1

The Mathematician, the Historian, and the History of Mathematics*

The historian's basic questions, whether he is a historian of mathematics or of political institutions, are: what was the past like? and how did the present come to be? The second question—how did the present come to be?—is the central one in the history of mathematics, whether done by historian or mathematician. But the historian's view of both past and present is quite different from that of the mathematician. The historian is interested in the past in its full richness, and sees any present fact as conditioned by a complex chain of causes in an almost unlimited past. The mathematician instead is oriented toward the present, and toward past mathematics chiefly insofar as it led to important present mathematics.[1]

I

What questions do mathematicians generally ask about the history of mathematics? "When was this concept first defined, and what problems led to its definition?" "Who first proved this theorem, and how did he do it?" "Is the proof correct by modern standards?" The mathematician begins with mathematics that is important now, and looks backwards for its antecedents. To a mathematician, all mathematics is contemporary; as Littlewood put it [A7, p. 81], the ancient Greeks were "Fellows of another College." True and significant mathematics is true and significant, whenever it may have been done.

The history of mathematics as written by mathematicians tends to be technical, to focus on the content of specific papers. It is written on a high mathematical level, and deals with significant mathematics. The title of E. T. Bell's *The Development of Mathematics* reflects the mathematician's view. The mathematician looks at the development of *mathematics*, as the result

---

* Reprinted with permission from *Historia Mathematica*, Vol. 2, 4(Nov. 1975) 439–447.

[1] By "historian" and "mathematician" I do not mean a classification according to the field of a person's Ph.D., but according to his general point of view. For our present purpose, Dirk Struik and Thomas Hawkins are historians; E. T. Bell and the authors of the *Encyclopädie der mathematischen Wissenschaften* [13] are mathematicians. In general mathematicians and historians have, while writing the history of mathematics, in fact taken the different approaches I describe, though there is no a priori reason they would necessarily have to do so.

of a chronologically and logically connected series of papers; he does not look at it as the work of people living in considerably different historical settings.

II

How is the historian different? First, it goes without saying that, in asking what the past was like, the historian will be more concerned than the mathematician about the non-mathematical, as well as the mathematical past. More surprising is that even when dealing with strictly technical questions, the historian may view things differently from the mathematician.

While the mathematician sees the past as part of the present, the historian sees the present as laden with archaeological relics from the past; he sees everything in the present as having many and diverse roots in the past, and as the end of long, complex processes. Many things we take for granted are neither logical nor natural. They might even appear arbitrary, but they are, instead, the products of particular historical situations. For example, the use of the letter "epsilon," as in delta-epsilon proofs, appears arbitrary, but it in fact records the origin of the use of inequalities in proofs in analysis. The origin was in the study of approximations, the notation "epsilon" is Cauchy's, and the letter seems to stand for "error."[2] Historians love this sort of explanation, and constantly search for ones like it. To take another kind of example, isn't it amazing that the standard proof that $\sqrt{2}$ is irrational is over two thousand years old?[3]

Another essential difference between the historian and the mathematician is this: the historian of mathematics will ask himself what the *total* mathematical past in some particular time-period was like. He will steep himself in many aspects of that mathematical past, not just those which have an obvious bearing on the antecedents of the particular mathematical development whose history he is tracing. [A2;A3;A4;A8] Let us see how this feature of the historian's approach can be of value to the mathematician interested in the history of mathematics. Obviously it is easier to find the antecedents of present ideas when one knows where to look. The better one knows the past, the wider the variety of places he can investigate. In addition, through familiarity with specific types of sources, the historian knows where to find the answers to particular types of questions. By contrast, someone relatively unfamiliar with the time period in question will not always understand what past mathematicians were trying to do, and will find that the terms used then did not always mean what they mean today.[4]

Let me give some examples to clarify this general point. There is a widespread impression that eighteenth-century mathematicians were very cavalier in their treatment of convergence, and it is sometimes even said that they assumed that once they had shown that the nth term of a series went to zero, the series converged. Did eighteenth-century mathematicians in fact make this error? Hadn't they ever heard of the harmonic series? My own sense of eighteenth-century mathematics says that eighteenth-century mathematicians weren't that incompetent. They knew

---

[2] In an approximation to the sum of an infinite series, an 18th century mathematician might take n terms and ask how large the "error"—the difference between the nth partial sum and the sum of the infinite series—might be. The series converges in Cauchy's sense when the difference can be made less than any assignable error. Compare his *Cours d'analyse* [B6(2), vol. 3] with his article in the *Comptes Rendus 37* (1853) [B6 (1), vol. 12, pp. 114–124].

[3] See Van der Waerden's *Science Awakening* [9, p. 110] and compare Aristotle's *Prior Analytics* i 23, 41a, 26–27.

[4] In my article [A5] in the *Am. Math. Mon.* 81, 354–365, I have treated this point at length, especially with respect to changing standards of proof in analysis.

the divergence of the harmonic series.[5] People like Euler, D'Alembert, Lagrange, and Laplace were not hopelessly confused. In fact, D'Alembert and Lagrange investigated the remainders of specific infinite series and tried to find bounds on the value of those remainders.[6] Why, then, did even these men sometimes say "the series converges" when they had shown only that the nth term goes to zero? Because, in the eighteenth century, the term "converge" was used in different ways; sometimes, it was used as we use it; often, however, it was used to say that the nth term went to zero or that the terms of the series got smaller.[7] This conclusion, which I reached after reading numerous eighteenth-century papers, can be reliably verified by looking at the Diderot-D'Alembert *Encyclopédie*.[8] The modern definition of convergent series—that the partial sums of the series have a limit—was established by Cauchy [B6 (2), vol. 3, p. 114]. It is hard to avoid reading this modern meaning back into eighteenth-century mathematics. But this linguistic point, once understood, makes sense out of much eighteenth-century work on infinite series.

Another advantage of knowing the mathematical past is that the historian can construct a total picture of the background of some specific modern achievement. Rather than just looking at the major papers on the same topic, he may find the antecedents of some modern theories in unlikely places. A well-known example is the way the general definition of function came, not merely out of attempts to describe the class of all known algebraic expressions, but, more importantly, from the attempts to characterize the solutions to the partial differential equation for the vibrating string.[9] Another example may be found in the way Thomas Hawkins in his *Lebesgue's theory of Integration* [E14] has presented the full nineteenth-century background, drawing on a wide variety of mathematical work besides earlier ideas on integration.

For another example, consider Cauchy's definition of the derivative and the proofs of theorems based on that definition. In looking at the introductory sections of eighteenth-century calculus books, one finds a long string of definitions of derivatives, and debates about their nature—debates stemming from the attack on the foundations of the calculus by Bishop Berkeley. One might well view each of these old definitions and polemics as major contributors to Cauchy's final formulation. However, what was more important than the explicit verbal definition Cauchy gave for the derivative are the associated in-equality proof-techniques he pioneered. And these techniques came from elsewhere. The basic inequality property Cauchy used to define the derivative came to him from Lagrange's work on the Lagrange remainder in the latter's lectures

---

[5] The divergence of the harmonic series was proved in the late 17th century by Johann and Jakob Bernoulli, and, for that matter, was shown in the 14th century by Nicole Oresme. For Oresme, see [3, p. 293]; for the Bernoullis, [16, pp. 320–24].

[6] See J. d'Alembert, "Reflexions sur les suites et sur les racines imaginaires," *Opuscules Mathématiques,* 1768, vol. 5, pp. 171–83, esp. p. 173. See also Lagrange, *Théorie des Fonctions Analytiques,* 2nd ed., 1813 in [B24, vol. IX, pp. 83–84].

[7] For examples of this usage, see d'Alembert, *op. cit.;* L. Euler, "De seriebus divergentibus" in [B11 (1), vol. 15, pp. 586, 588]; and Lagrange, "Sur la résolution des équations numériques," 1772 [B24, II, p. 541].

[8] See also d'Alembert, *Dictionnaire raisonné des mathématiques*, which collects the mathematical articles from the *Encyclopédie,* articles "Convergence" and "Serié ou suite." Compare G. S. Kluegel, *Mathematisches Woerterbuch,* 1803, article "Convergirend, Annaehernd."

[9] See [B23]; compare [16, pp. 351–68]; and C. Truesdell, "The rational mechanics of flexible or elastic bodies, 1638–1788" in [B11 (2), 11, Sect. 2 (1960)].

on the calculus at the *Ecole polytechnique*.[10] The inequality proof-techniques themselves were developed largely in the study of algebraic approximations in the eighteenth century.[11]

Let us now take up a characteristic of the historian which we have not yet considered. We expect the historian to know the general history of a particular time as well as the mathematics of that time. He should have a sense of what it was like to be a person, not just a mathematician, at that time. Sometimes such knowledge has great explanatory value. One would not want to treat the history of the foundations of the calculus without knowing about the attacks on the calculus by the theologian Berkeley [B3, ch. VI]; or the flowering of seventeenth-and eighteenth-century mathematics without reference to the contemporary explosion in the natural sciences, especially Newtonian physics [A6]. One cannot treat the growth of the French school of mathematics in the nineteenth century without mentioning a major cause—the founding by the French revolutionary government of the *Ecole polytechnique*, providing employment and a first-rate mathematical community for its faculty, and an excellent mathematical education for its students.[12] One would not want to explain the relative absence of women in the ranks of nineteenth-century mathematicians without referring to the lack of access to higher education for women in Europe at a time when mathematics was so specialized and advanced that formal training was essential.[13]

By virtue of his training, the historian has been exposed to a number of general historiographical questions and is used to hearing them asked. For instance, there are theories of the nature of scientific change, like Thomas Kuhn's theory of scientific revolutions [A9]. Again, there are sociologically based theories like Robert Merton's analysis of priority controversies in science [A13]. The historian of mathematics, without having to become a disciple of Kuhn or Merton, can use such theories to help ask fruitful questions about the past of mathematics, and about the time period in which the mathematics occurred. He has the questions already at hand, and need not figure them out from first principles.

III

Our description of the possible contributions of historians and mathematicians to the writing of the history of mathematics has required, as well, some description of what the history of mathematics is like. Let us know turn to a different, but related question. What value has the history of mathematics, whether done by historian or mathematician? Of course, there is an inherent fascination in any history, and work in the history of mathematics certainly should be an element in the history of human culture in general. But another essential use exists—for the mathematician—in teaching and understanding mathematics.

Historical background can help teach mathematics in three ways. First, the history can help the teacher understand the inherent difficulty of certain concepts. A concept which took hundreds of years to develop is probably hard, and the historical difficulties may well resemble student difficulties.

---

[10] See [A5 pp. 361–63] and *Leçons sur le calcul des fonctions* (2nd. ed. 1806) in [B24], vol. X, p. 87].

[11] See Lagrange, *Traité de la résolution des équations numériques de tous les degrés,* 2nd ed., 1808, in [B24, vol. VIII, pp. 46–7, 163]. Compare d'Alembert, *Opuscules mathématiques,* 1768, vol. 5, pp. 171–83.

[12] J. T. Merz, *History of European Thought in the Nineteenth Century,* vol. I (1904), Dover reprint, 1965.

[13] See [B29]; L. Osen, *Women in Mathematics,* M.I.T. Press, 1974; J. L. Coolidge, "Six female mathematicians," *scripta Math.* 7 (1951), pp. 20–31.

Second, understanding how a mathematical idea arose can help motivate students. It helps answer questions like, "Why might somebody want to think about it in this particular way?".[14] Isolated historical comments, of course, do not make history. What one would like to do for one's students—and for that matter, for oneself—is to give them a sense of how the whole subject developed, and how the whole background of the subject fits together. Such a sense would motivate not just one concept or proof, but the entire subject.

Third, the historical background can help the student—or the mathematician—see how mathematics fits in with the rest of human thought; how Descartes the mathematician relates to Descartes the philosopher; how the rise of German mathematics in the mid-nineteenth century fits into the rise of German science, technology, and national power at that time. To see past mathematics in its historical context helps to see present mathematics in its philosophical, scientific, and social context, and to have a better understanding of the place of mathematics in the world.

IV

If the history of mathematics is indeed to be used in these ways, we need more of it. There now exists a technical literature which has established what the important results and their major antecedents are in many areas.[15] There is a need now for more studies on the full historical background of many subjects in modern mathematics—the theory of functions of a complex variable, for instance, or the rise of abstract algebra, or the philosophical and mathematical impact of non-Euclidean geometry. And the existing work needs to be made more available through the offering of full-scale courses in the history of mathematics, placing the monographs of Boyer and Hawkins along with the general histories of mathematics on the library shelves, and ordering *Archive for the History of the Exact Sciences* and *Historia Mathematica* for the departmental library along with the *Bulletin* and the *Monthly*. Finally, there are now too few historians of mathematics. The path for the historian of mathematics is difficult; he needs the historian's training, but also needs to know a great deal of mathematics. The history of science is itself a young and relatively small profession; the number of historians of mathematics, because of the types of knowledge needed, is even smaller. Still the need for such people is apparent.

Even if historians of mathematics were legion, however, the contribution of mathematicians to the history of mathematics would remain crucial. Mathematicians, of course, bring a higher level of mathematical knowledge to any historical task. Historians of mathematics should certainly know the mathematics whose history they are writing. But mathematicians are still needed—and not just because they know the mathematics better. Historians need the mathematician's point of view about what is mathematically important. The mathematician's work determines what it is that most needs a historical explanation. Only the mathematician can tell us which of a half-dozen contemporary concepts is really the crucial one, and which older concepts are worth looking into again—infinitesimals are one example (See Abraham Robinson's *Non-Standard Analysis* [C29], esp. ch. 10). Furthermore, the mathematician has a better idea of the logical relationship between

---

[14] A view championed by Lebesgue. See May's biography in H. Lebesgue, *Measure and the Integral* [B27, p. 5].

[15] A good introduction to the literature up to 1936, with extensive and humane annotations, is Sarton's *Study of the History of Mathematics* [7]. For a detailed, up-to-date, and extremely valuable guide see May's *Bibliography* [6].

mathematical ideas, and can suggest connections to the historian which might not be apparent from the historical record alone.

We have seen that the mathematician and the historian bring different skills and different perspectives to their common task of explaining the mathematical present by means of the past. Therefore, as this conference by its existence declares, collaboration between mathematicians and historians can be fruitful. The value of such a collaboration will be enhanced if each collaborator understands the unique contributions which can be made by the other. The importance of the common task, I think, makes it well worth the collective efforts.

Discussion

The discussion began with three comments by Dieudonné. First, he raised a technical point regarding convergence in the eighteenth century. He claimed that there are many examples to be found among the works of eighteenth-century mathematicians where divergent series are given a sum. Secondly, Dieudonné commented regarding the relationship between mathematics and factors external to its development. He stated that, for example, the general history of the seventeenth century had no connection with Fermat's theory of numbers. Further, in spite of the fact that Descartes, Leibniz, and Cantor were all philosophers and mathematicians, mathematicians in general are *not* philosophers; they *do* mathematics. And thirdly, Dieudonné asserted that simply because the choice of notation may be motivated by circumstances external to mathematical reasoning, does not, however, make this fact significant. He firmly held that such motivating circumstances have no significance.

In her response to Dieudonné's first point, Grabiner emphazized her basic agreement with Dieudonné. Of course it is true, she agreed, that mathematicians in the eighteenth century used divergent series; she had not intended to dispute that fact (See for instance "De seriebus divergentibus" in [B11 (1)vol. 15, pp. 586–588). She re-emphasized her claim that *some* (and not all) historical discussions of infinite series can be illuminated by considering the *two* definitions of convergence used in the eighteenth century.

Kahane emphasized the difference between *definitions* of convergence and *proofs* of convergence. He claimed that already in 1807, Fourier had in mind a definition of convergence based on partial sums, i.e., before Cauchy. [This definition is to be found in Fourier's basic paper on heat conduction submitted to the Academy of Sciences of Paris in 1807. The manuscript is in the library of the Ecole des Ponts et Chaussées and was later published in 1822 as part of his *Théorie analytique de la chaleur—Ed.*]. It would have been possible, stated Kahane, for Fourier to give proof of convergence, but he did not do so. Actually, when Cauchy attempted to prove the convergence of Fourier series, he committed several notable errors, one of which was to mistake absolute convergence for conditional convergence. Ironically, this proved to be helpful, by providing a motivation for Dirichlet to press on, concluded Kahane.

In regard to Dieudonné's second comment, Edwards began by pointing out that precisely in Fermat's time, the rediscovery and translation of the texts of Diophantus occurred. He believed that the events in the general cultural history of the era were extremely relevant to the development of the theory of numbers by Fermat. Hawkins further noted that at least *in the past* many mathematicians were interested in philosophy.

Dou continued in this vein. To him, mathematics is becoming separated from the world, and this is dangerous for mathematics. Even with the Greeks, Dou claimed, mathematics was a real part of life, something to live with. In recent times, there has been a move to separate mathematics and philosophy, and consequently there has arisen a severe need to bridge the gap between the two. Whereas the choice of notation may be irrelevant to mathematics, its philosophy is not.

Putnam concluded this line of thought by noting that the philosophy of mathematics from Plato on has always been of great importance not only for mathematics, but for epistemology and metaphysics as well. Mackey concluded the discussion with a comment on the pedagogical value of the history and philosophy of mathematics to students of mathematics. In Mackey's opinion, both of these approaches to mathematics can be illuminating, but one must question the level of accuracy required. Neither the historian's nor the philosopher's detailed accuracy is beneficial pedagogically. Mackey expressed interest in history, but felt that because of the pressures of his discipline, he could not be interested in too detailed a history.

2

Who Gave You the Epsilon? Cauchy and the Origins of Rigorous Calculus*

Student: The car has a speed of 50 miles an hour. What does that mean?
Teacher: Given any $\varepsilon > 0$, there exists a δ such that if $|t_2 - t_l| < \delta$, then $\left|\frac{s_2-s_1}{t_2-t_1} - 50\right| < \varepsilon$.
Student: How in the world did anybody ever think of such an answer?

Perhaps this exchange will remind us that the rigorous basis for the calculus is not at all intuitive—in fact, quite the contrary. The calculus is a subject dealing with speeds and distances, with tangents and areas—not inequalities. When Newton and Leibniz invented the calculus in the late seventeenth century, they did not use delta-epsilon proofs. It took a hundred and fifty years to develop them. This means that it was probably very hard, and it is no wonder that a modern student finds the rigorous basis of the calculus difficult. How, then, did the calculus get a rigorous basis in terms of the algebra of inequalities?

Delta-epsilon proofs are first found in the works of Augustin-Louis Cauchy (1789–1867). This is not always recognized, since Cauchy gave a purely verbal definition of limit, which at first glance does not resemble modern definitions ([6], p. 19):

> When the successively attributed values of the same variable indefinitely approach a fixed value, so that finally they differ from it by as little as desired, the last is called the *limit* of all the others.

Cauchy also gave a purely verbal definition of the derivative of $f(x)$ as the limit, when it exists, of the quotient of differences $(f(x + h) - f(x))/h$ when h goes to zero, a statement much like those that had already been made by Newton, Leibniz, d'Alembert, Maclaurin, and Euler. But what is significant is that Cauchy translated such verbal statements into the precise language of inequalities when he needed them in his proofs. For instance, for the derivative Cauchy ([8], p. 44) makes the following statement:

> Let δ, ε be two very small numbers; the first is chosen so that for all numerical [i.e., absolute] values of h less than δ, and for any value of x included [in the interval of definition], the ratio $(f(x + h) - f(x))/h$ will always be greater than $f'(x) - \varepsilon$ and less than $f'(x) + \varepsilon$. (*)

* Reprinted from *Amer Math Monthly* **90** (1983), 185–194.

This one example will be enough to indicate how Cauchy did the calculus, because the question to be answered in the present paper is not, "how is a rigorous delta-epsilon proof constructed?" As Cauchy's intellectual heirs we all know this. The central question is, how and why was Cauchy able to put the calculus on a rigorous basis, when his predecessors were not?

The answers to this historical question cannot be found by reflecting on the logical relations between the concepts, but by looking in detail at the past and seeing how the existing state of affairs in fact developed from that past. Thus we will examine the mathematical situation in the seventeenth and eighteenth centuries—the background against which we can appreciate Cauchy's innovation. We will describe the powerful techniques of the calculus of this earlier period and the relatively unimpressive views put forth to justify them. We will then discuss how a sense of urgency about rigorizing analysis gradually developed in the eighteenth century. Most important, we will explain the development of the mathematical techniques necessary for the new rigor from the work of men like Euler, d'Alembert, Poisson, and especially Lagrange. Finally, we will show how these mathematical results, though often developed for purposes far removed from establishing foundations for the calculus, were used by Cauchy in constructing his new rigorous analysis.

The Practice of Analysis: From Newton to Euler

In the late seventeenth century, Newton and Leibniz, almost simultaneously, independently invented the calculus. This invention involved three things. First, they invented the general concepts of differential quotient and integral (these are Leibniz's terms; Newton called the concepts "fluxion" and "fluent"). Second, they devised a notation for these concepts which made the calculus an algorithm: the methods not only worked, but were easy to use. Their notations had great heuristic power, and we still use Leibniz's dy/dx and $\int y\,dx$, and Newton's $\dot{x}$, today. Third, both men realized that the basic processes of finding tangents and areas, that is, differentiating and integrating, are mutually inverse—what we now call the Fundamental Theorem of Calculus.

Once the calculus had been invented, mathematicians possessed an extremely powerful set of methods for solving problems in geometry, in physics, and in pure analysis. But what was the nature of the basic concepts? For Leibniz, the differential quotient was a ratio of infinitesimal differences, and the integral was a sum of infinitesimals. For Newton, the derivative, or fluxion, was described as a rate of change; the integral, or fluent, was its inverse. In fact, throughout the eighteenth century, the integral was generally thought of as the inverse of the differential. One might imagine asking Leibniz exactly what an infinitesimal was, or Newton what a rate of change might be. Newton's answer, the best of the eighteenth century, is instructive. Consider a ratio of finite quantities (in modern notation, $(f(x+h)-f(x))/h$ as h goes to zero). The ratio eventually becomes what Newton called an "ultimate ratio". Ultimate ratios are ([26], p. 39)

> limits to which the ratios of quantities decreasing without limit do always converge, and to which they approach nearer than by any given difference, but never go beyond, nor ever reach until the quantities vanish.

Except for "reaching" the limit when the quantities vanish, we can translate Newton's words into our algebraic language. Newton himself, however, did not do this, nor did most of his followers in the eighteenth century. Moreover, "never go beyond" does not allow a variable to oscillate about its limit. Thus, though Newton's is an intuitively pleasing picture, as it stands it was not and could not be used for proofs about limits. The definition sounds good, but it was not understood or applied in algebraic terms.

But most eighteenth-century mathematicians would object, "Why worry about foundations?" In the eighteenth century, the calculus, intuitively understood and algorithmically executed, was applied to a wide range of problems. For instance, the partial differential equation for vibrating strings was solved; the equations of motion for the solar system were solved; the Laplace transform and the calculus of variations and the gamma function were invented and applied; all of mechanics was worked out in the language of the calculus. These were great achievements on the part of eighteenth-century mathematicians. Who would be greatly concerned about foundations when such important problems could be successfully treated by the calculus? Results were what counted.

This point will be better appreciated by looking at an example which illustrates both the "uncritical" approach to concepts of the eighteenth century and the immense power of eighteenth-century techniques, from the work of the great master of such techniques: Leonhard Euler. The problem is to find the sum of the series

$$\frac{1}{1}+\frac{1}{4}+\frac{1}{9}+\cdots+\frac{1}{k^2}+\cdots.$$

It clearly has a finite sum since it is bounded above by the series

$$1+\frac{1}{1\cdot 2}+\frac{1}{2\cdot 3}+\frac{1}{3\cdot 4}+\cdots+\frac{1}{(k-1)\cdot k}+\cdots,$$

whose sum was known to be 2; Johann Bernoulli had found this sum by treating

$$1+\frac{1}{1\cdot 2}+\frac{1}{2\cdot 3}+\frac{1}{3\cdot 4}+\cdots$$

as the difference between the series $\frac{1}{1}+\frac{1}{2}+\frac{1}{3}+\cdots$ and the series $\frac{1}{2}+\frac{1}{3}+\frac{1}{4}+\cdots$, and observing that this difference telescopes ([2], IV, 8; cited in [28], p. 321).

Euler's summation of $\sum 1/k^2$ makes use of a lemma from the theory of equations: given a polynomial equation whose constant term is 1, the coefficient of the linear term is the sum of the reciprocals of the roots with the signs changed. This result was both discovered and demonstrated by considering the equation $(x-a)(x-b)=0$, having roots a and b. Multiplying and then dividing out ab, we obtain

$$\frac{1}{ab}x^2-\left(\frac{1}{a}+\frac{1}{b}\right)x+1=0;$$

the result is now obvious, as is the extension to equations of higher degree. Euler's solution then considers the equation $\sin x=0$.

Expanding this as an infinite series, Euler obtained

$$x-\frac{x^3}{3!}+\frac{x^5}{5!}-\cdots=0.$$

Dividing by x yields

$$1-\frac{x^2}{3!}+\frac{x^4}{5!}-\cdots=0.$$

Finally, substituting $x^2=u$ produces

$$1-\frac{u}{3!}+\frac{u^2}{5!}-\cdots=0.$$

But Euler thought that power series could be manipulated just like polynomials. Thus, we now have a polynomial equation in u, whose constant term is 1. Applying the lemma to it, the coefficient of the linear term with the sign changed is $1/3! = 1/6$. The roots of the equation in u are the roots of $\sin x = 0$ with the substitution $u = x^2$, namely $\pi^2, 4\pi^2, 9\pi^2, \ldots$. Thus the lemma implies

$$\frac{1}{6} = \frac{1}{\pi^2} + \frac{1}{4\pi^2} + \frac{1}{9\pi^2} + \cdots .$$

Multiplying by π^2 yields the sum of the original series ([4], p. 487; [13], (2) Vol. 14, pp. 73–86):

$$\frac{1}{1} + \frac{1}{4} + \frac{1}{9} + \cdots + \frac{1}{k^2} + \cdots = \frac{\pi^2}{6}.$$

Though it is easy to criticize eighteenth-century arguments like this for their lack of rigor, it is also unfair. Foundations, precise specifications of the conditions under which such manipulations with infinities or infinitesimals were admissible, were not very important to men like Euler, because without such specifications they made important new discoveries, whose results in cases like this could readily be verified. When the foundations of the calculus were discussed in the eighteenth century, they were treated as secondary. Discussions of foundations appeared in the introductions to books, in popularizations, and in philosophical writings, and were not—as they are now and have been since Cauchy's time—the subject of articles in research-oriented journals.

Thus, where we once had one question to answer, we now have two. The first remains, where do Cauchy's rigorous techniques come from? Second, one must now ask, why rigorize the calculus in the first place? If few mathematicians were very interested in foundations in the eighteenth century ([16], Chapter 6), then when, and why, were attitudes changed?

Of course, to establish rigor, it is necessary—though not sufficient—to think rigor is significant. But more important, to establish rigor, it is necessary (though also not sufficient) to have a set of techniques in existence which are suitable for that purpose. In particular, if the calculus is to be made rigorous by being reduced to the algebra of inequalities, one must have both the algebra of inequalities, and facts about the concepts of the calculus that can be expressed in terms of the algebra of inequalities.

In the early nineteenth century, three conditions held for the first time: Rigor was considered important; there was a well-developed algebra of inequalities; and, certain properties were known about the basic concepts of analysis—limits, convergence, continuity, derivatives, integrals—properties which could be expressed in the language of inequalities if desired. Cauchy, followed by Riemann and Weierstrass, gave the calculus a rigorous basis, using the already-existing algebra of inequalities, and built a logically-connected structure of theorems about the concepts of the calculus. It is our task to explain how these three conditions—the developed algebra of inequalities, the importance of rigor, the appropriate properties of the concepts of the calculus—came to be.

The Algebra of Inequalities

Today, the algebra of inequalities is studied in calculus courses because of its use as a basis for the calculus, but why should it have been studied in the eighteenth century when this application was unknown? In the eighteenth century, inequalities were important in the study of a major class of results: approximations.

For example, consider an equation such as $(x+1)^{\mu} = a$, for μ not an integer. Usually a cannot be found exactly, but it can be approximated by an infinite series. In general, given some number n of terms of such an approximating series, eighteenth-century mathematicians sought to compute an upper bound on the error in the approximation—that is, the difference between the sum of the series and the nth partial sum. This computation was a problem in the algebra of inequalities. Jean d'Alembert [10] solved it for the important case of the binomial series; given the number of terms of the series n, and assuming implicitly that the series converges to its sum, he could find the bounds on the error—that is, on the remainder of the series after the nth term—by bounding the series above and below with convergent geometric progressions. Similarly, Joseph-Louis Lagrange [24] invented a new approximation method using continued fractions and, by extremely intricate inequality calculations, gave necessary and sufficient conditions for a given iteration of the approximation to be closer to the result than the previous iteration. Lagrange also derived the Lagrange remainder of the Taylor series [23], using an inequality which bounded the remainder above and below by the maximum and minimum values of the nth derivative and then applying the intermediate-value theorem for continuous functions. Thus, through such eighteenth-century work ([16], pp. 56–68; [14], Chapters 2–4), there was by the end of the eighteenth century a developed algebra of inequalities, and people used to working with it. Given an n, these people are used to finding an error—that is, an epsilon.

Changing Attitudes Toward Rigor

Mathematicians were much more interested in finding rigorous foundations for the calculus in 1800 than they had been a hundred years before. There are many reasons for this: no one enough by itself, but apparently sufficient when acting together. Of course one might think that eighteenth-century mathematicians were always making errors because of the lack of an explicitly-formulated rigorous foundation. But this did not occur. They were usually right, and for two reasons. One is that if one deals with real variables, functions of one variable, series which are power series, and functions arising from physical problems, errors will not occur too often. A second reason is that mathematicians like Euler and Laplace had a deep insight into the basic properties of the concepts of the calculus, and were able to choose fruitful methods and evade pitfalls. The only "error" they committed was to use methods that shocked mathematicians of later ages who had grown up with the rigor of the nineteenth century.

What then were the reasons for the deepened interest in rigor? One set of reasons was philosophical. In 1734, the British philosopher Bishop Berkeley ([1], Section 35) had attacked the calculus on the ground that it was not rigorous. In *The Analyst, or a Discourse Addressed to an Infidel Mathematician*, he said that mathematicians had no business attacking the unreasonableness of religion, given the way they themselves reasoned. He ridiculed fluxions—"velocities of evanescent increments"—calling the evanescent increments "ghosts of departed quantities". Even more to the point, he correctly criticized a number of specific arguments from the writings of his mathematical contemporaries. For instance, he attacked the process of finding the fluxion (our derivative) by reviewing the steps of the process: if we consider $y = x^2$, taking the ratio of the differences $((x+h)^2 - x^2)/h$, then simplifying to $2x + h$, then letting h vanish, we obtain $2x$. But is h zero? If it is, we cannot meaningfully divide by it; if it is not zero, we have no right to throw it away. As Berkeley put it ([1], Section 15), the quantity we have called h "might have signified either an increment or nothing. But then, which of these soever you make it signify, you must argue consistently with such its signification".

Since an adequate response to Berkeley's objections would have involved recognizing that an equation involving limits is a shorthand expression for a sequence of inequalities—a subtle and difficult idea—no eighteenth-century analyst gave a fully adequate answer to Berkeley. However, many tried. Maclaurin, d'Alembert, Lagrange, Lazare Carnot, and possibly Euler, all knew about Berkeley's work, and all wrote something about foundations. So Berkeley did call attention to the question. However, except for Maclaurin, no leading mathematician spent much time on the question because of Berkeley's work, and even Maclaurin's influence lay in other fields.

Another factor contributing to the new interest in rigor was that there was a limit to the number of results that could be reached by eighteenth-century methods. Near the end of the century, some leading mathematicians had begun to feel that this limit was at hand. D'Alembert and Lagrange indicate this in their correspondence, with Lagrange calling higher mathematics "decadent" ([2], Vol. 13, p. 229). The philosopher Diderot went so far as to claim that the mathematicians of the eighteenth century had "erected the pillars of Hercules" beyond which it was impossible to go ([11], pp. 180–181). Thus, there was a perceived need to consolidate the gains of the past century.

Another factor was Lagrange, who became increasingly interested in foundations, and, through his activities, interested other mathematicians. In the eighteenth century, scientific academies offered prizes for solving major outstanding problems. In 1784, Lagrange and his colleagues posed the problem of the foundations of the calculus as the Berlin Academy's prize problem. Nobody solved it to Lagrange's satisfaction, but two of the entries in the competition were later expanded into full-length books, the first on the Continent, on foundations: Simon L'Huilier's *Exposition Elémentaire des Principes des Calculs Supérieurs*, Berlin, 1787, and Lazare Carnot's *Réflexions sur la Métaphysique du Calcul Infinitésimal*, Paris, 1797. Thus Lagrange clearly helped revive interest in the problem.

Lagrange's interest stemmed in part from his respect for the power and generality of algebra; he wanted to gain for the calculus the certainty he believed algebra to possess. But there was another factor increasing interest in foundations, not only for Lagrange, but for many other mathematicians by the end of the eighteenth century: the need to teach. Teaching forces one's attention to basic questions. Yet before the mid-eighteenth century, mathematicians had often made their living by being attached to royal courts. But royal courts declined; the number of mathematicians increased; and mathematics began to look useful. First in military schools and later on at the Ecole Polytechnique in Paris, another line of work became available: teaching mathematics to students of science and engineering. The Ecole Polytechnique was founded by the French revolutionary government to train scientists, who, the government believed, might prove useful to a modern state. And it was as a lecturer in analysis at the Ecole Polytechnique that Lagrange wrote his two major works on the calculus which treated foundations; similarly, it was 40 years earlier, teaching the calculus at the Military Academy at Turin, that Lagrange had first set out to work on the problem of foundations. Because teaching forces one to ask basic questions about the nature of the most important concepts, the change in the economic circumstances of mathematicians—the need to teach—provided a catalyst for the crystallization of the foundations of the calculus out of the historical and mathematical background. In fact, even well into the nineteenth century, much of foundations was born in the teaching situation: Weierstrass's foundations come from his lectures at Berlin; Dedekind first thought of the problem of continuity while teaching at Zurich; Dini and Landau turned to foundations while teaching analysis; and, most important for our present purposes, so did Cauchy. Cauchy's foundations of analysis appear in the books based on his lectures at the Ecole Polytechnique; his book of 1821 was the first example of the great French tradition of *Cours d'Analyse*.

The Concepts of the Calculus

Arising from algebra, the algebra of inequalities was now there for the calculus to be reduced to; the desire to make the calculus rigorous had arisen through consolidation, through philosophy, through teaching, through Lagrange. Now let us turn to the mathematical substance of eighteenth-century analysis, to see what was known about the concepts of the calculus before Cauchy, and what he had to work out for himself, in order to define, and prove theorems about, limit, convergence, continuity, derivatives, and integrals.

First, consider the concept of limit. As we have already pointed out, since Newton the limit had been thought of as a bound which could be approached closer and closer, though not surpassed. By 1800, with the work of L'Huilier and Lacroix on alternating series, the restriction that the limit be one-sided had been abandoned. Cauchy systematically translated this refined limit-concept into the algebra of inequalities, and used it in proofs once it had been so translated; thus he gave reality to the oft-repeated eighteenth-century statement that the calculus could be based on limits.

For example, consider the concept of convergence. Maclaurin had said already that the sum of a series was the limit of the partial sums. For Cauchy, this meant something precise. It meant that, given an ε, one could find n such that, for more than n terms, the sum of the infinite series is within ε of the nth partial sum. That is the reverse of the error-estimating procedure that d'Alembert had used. From his definition of a series having a sum, Cauchy could prove that a geometric progression with radius less in absolute value than 1 converged to its usual sum. As we have said, d'Alembert had shown that the binomial series for, say, $(1+x)^{p/q}$ could be bounded above and below by convergent geometric progressions. Cauchy assumed that if a series of positive terms is bounded above, term-by-term, by a convergent geometric progression, then it converges; he then used such comparisons to prove a number of tests for convergence: the root test, the ratio test, the logarithm test. The treatment is quite elegant ([6], (2), Vol. 3, pp. 114–138). Taking a technique used a few times by men like d'Alembert and Lagrange on an *ad hoc* basis in approximations, and using the definition of the sum of a series based on the limit concept, Cauchy created the first rigorous theory of convergence.

Let us now turn to the concept of continuity. Cauchy ([6], p. 43) gave essentially the modern definition of continuous function, saying that the function $f(x)$ is continuous on a given interval if for each x in that interval "the numerical [i.e., absolute] value of the difference $f(x+a)-f(x)$ decreases indefinitely with a". He used this definition in proving the intermediate-value theorem for continuous functions ([6], pp. 378–380; English translation in [16], pp. 167–168). The proof proceeds by examining a function $f(x)$ on an interval, say $[b, c]$, where $f(b)$ is negative, $f(c)$ positive, and dividing the interval $[b, c]$ into m parts of width $h=(c-b)/m$. Cauchy considered the sign of the function at the points

$$f(b), f(b+h), \ldots, f(b+(m-1)h), f(c);$$

unless one of the values of f is 0, there are two values of x differing by h such that f is negative at one, positive at the other. Repeating this process for new intervals of width

$$(c - b)/m, (c - b)/m^2, \ldots,$$

gives an increasing sequence of values of x: $b, b_1, b_2, \ldots$ for which f is negative, and a decreasing sequence of values of x: $c, c_1, c_2, \ldots$ for which f is positive, and such that the difference between b_k and c_k goes to 0. Cauchy asserted that these two sequences must have a common limit a. He then argued that since $f(x)$ is continuous, the sequence of negative values $f(b_k)$ and of positive values $f(c_k)$ both converge toward the common limit $f(a)$, which must therefore be 0.

Cauchy's proof involves an already existing technique, which Lagrange had applied in approximating real roots of polynomial equations. If a polynomial was negative for one value of the variable, positive for another, there was a root in between, and the difference between those two values of the variable bounded the error made in taking either as an approximation to the root ([20], p. 87, [21]). Thus again we have the algebra of inequalities providing a technique which Cauchy transformed from a tool of approximation to a tool of rigor.

It is worth remarking at this point that Cauchy, in his treatment both of convergence and of continuity, implicitly assumed various forms of the completeness property for the real numbers. For instance, he treated as obvious that a series of positive terms, bounded above by a convergent geometric progression, converges: also, his proof of the intermediate-value theorem assumes that a bounded monotone sequence has a limit.

While Cauchy was the first systematically to exploit inequality proof techniques to prove theorems in analysis, he did not identify all the implicit assumptions about the real numbers that such inequality techniques involve. Similarly, as the reader may have already noticed, Cauchy's definition of continuous function does not distinguish between what we now call point-wise and uniform continuity; also, in treating series of functions, Cauchy did not distinguish between point-wise and uniform convergence. The verbal formulations like "for all" that are involved in choosing deltas did not distinguish between "for any epsilon and for all x" and "for any x, given any epsilon" (Grattan-Guinness [18] puts it well: "Uniform convergence was tucked away in the work "always," with no reference to the variable at all."). Nor was it at all clear in the 1820s how much depended on this distinction, since proofs about continuity and convergence were in themselves so novel. We shall see the same confusion between uniform and point-wise convergence as we turn now to Cauchy's theory of the derivative.

Again we begin with an approximation. Lagrange gave the following inequality about the derivative:

$$(**) \qquad f(x+h) = f(x) + hf'(x) + hV,$$

where V goes to 0 with h. He interpreted this to mean that, given any D, one can find h sufficiently small so that V is between $-D$ and $+D$ ([21], p. 87; [23], p. 77). Clearly this interpretation is equivalent to statement (*) above, Cauchy's delta-epsilon characterization of the derivative. But how did Lagrange obtain this result? The answer is surprising; for Lagrange, formula (**) was a consequence of Taylor's theorem. Lagrange believed that any function (that is, any analytic expression, whether finite or infinite, involving the variable) had a unique power-series expansion (except possibly at a finite number of isolated points). This is because he believed that there was an "algebra of infinite series", an algebra exemplified by work of Euler such as the example we gave above. And Lagrange said that the way to make the calculus rigorous was to reduce it to algebra. Although there is no "algebra" of infinite series that gives power-series expansions without any consideration of convergence and limits, this assumption led Lagrange to define $f'(x)$ without reference to limits, as the coefficient of the linear term in h in the Taylor series expansion for $f(x+h)$. Following Euler, Lagrange then said that, for any power series in h, one could take h sufficiently small so that any given term of the series exceeded the sum of all the rest of the terms following it; this approximation, said Lagrange, is assumed in applications of the calculus to geometry and mechanics ([23], p. 29; [21], p. 101; [12]). Applying this approximation to the linear term in the Taylor series produces (**), which I call the Lagrange property of the derivative. (Like Cauchy's (*), the inequality-translation Lagrange gives for (**) assumes that, given any D, one finds h sufficiently small so $|V| \leq D$ with no mention whatever of x.)

Not only did Lagrange state property (**) and the associated inequalities, he used them as a basis for a number of proofs about derivatives: for instance, to prove that a function with positive derivative on an interval is increasing there, to prove the mean-value theorem for derivatives, and to obtain the Lagrange remainder for the Taylor series. (Details may be found in the works cited in [16] and [17].) Lagrange also applied his results to characterize the properties of maxima and minima, and orders of contact between curves.

With a few modifications, Lagrange's proofs are valid—provided that property (**) can be justified. Cauchy borrowed and simplified what are in effect Lagrange's inequality proofs about derivatives, with a few improvements, basing them on his own (*). But Cauchy made these proofs legitimate because Cauchy defined the derivative precisely to satisfy the relevant inequalities. Once again, the key properties come from an approximation. For Lagrange, the derivative was *exactly*—no epsilons needed—the coefficient of the linear term in the Taylor series; formula (**), and the corresponding inequality that $f(x+h) - f(x)$ lies between $h(f'(x) \pm D)$, were approximations. Cauchy ([16], Chapter 5, [17]) brought Lagrange's inequality properties and proofs together with a definition of derivative devised to make those techniques rigorously founded.

The last of the concepts we shall consider, the integral, followed an analogous development. In the eighteenth century, the integral was usually thought of as the inverse of the differential. But sometimes the inverse could not be computed exactly, so men like Euler remarked that the integral could be approximated as closely as one liked by a sum. Of course, the geometric picture of an area being approximated by rectangles, or the Leibnizian definition of the integral as a sum, suggests this immediately. But what is important for our purposes is that much work was done on approximating the values of definite integrals in the eighteenth century, including considerations of how small the subintervals used in the sums should be when the function oscillates to a greater or lesser extent. For instance, Euler [23] treated sums of the form

$$\sum_{k=0}^{n} f(x_k)(x_{k+1} - x_k)$$

as approximations to the integral $\int_{x_0}^{x_n} f(x)\,dx$.

In 1820, S.-D. Poisson, who was interested in complex integration and therefore more concerned than most people about the existence and behavior of integrals, asked the following question. If the integral F is defined as the anti-derivative of f, and if $b - a = nh$, can it be proved that $F(b) - F(a) = \int_a^b f(x)\,dx$ is the limit of the sum

$$S = hf(a) + hf(a+h) + \cdots + hf(a+(n-1)h)$$

as h gets small? (S is an approximating sum of the eighteenth-century sort.) Poisson called this result "the fundamental proposition of the theory of definite integrals." He proved it by using another inequality-result: the Taylor series with remainder. First, he wrote $F(b) - F(a)$ as the telescoping sum

$$(***) \quad F(a+h) - F(a) + F(a+2h) - F(a+h) + \cdots + F(b) - F(a+(n-1)h).$$

Then, for each of the terms of the form $F(a+kh) - F(a+(k-1)h)$, Taylor's series with remainder gives, since by definition $F' = f$,

$$F(a+kh) - F(a+(k-1)h) = hf(a+(k-1)h) + R_k h^{1+w},$$

where $w > 0$, for some R_k. Thus the telescoping sum (2) becomes

$$hf(a) + hf(a+h) + \cdots + hf(a+(n-1)h) + (R_1 + \cdots + R_n)h^{1+w}.$$

So $F(b) - F(a)$ and the sum S differ by $(R_1 + \cdots + R_n)h^{1+w}$. Letting R be the maximum value for the R_k,

$$(R_1 + \cdots + R_n)h^{1+w} \leq n \cdot R(h^{1+w}) = R \cdot nh \cdot h^w = R(b-a)h^w.$$

Therefore, if h is taken sufficiently small, $F(b) - F(a)$ differs from S by less than any given quantity [27].

Poisson's was the first attempt to prove the equivalence of the antiderivative and limit-of-sums conceptions of the integral. However, besides the implicit assumptions of the existence of antiderivatives and bounded first derivatives for f on the given interval, the proof assumes that the subintervals on which the sum is taken are all equal. Should the result not hold for unequal divisions also? Poisson thought so, and justified it by saying ([27], pp. 329–330),

> if the integral is represented by the area of a curve, this area will be the same, if we divide the difference ... into an infinite number of equal parts, or an infinite number of unequal parts following any law.

This, however, is an assertion, not a proof. And Cauchy saw that a proof was needed.

Cauchy ([6], (2) Vol. 3, p. iii) did not like formalistic arguments in supposedly rigorous subjects, saying that most algebraic formulas hold "only under certain conditions, and for certain values of the quantities they contain" . In particular, one could not assume that what worked for finite expressions automatically worked for infinite ones. Thus, Cauchy showed that the sum of the series $\frac{1}{1} + \frac{1}{4} + \frac{1}{9} + \cdots$ was $\pi^2/6$ by actually calculating the difference between the nth partial sum and $\pi^2/6$ and showing that it was arbitrarily small ([6], (2) Vol. 3, pp. 456–457).

Similarly, just because there was an operation called taking a derivative did not mean that the inverse of that operation always produced a result. The existence of the definite integral had to be proved. And how does one prove existence in the 1820s? One constructs the mathematical object in question by using an eighteenth-century approximation that converges to it. Cauchy defined the integral as the limit of Euler-style sums $\sum f(x_k)(x_{k+1} - x_k)$ for $x_{k+1} - x_k$ sufficiently small. Assuming explicitly that $f(x)$ was continuous on the given interval (and implicitly that it was uniformly continuous), Cauchy was able to show that all sums of that form approach a fixed value, called by definition the integral of the function on that interval. This is an extremely hard proof ([5], pp. 122–125; [16], pp. 171–175 in English translation). Finally, borrowing from Lagrange the mean-value theorem for integrals, Cauchy proved the Fundamental Theorem of Calculus ([5], p. 151–152).

Conclusion

Here are all the pieces of the puzzle we originally set out to solve. Algebraic approximations produced the algebra of inequalities; eighteenth-century approximations in the calculus produced the useful properties of the concepts of analysis: d'Alembert's error-bounds for series, Lagrange's inequalities about derivatives, Euler's approximations to integrals. There was a new interest in foundations. All that was needed was a sufficiently great genius to build the new foundation.

Two men came close. In 1816, Carl Friedrich Gauss gave a rigorous treatment of the convergence of the hypergeometric series, using the technique of comparing a series with convergent geometric progressions; however, Gauss did not give a general foundation for all of analysis. Bernhard Bolzano, whose work was little known until the 1860s, echoing Lagrange's call to reduce the calculus to algebra, gave in 1817 a definition of continuous function like Cauchy's and then proved—by a different technique from Cauchy's—the intermediate-value theorem [3]. But it was Cauchy who gave rigorous definitions and proofs for all the basic concepts; it was he who realized the far-reaching power of the inequality-based limit concept; and it was he who gave us—except for a few implicit assumptions about uniformity and about completeness—the modern rigorous approach to calculus.

Mathematicians are used to taking the rigorous foundations of the calculus as a completed whole. What I have tried to do as a historian is to reveal what went into making up that great achievement. This needs to be done, because completed wholes by their nature do not reveal the separate strands that go into weaving them—especially when the strands have been considerably transformed. In Cauchy's work, though, one trace indeed was left of the origin of rigorous calculus in approximations—the letter epsilon. The ε corresponds to the initial letter in the word 'erreur' (or "error"), and Cauchy in fact used ε for "error" in some of his work on probability [9]. It is both amusing and historically appropriate that the ε, once used to designate the 'error' in approximations, has become transformed into the characteristic symbol of precision and rigor in the calculus. As Cauchy transformed the algebra of inequalities from a tool of approximation to a tool of rigor, so he transformed the calculus from a powerful method of generating results to the rigorous subject we know today.

References

1. George Berkeley, *The Analyst, or a Discourse Addressed to an Infidel Mathematician*, 1736.
2. Johann Bernoulli, *Opera Omnia*, 1742.
3. B. Bolzano, *Rein analytischer Beweis des Lehrsatzes dass zwischen je zwey Werthen, die ein entgegengesetztes Resultat gewaehren, wenigstens cine reele Wurzel der Gleichung liege,* Prague, 1817; English version, S. B. Russ, A translation of Bolzano's paper on the intermediate value theorem, *Hist. Math.* **7** (1980), 156–185.
4. C. Boyer, *History of Mathematics*, John Wiley, 1968.
5. A.-L. Cauchy, *Calcul Infinitesimal*, in [7], (2), Vol. 4.
6. A.-L. Cauchy, *Cours d'Analyse*, Paris, 1821; in [7], (2), Vol. 3.
7. A.-L. Cauchy, *Oeuvres Complètes d'Augustin Cauchy*, Gauthier-Villars, 1899.
8. A.-L. Cauchy, Résumé des leçons données à l'école royale polytechnique sur le calcul infinitésimal, 1823; in [7], (2), Vol. 4.

9. A.-L. Cauchy, Sur la plus grande erreur à craindre dans un résultat moyen, et sur le système de facteurs qui rend cette plus grande erreur un minimum, *C. R. Acad. Sci. Paris* **37**, 1853; in [7], (1), Vol. 12.
10. J. d'Alembert, Réflexions sur les suites et sur les racines imaginaires, in *Opuscules Mathématiques*, Vol. 5, Briasson, 1768, pp. 171–215.
11. D. Diderot, De l'interprétation de la nature, in *Oeuvres Philosophiques* (ed. P. Vernière), Garnier, 1961, pp. 180–181.
12. L. Euler, *Institutiones Calculi Integralis*, 3 vols, St. Petersburg, 1768–1770; in [13], (1), Vol. II.
13. L. Euler, *Opera Omnia*, Leipzig, 1912.
14. H. Goldstine, *A History of Numerical Analysis from the 16th through the 19th Century*, Springer-Verlag, 1977.
15. J. V. Grabiner, Cauchy and Bolzano: tradition and transformation in the history of mathematics, in [25].
16. J. V. Grabiner, *The Origins of Cauchy's Rigorous Calculus*, MIT Press, 1981.
17. J. V. Grabiner, The origins of Cauchy's theory of the derivative, *Hist. Math.* **5** (1978), 379–409.
18. I. Grattan-Guinness, *Development of the Foundations of Mathematical Analysis from Euler to Riemann*, MIT Press, 1970.
19. A. P. Iushkevich, O vozniknoveniya poiyatiya ob opredelennom integrale Koshi, *Trudy Instituta Istorii Estestvoznaniya, Akademia Nauk SSSR* **1** (1947), 373–411.
20. J.-L. Lagrange, *Equations Numériques*, in [22], Vol. 8.
21. J.-L. Lagrange, *Leçons sur le Calcul des Fonctions*, Paris, 1806, in [22], Vol. 10.
22. J.-L. Lagrange, *Oeuvres de Lagrange*, Gauthier-Villars, 1867–1892.
23. J.-L. Lagrange, *Théorie des Fonctions Analytiques*, 2nd ed., Paris, 1813, in [22], Vol. 9.
24. J.-L. Lagrange, *Traité de la Résolution des Équations Numériques de Tous les Degrés*, 2nd ed., Courcier, 1808, in [22], Vol. 8.
25. E. Mendelsohn, *Transformation and Tradition in the Sciences*, Cambridge, 1984.
26. Isaac Newton, *Mathematical Principles of Natural Philosophy* (transl. A. Motte, revised by Florian Cajori), 3rd ed., 1726, Univ. of California Press, 1934.
27. S. D. Poisson, Suite du mémoire sur les integrales définies, *J. de l'Ecole polytechnique*, Cah. **18** II (1820), 295–341.
28. D. J. Struik, *A Source Book in Mathematics, 1200–1800*, Harvard, 1969.

3

The Changing Concept of Change: The Derivative from Fermat to Weierstrass*

First the derivative was used, then discovered, explored and developed, and only then, defined.

Some years ago while teaching the history of mathematics, I asked my students to read a discussion of maxima and minima by the seventeenth-century mathematician, Pierre Fermat. To start the discussion, I asked them, "Would you please define a relative maximum?" They told me it was a place where the derivative was zero. "If that's so," I asked, "then what is the definition of a relative minimum?" They told me, *that's* a place where the derivative is zero. "Well, in that case," I asked, "what is the difference between a maximum and a minimum?" They replied that in the case of a maximum, the second derivative is negative.

What can we learn from this apparent victory of calculus over common sense?

I used to think that this story showed that these students did not understand the calculus, but I have come to think the opposite: they understood it very well. The students' answers are a tribute to the power of the calculus in general, and the power of the concept of derivative in particular. Once one has been initiated into the calculus, it is hard to remember what it was like *not* to know what a derivative is and how to use it, and to realize that people like Fermat once had to cope with finding maxima and minima without knowing about derivatives at all.

Historically speaking, there were four steps in the development of today's concept of the derivative, which I list here in chronological order. The derivative was first *used*; it was then *discovered*; it was then *explored and developed*; and it was finally *defined.* That is, examples of what we now recognize as derivatives first were used on an ad hoc basis in solving particular problems; then the general concept lying behind these uses was identified (as part of the invention of the calculus); then many properties of the derivative were explained and developed in applications both to mathematics and to physics; and finally, a rigorous definition was given and the concept of derivative was embedded in a rigorous theory. I will describe the steps, and give one detailed mathematical example from each. We will then reflect on what it all means—for the teacher, for the historian, and for the mathematician.

---

* Reprinted from Math. Mag. 56, 4(Sept. 1983) 195–206.

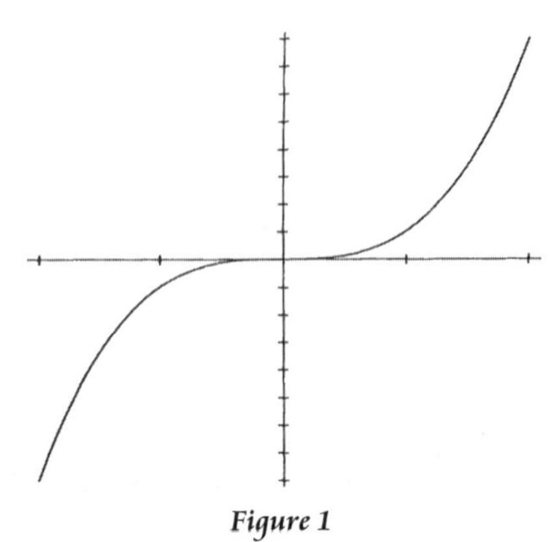

Figure 1

The Seventeenth-Century Background

Our story begins shortly after European mathematicians had become familiar once more with Greek mathematics, learned Islamic algebra, synthesized the two traditions, and struck out on their own. François Vieta invented symbolic algebra in 1591; Descartes and Fermat independently invented analytic geometry in the 1630's. Analytic geometry meant, first, that curves could be represented by equations; conversely, it meant also that every equation determined a curve. The Greeks and Muslims had studied curves, but not that many—principally the circle and the conic sections plus a few more defined as loci. Many problems had been solved for these, including finding their tangents and areas. But since any equation could now produce a new curve, students of the geometry of curves in the early seventeenth century were suddenly confronted with an explosion of curves to consider. With these new curves, the old Greek methods of synthetic geometry were no longer sufficient. The Greeks, of course, had known how to find the tangents to circles, conic sections, and some more sophisticated curves such as the spiral of Archimedes, using the methods of synthetic geometry. But how could one describe the properties of the tangent at an arbitrary point on a curve defined by a ninety-sixth degree polynomial? The Greeks had defined a tangent as a line which touches a curve without cutting it, and usually expected it to have only one point in common with the curve. How then was the tangent to be defined at the point (0,0) for a curve like $y = x^3$ (Figure 1), or to a point on a curve with many turning points (Figure 2)?

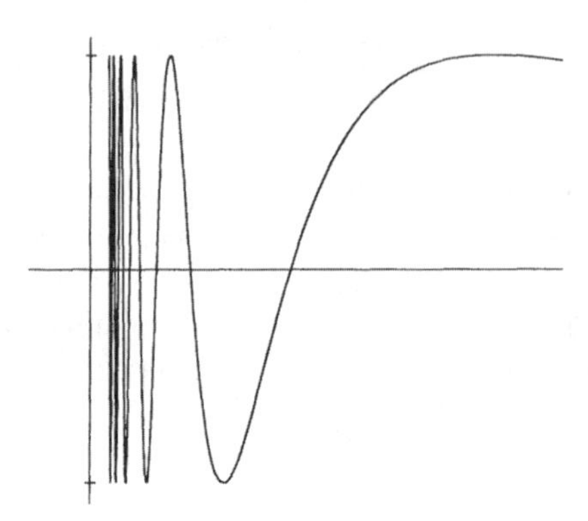

Figure 2

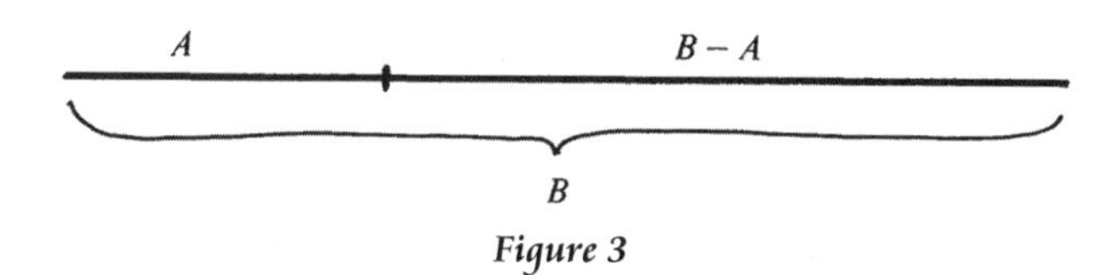

Figure 3

The same new curves presented new problems to the student of areas and arc lengths. The Greeks had also studied a few cases of what they called "isoperimetric" problems. For example, they asked: of all plane figures with the same perimeter, which one has the greatest area? The circle, of course, but the Greeks had no general method for solving all such problems. Seventeenth-century mathematicians hoped that the new symbolic algebra might somehow help solve all problems of maxima and minima.

Thus, though a major part of the agenda for seventeenth-century mathematicians—tangents, areas, extrema—came from the Greeks, the subject matter had been vastly extended, and the solutions would come from using the new tools: symbolic algebra and analytic geometry.

Finding Maxima, Minima, and Tangents

We turn to the first of our four steps in the history of the derivative: its *use*, and we also illustrate some of the general statements we have made. We shall look at Pierre Fermat's method of finding maxima and minima, which dates from the 1630's **[8]**. Fermat illustrated his method first in solving a simple problem, whose solution was well known: *Given a line, to divide it into two parts so that the product of the parts will be a maximum.* Let the length of the line be designated B and the first part A (Figure 3). Then the second part is $B - A$ and the product of the two parts is

$$A(B - A) = AB - A^2. \tag{1}$$

Fermat had read in the writings of the Greek mathematician Pappus of Alexandria that a problem which has, in general, two solutions will have only one solution in the case of a maximum. This remark led him to his method of finding maxima and minima. Suppose in the problem just stated there is a second solution. For this solution, let the first part of the line be designated as $A + E$; the second part is then $B - (A + E) = B - A - E$. Multiplying the two parts together, we obtain for the product

$$BA + BE + A^2 - AE - EA - E^2 = AB - A^2 - 2AE + BE - E^2. \tag{2}$$

Following Pappus' principle for the maximum, instead of two solutions, there is only one. So we set the two products (1) and (2) "sort of" equal; that is, we formulate what Fermat called the pseudo-equality:

$$AB - A^2 = AB - A^2 - 2AE + BE - E^2.$$

Simplifying, we obtain

$$2AE + E^2 = BE$$

and

$$2A + E = B$$

Now Fermat said, with no justification and no ceremony, "suppress E." Thus he obtained

$$A = B/2,$$

which indeed gives the maximum sought. He concluded, "We can hardly expect a more general method." And, of course, he was right.

Notice that Fermat did not call E infinitely small, or vanishing, or a limit; he did not explain why he could first divide by E (treating it as nonzero) and then throw it out (treating it as zero). Furthermore, he did not explain what he was doing as a special case of a more general concept, be it derivative, rate of change, or even slope of tangent. He did not even understand the relationship between his maximum-minimum method and the way one found tangents; in fact he followed his treatment of maxima and minima by saying that the same method—that is, adding E, doing the algebra, then suppressing E—could be used to find tangents [**8**, p. 223].

Though the considerations that led Fermat to his method may seem surprising to us, he did devise a method of finding extrema that worked, and it gave results that were far from trivial. For instance, Fermat applied his method to optics. Assuming that a ray of light which goes from one medium to another always takes the quickest path (what we now call the Fermat least-time principle), he used his method to compute the path taking minimal time. Thus he showed that his least-time principle yields Snell's law of refraction [**7**] [**12**, pp. 387–390].

Though Fermat did not publish his method of maxima and minima, it became well known through correspondence and was widely used. After mathematicians had become familiar with a variety of examples, a pattern emerged from the solutions by Fermat's method to maximum-minimum problems. In 1659, Johann Hudde gave a general verbal formulation of this pattern [**3**, p. 186], which, in modern notation, states that, *given a polynomial of the form*

$$y = \sum_{k=0}^{n} a_k x^k,$$

there is a maximum or minimum when

$$\sum_{k=1}^{n} k a_k x^{k-1} = 0.$$

Of even greater interest than the problem of extrema in the seventeenth century was the finding of tangents. Here the tangent was usually thought of as a secant for which the two points came closer and closer together until they coincided. Precisely what it meant for a secant to "become" a tangent was never completely explained. Nevertheless, methods based on this approach worked. Given the equation of a curve

$$y = f(x),$$

Fermat, Descartes, John Wallis, Isaac Barrow, and many other seventeenth-century mathematicians were able to find the tangent. The method involves considering, and computing, the slope of the secant,

$$\frac{f(x+h) - f(x)}{h},$$

doing the algebra required by the formula for $f(x+h)$ in the numerator, then dividing by h. The diagram in Figure 4 then suggests that when the quantity h vanishes, the secant becomes the

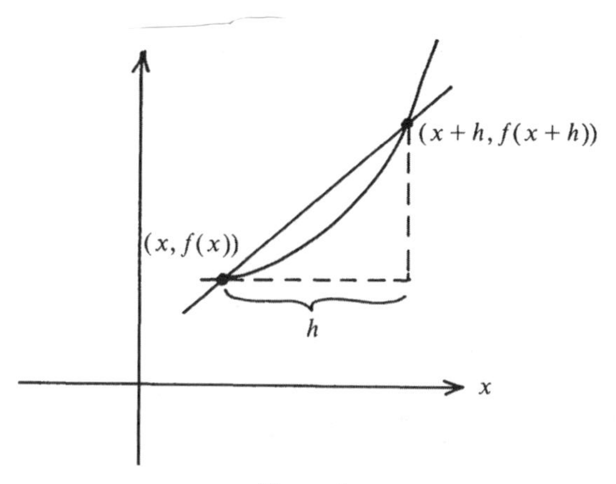

Figure 4

tangent, so that neglecting h in the expression for the slope of the secant gives the slope of the tangent. Again, a general pattern for the equations of slopes of tangents soon became apparent, and a rule analogous to Hudde's rule for maxima and minima was stated by several people, including René Sluse, Hudde, and Christiaan Huygens [**3**, pp. 185–186].

By the year 1660, both the computational and the geometric relationships between the problem of extrema and the problem of tangents were clearly understood; that is, a maximum was found by computing the slope of the tangent, according to the rule, and asking when it was zero. While in 1660 there was not yet a general concept of derivative, there was a general method for solving one type of geometric problem. However, the relationship of the tangent to other geometric concepts—area, for instance—was not understood, and there was no completely satisfactory definition of tangent. Nevertheless, there was a wealth of methods for solving problems that we now solve by using the calculus, and in retrospect, it would seem to be possible to generalize those methods. Thus in this context it is natural to ask, how did the derivative as we know it come to be?

It is sometimes said that the idea of the derivative was motivated chiefly by physics. Newton, after all, invented both the calculus and a great deal of the physics of motion. Indeed, already in the Middle Ages, physicists, following Aristotle who had made "change" the central concept in his physics, logically analyzed and classified the different ways a variable could change. In particular, something could change uniformly or nonuniformly; if nonuniformly, it could change uniformly-nonuniformly or nonuniformly-nonuniformly, etc. [**3**, pp. 73–74]. These medieval classifications of variation helped to lead Galileo in 1638, without benefit of calculus, to his successful treatment of uniformly accelerated motion. Motion, then, could be studied scientifically. Were such studies the origin and purpose of the calculus? The answer is no. However plausible this suggestion may sound, and however important physics was in the later development of the calculus, physical questions were in fact neither the immediate motivation nor the first application of the calculus. Certainly they prepared people's thoughts for some of the properties of the derivative, and for the introduction into mathematics of the concept of change. But the immediate motivation for the general concept of derivative—as opposed to specific examples like speed or slope of tangent—did not come from physics. The first problems to be solved, as well as the first applications, occurred in mathematics, especially geometry (see [**1**, chapter 7]; see also [**3**; chapters 4–5], and, for Newton, [**17**]). The concept of derivative then developed gradually, together with the ideas of

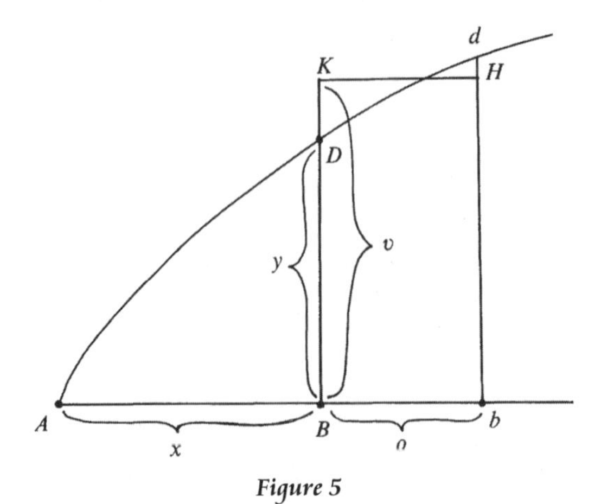

Figure 5

extrema, tangent, area, limit, continuity, and function, and it interacted with these ideas in some unexpected ways.

Tangents, Areas, and Rates of Change

In the latter third of the seventeenth century, Newton and Leibniz, each independently, invented the calculus. By "inventing the calculus" I mean that they did three things. First, they took the wealth of methods that already existed for finding tangents, extrema, and areas, and they subsumed all these methods under the heading of two general concepts, the concepts which we now call **derivative** and **integral.** Second, Newton and Leibniz each worked out a notation which made it easy, almost automatic, to use these general concepts. (We still use Newton's $\dot{x}$ and we still use Leibniz's dy/dx and $\int y\,dx$.) Third, Newton and Leibniz each gave an argument to prove what we now call the Fundamental Theorem of Calculus: the derivative and the integral are mutually inverse. Newton called our "derivative" a *fluxion*—a rate of flux or change; Leibniz saw the derivative as a ratio of infinitesimal differences and called it the *differential quotient.* But whatever terms were used, the concept of derivative was now embedded in a general subject—the calculus—and its relationship to the other basic concept, which Leibniz called the integral, was now understood. Thus we have reached the stage I have called *discovery.*

Let us look at an early Newtonian version of the Fundamental Theorem [**13**, sections 54–5, p. 23]. This will illustrate how Newton presented the calculus in 1669, and also illustrate both the strengths and weaknesses of the understanding of the derivative in this period.

Consider with Newton a curve under which the area up to the point $D = (x, y)$ is given by z (see Figure 5). His argument is general: "Assume any relation betwixt x and z that you please;" he then proceeded to find y. The example he used is

$$z = \frac{n}{m+n} a x^{(m+n)/n};$$

however, it will be sufficient to use $z = x^3$ to illustrate his argument.

In the diagram in Figure 5, the auxiliary line bd is chosen so that $Bb = o$, where o is not zero. Newton then specified that $BK = v$ should be chosen so that area $BbHK$ = area $BbdD$.

Thus $ov =$ area $BbdD$. Now, as x increases to $x + o$, the change in the area z is given by

$$z(x+o) - z(x) = x^3 + 3x^2o + 3xo^2 + o^3 - x^3 = 3x^2o + 3xo^2 + o^3,$$

which, by the definition of v, is equal to ov. Now since $3x^2o + 3xo^2 + o^3 = ov$, dividing by o produces $3x^2 + 3ox + o^2 = v$. Now, said Newton, "If we suppose Bb to be diminished infinitely and to vanish, or o to be nothing, v and y in that case will be equal and the terms which are multiplied by o will vanish: so that there will remain..."

$$3x^2 = y.$$

What has he shown? Since $(z(x+o) - z(x))/o$ is the rate at which the area z changes, that rate is given by the ordinate y. Moreover, we recognize that $3x^2$ would be the slope of the tangent to the curve $z = x^3$. Newton went on to say that the argument can be reversed; thus the converse holds too. We see that derivatives are fundamentally involved in areas as well as tangents, so the concept of derivative helps us to see that these two problems are mutually inverse. Leibniz gave analogous arguments on this same point (see, e.g. [**16**, pp. 282–284]).

Newton and Leibniz did not, of course, have the last word on the concept of derivative. Though each man had the most useful properties of the concept, there were still many unanswered questions. In particular, what, exactly, is a differential quotient? Some disciples of Leibniz, notably Johann Bernoulli and his pupil the Marquis de l'Hospital, said a differential quotient was a ratio of infinitesimals; after all, that is the way it was calculated. But infinitesimals, as seventeenth-century mathematicians were well aware, do not obey the Archimedean axiom. Since the Archimedean axiom was the basis for the Greek theory of ratios, which was, in turn, the basis of arithmetic, algebra, and geometry for seventeenth-century mathematicians, non-Archimedean objects were viewed with some suspicion. Again, what is a fluxion? Though it can be understood intuitively as a velocity, the proofs Newton gave in his 1671 *Method of Fluxions* all involved an "indefinitely small quantity o," [**14**, pp. 32–33] which raises many of the same problems that the o which "vanishes" raised in the Newtonian example of 1669 we saw above. In particular, what is the status of that little o? Is it zero? If so, how can we divide by it? If it is not zero, aren't we making an error when we throw it away? These questions had already been posed in Newton's and Leibniz's time. To avoid such problems, Newton said in 1687 that quantities defined in the way that $3x^2$ was defined in our example were the *limit* of the ratio of vanishing increments. This sounds good, but Newton's understanding of the term "limit" was not ours. Newton in his *Principia* (1687) described limits as "ultimate ratios"—that is, the value of the ratio of those vanishing quantities just when they are vanishing. He said, "Those ultimate ratios with which quantities vanish are not truly the ratios of ultimate quantities, but limits towards which the ratios of quantities decreasing without limit do always converge; and to which they approach nearer than by any given difference, but never go beyond, nor in effect attain to, till the quantities are diminished in infinitum" [**15**, Book I, Scholium to Lemma XI, p. 39].

Notice the phrase "but never go beyond"—so a variable cannot oscillate about its limit. By "limit" Newton seems to have had in mind "bound," and mathematicians of his time often cite the particular example of the circle as the limit of inscribed polygons. Also, Newton said, "nor... attain to, till the quantities are diminished in infinitum." This raises a central issue: it was often asked whether a variable quantity ever actually reached its limit. If it did not, wasn't there an error? Newton did not help clarify this when he stated as a theorem that "Quantities and the ratios of quantities which in any finite time converge continually to equality, and before the end of that time approach nearer to each other than by any given difference, become ultimately equal" [15, Book I, Lemma I, p. 29]. What does "become ultimately equal" mean? It was not really clear in the eighteenth century, let alone the seventeenth.

In 1734, George Berkeley, Bishop of Cloyne, attacked the calculus on precisely this point. Scientists, he said, attack religion for being unreasonable; well, let them improve their own reasoning first. A quantity is either zero or not; there is nothing in between. And Berkeley characterized the mathematicians of his time as men "rather accustomed to compute, than to think" [2].

Perhaps Berkeley was right, but most mathematicians were not greatly concerned. The concepts of differential quotient and integral, concepts made more effective by Leibniz's notation and by the Fundamental Theorem, had enormous power. For eighteenth-century mathematicians, especially those on the Continent where the greatest achievements occurred, it was enough that the concepts of the calculus were understood sufficiently well to be applied to solve a large number of problems, both in mathematics and in physics. So, we come to our third stage: *exploration and development.*

Differential Equations, Taylor Series, and Functions

Newton had stated his three laws of motion in words, and derived his physics from those laws by means of synthetic geometry [15]. Newton's second law stated: "*The change of motion* [*our 'momentum'*] *is proportional to the motive force impressed, and is made in the direction of the* [*straight*] *line in which that force is impressed*" [15, p. 13]. Once translated into the language of the calculus, this law provided physicists with an instrument of physical discovery of tremendous power—because of the power of the concept of the derivative.

To illustrate, if F is force and x distance (so $m\dot{x}$ is momentum and, for constant mass, $m\ddot{x}$ the rate of change of momentum), then Newton's second law takes the form $F = m\ddot{x}$. Hooke's law of elasticity (when an elastic body is distorted the restoring force is proportional to the distance [in the opposite direction] of the distortion) takes the algebraic form $F = -kx$. By equating these expressions for force, Euler in 1739 could easily both state and solve the differential equation $m\ddot{x} + kx = 0$ which describes the motion of a vibrating spring [10, p. 482]. It was mathematically surprising, and physically interesting, that the solution to that differential equation involves sines and cosines.

An analogous, but considerably more sophisticated problem, was the statement and solution of the partial differential equation for the vibrating string. In modern notation, this is

$$\frac{\partial^2 y}{\partial t^2} = \frac{T \partial^2 y}{\mu \partial x^2},$$

where T is the tension in the string and μ is its mass per unit length. The question of how the solutions to this partial differential equation behaved was investigated by such men as d'Alembert, Daniel Bernoulli, and Leonhard Euler, and led to extensive discussions about the nature of continuity, and to an expansion of the notion of function from formulas to more general dependence relations [10, pp. 502–514], [16, pp. 367–368]. Discussions surrounding the problem of the vibrating string illustrate the unexpected ways that discoveries in mathematics and physics can interact ([16, pp. 351–368] has good selections from the original papers). Numerous other examples could be cited, from the use of infinite-series approximations in celestial mechanics to the dynamics of rigid bodies, to show that by the mid-eighteenth century the differential equation had become the most useful mathematical tool in the history of physics.

Another useful tool was the Taylor series, developed in part to help solve differential equations. In 1715, Brook Taylor, arguing from the properties of finite differences, wrote an equation expressing what we would write as $f(x+h)$ in terms of $f(x)$ and its quotients of differences of various orders. He then let the differences get small, passed to the limit, and gave the formula that still bears his name: the Taylor series. (Actually, James Gregory and Newton had anticipated this discovery, but Taylor's work was more directly influential.) The importance of this property of derivatives was soon recognized, notably by Colin Maclaurin (who has a special case of it named after him), by Euler, and by Joseph-Louis Lagrange. In their hands, the Taylor series became a powerful tool in studying functions and in approximating the solution of equations.

But beyond this, the study of Taylor series provided new insights into the nature of the derivative. In 1755, Euler, in his study of power series, had said that for any power series,

$$a + bx + cx^2 + dx^3 + \cdots,$$

one could find x sufficiently small so that if one broke off the series after some particular term—say x^2—the x^2 term would exceed, in absolute value, the sum of the entire remainder of the series [**6**, section 122]. Though Euler did not prove this—he must have thought it obvious since he usually worked with series with finite coefficients—he applied it to great advantage. For instance, he could use it to analyze the nature of maxima and minima. Consider, for definiteness, the case of maxima. If $f(x)$ is a relative maximum, then by definition, for small h,

$$f(x-h) < f(x) \quad \text{and} \quad f(x+h) < f(x).$$

Taylor's theorem gives, for these inequalities,

$$f(x-h) = f(x) - h\frac{df(x)}{dx} + h^2\frac{d^2f(x)}{dx^2} - \cdots < f(x) \tag{3}$$

$$f(x+h) = f(x) - h\frac{df(x)}{dx} + h^2\frac{d^2f(x)}{dx^2} + \cdots < f(x). \tag{4}$$

Now if h is so small that $h\,df(x)/dx$ dominates the rest of the terms, the only way that both of the inequalities (3) and (4) can be satisfied is for $df(x)/dx$ to be zero. Thus the differential quotient is zero for a relative maximum. Furthermore, Euler argued, since h^2 is always positive, if $d^2f(x)/dx^2 \neq 0$, the only way both inequalities can be satisfied is for $d^2f(x)/dx^2$ to be negative. This is because the h^2 term dominates the rest of the series—unless $d^2f(x)/dx^2$ is itself zero, in which case we must go on and think about even higher-order differential quotients. This analysis, first given and demonstrated geometrically by Maclaurin, was worked out in full analytic detail by Euler [**6**, sections 253–254], [**9**, pp. 117–118]. It is typical of Euler's ability to choose computations that produce insight into fundamental concepts. It assumes, of course, that the function in question has a Taylor series, an assumption which Euler made without proof for many functions; it assumes also that the function is uniquely the sum of its Taylor series, which Euler took for granted. Nevertheless, this analysis is a beautiful example of the exploration and development of the concept of the differential quotient of first, second, and nth orders—a development which completely solves the problem of characterizing maxima and minima, a problem which goes back to the Greeks.

Lagrange and the Derivative as a Function

Though Euler did a good job analyzing maxima and minima, he brought little further understanding of the nature of the differential quotient. The new importance given to Taylor series meant that one had to be concerned not only about first and second differential quotients, but about differential quotients of any order.

The first person to take these questions seriously was Lagrange. In the 1770's, Lagrange was impressed with what Euler had been able to achieve by Taylor-series manipulations with differential quotients, but Lagrange soon became concerned about the logical inadequacy of all the existing justifications for the calculus. In particular, Lagrange wrote in 1797 that the Newtonian limit-concept was not clear enough to be the foundation for a branch of mathematics. Moreover, in not allowing variables to surpass their limits, Lagrange thought the limit-concept too restrictive. Instead, he said, the calculus should be reduced to algebra, a subject whose foundations in the eighteenth century were generally thought to be sound [**11**, pp. 15–16].

The algebra Lagrange had in mind was what he called the algebra of infinite series, because Lagrange was convinced that infinite series were part of algebra. Just as arithmetic deals with infinite decimal fractions without ceasing to be arithmetic, Lagrange thought, so algebra deals with infinite algebraic expressions without ceasing to be algebra. Lagrange believed that expanding $f(x+h)$ into a power series in h was always an algebraic process. It is obviously algebraic when one turns $1/(1-x)$ into a power series by dividing. And Euler had found, by manipulating formulas, infinite power-series expansions for functions like $\sin x$, $\cos x$, e^x. If functions like those have power-series expansions, perhaps everything could be reduced to algebra. Euler, in his book *Introduction to the analysis of the infinite* (*Introductio in analysin infinitorum*, 1748), had studied infinite series, infinite products, and infinite continued fractions by what he thought of as purely algebraic methods. For instance, he converted infinite series into infinite products by treating a series as a very long polynomial. Euler thought that this work was purely algebraic, and—what is crucial here—Lagrange also thought Euler's methods were purely algebraic. So Lagrange tried to make the calculus rigorous by reducing it to the algebra of infinite series.

Lagrange stated in 1797, and thought he had proved, that any function (that is, any analytic expression, finite or infinite) had a power-series expansion:

$$f(x+h) = f(x) + p(x)h + q(x)h^2 + r(x)h^3 + \cdots, \tag{5}$$

except, possibly, for a finite number of isolated values of x. He then defined a new function, the coefficient of the linear term in h which is $p(x)$ in the expansion shown in (5)) and called it the **first derived function** of $f(x)$. Lagrange's term "derived function" (*fonction dérivée*) is the origin of our term "derivative." Lagrange introduced a new notation, $f'(x)$, for that function. He defined $f''(x)$ to be the first derived function of $f'(x)$, and so on, recursively. Finally, using these definitions, he proved that, in the expansion (5) above, $q(x) = f''(x)/2$, $r(x) = f'''(x)/6$, and so on [**11**, chapter 2].

What was new about Lagrange's definition? The concept of *function*—whether simply an algebraic expression (possibly infinite) or, more generally, any dependence relation—helps free the concept of derivative from the earlier ill-defined notions. Newton's explanation of a fluxion as a rate of change appeared to involve the concept of motion in mathematics; moreover, a fluxion seemed to be a different kind of object than the flowing quantity whose fluxion it was. For Leibniz, the differential quotient had been the quotient of vanishingly small differences; the second differential quotient, of even smaller differences. Bishop Berkeley, in his attack on the calculus, had made fun of these earlier concepts, calling vanishing increments "ghosts of departed quantities" [**2**, section 35]. But since, for Lagrange, the derivative was a function, it was now the same sort of object as the original function. The second derivative is precisely the same sort of object as the first derivative; even the nth derivative is simply another function, defined as the coefficient of h in the Taylor series for $f^{(n-1)}(x+h)$. Lagrange's notation $f'(x)$ was designed precisely to make this point.

We cannot fully accept Lagrange's definition of the derivative, since it assumes that every differentiable function is the sum of a Taylor series and thus has infinitely many derivatives. Nevertheless, that definition led Lagrange to a number of important properties of the derivative. He used his definition together with Euler's criterion for using truncated power series in approximations to give a most useful characterization of the derivative of a function [**9**, p. 116, pp. 118–121]:

$$f(x+h) = f(x) + hf'(x) + hH, \text{ where } H \text{ goes to zero with } h.$$

(I call this the *Lagrange property of the derivative.*) Lagrange interpreted the phrase "H goes to zero with h" in terms of inequalities. That is, he wrote that,

$$\begin{aligned} &\textit{Given } D, h \textit{ can be chosen so that } f(x+h) - f(x) \\ &\textit{lies between } h(f'(x) - D) \textit{ and } h(f'(x) + D). \end{aligned} \tag{6}$$

Formula (6) is recognizably close to the modern delta-epsilon definition of the derivative.

Lagrange used inequality (6) to prove theorems. For instance, he proved that a function with positive derivative on an interval is increasing there, and used that theorem to derive the Lagrange remainder of the Taylor series [**9**, pp. 122–127], [**11**, pp. 78–85]. Furthermore, he said, considerations like inequality (6) are what make possible applications of the differential calculus to a whole range of problems in mechanics, in geometry, and, as we have described, the problem of maxima and minima (which Lagrange solved using the Taylor series remainder which bears his name [**11**, pp. 233–237]).

In Lagrange's 1797 work, then, the derivative is defined by its position in the Taylor series—a strange definition to us. But the derivative is also *described* as satisfying what we recognize as the appropriate delta-epsilon inequality, and Lagrange applied this inequality and its nth-order analogue, the Lagrange remainder, to solve problems about tangents, orders of contact between curves, and extrema. Here the derivative was clearly a function, rather than a ratio or a speed.

Still, it is a lot to assume that a function has a Taylor series if one wants to define only *one* derivative. Further, Lagrange was wrong about the algebra of infinite series. As Cauchy pointed out in 1821, the algebra of finite quantities cannot automatically be extended to infinite processes. And, as Cauchy also pointed out, manipulating Taylor series is not foolproof. For instance, e^{-1/x^2} has a zero Taylor series about $x = 0$, but the function is not identically zero. For these reasons, Cauchy rejected Lagrange's definition of derivative and substituted his own.

Definitions, Rigor, and Proofs

Now we come to the last stage in our chronological list: *definition.* In 1823, Cauchy defined the derivative of $f(x)$ as the limit, when it exists, of the quotient of differences $(f(x+h) - f(x))/h$ as h goes to zero [**4**, pp. 22–23]. But Cauchy understood "limit" differently than had his

Dates refer to these mathematician's major works which

predecessors. Cauchy entirely avoided the question of whether a variable ever reached its limit; he just didn't discuss it. Also, knowing an absolute value when he saw one, Cauchy followed Simon l'Huilier and S.-F. Lacroix in abandoning the restriction that variables never surpass their limits. Finally, though Cauchy, like Newton and d'Alembert before him, gave his definition of limit in words, Cauchy's understanding of limit (most of the time, at least) was algebraic. By this, I mean that when Cauchy needed a limit property in a proof, he used the algebraic inequality-characterization of limit. Cauchy's proof of the mean value theorem for derivatives illustrates this. First he proved a theorem which states: *if* $f(x)$ *is continuous on* $[x, x+a]$, *then*

$$\min_{[x,x+a]} f'(x) \leq \frac{f(x+a)-f(x)}{a} \leq \max_{[x,x+a]} f'(x). \tag{7}$$

The first step in his proof is [**4**, p. **44**]:

> Let δ, ε be two very small numbers; the first is chosen so that for all [absolute] values of h less than δ, and for any value of x [on the given interval], the ratio $(f(x+h)-f(x))/h$ will always be greater than $f'(x)-\varepsilon$ and less than $f'(x)+\varepsilon$.

(The notation in this quote is Cauchy's, except that I have substituted h for the i he used for the increment.) Assuming the intermediate-value theorem for continuous functions, which Cauchy had proved in 1821, the mean-value theorem is an easy corollary of (7) [**4**, pp. 44–45], [**9**, pp. 168–170].

Cauchy took the inequality-characterization of the derivative from Lagrange (possibly via an 1806 paper of A.-M. Ampère [**9**, pp. 127–132]). But Cauchy made that characterization into a definition of derivative. Cauchy also took from Lagrange the name derivative and the notation $f'(x)$, emphasizing the functional nature of the derivative. And, as I have shown in detail elsewhere [**9**, chapter 5], Cauchy adapted and improved Lagrange's inequality proof-methods to prove results like the mean-value theorem, proof-methods now justified by Cauchy's definition of derivative.

But of course, with the new and more rigorous definition, Cauchy went far beyond Lagrange. For instance, using his concept of limit to define the integral as the limit of sums, Cauchy made a good first approximation to a real proof of the Fundamental Theorem of Calculus [**9**, pp. 171–175], [**4**, pp. 122–125, 151–152]. And it was Cauchy who not only raised the question, but gave the first proof, of the existence of a solution to a differential equation [**9**, pp. 158–159].

contributed to the evolution of the concept of the derivative.

After Cauchy, the calculus itself was viewed differently. It was seen as a rigorous subject, with good definitions and with theorems whose proofs were based on those definitions, rather than merely as a set of powerful methods. Not only did Cauchy's new rigor establish the earlier results on a firm foundation, but it also provided a framework for a wealth of new results, some of which could not even be formulated before Cauchy's work.

Of course, Cauchy did not himself solve all the problems occasioned by his work. In particular, Cauchy's definition of the derivative suffers from one deficiency of which he was unaware. Given an ε, he chose a δ which he assumed would work for any x. That is, he assumed that the quotient of differences converged uniformly to its limit. It was not until the 1840's that G. G. Stokes, V. Seidel, K. Weierstrass, and Cauchy himself worked out the distinction between convergence and uniform convergence. After all, in order to make this distinction, one first needs a clear and algebraic understanding of what a limit is—the understanding Cauchy himself had provided.

In the 1850's, Karl Weierstrass began to lecture at the University of Berlin. In his lectures, Weierstrass made algebraic inequalities replace words in theorems in analysis, and used his own clear distinction between pointwise and uniform convergence along with Cauchy's delta-epsilon techniques to present a systematic and thoroughly rigorous treatment of the calculus. Though Weierstrass did not publish his lectures, his students—H. A. Schwarz, G. Mittag-Leffler, E. Heine, S. Pincherle, Sonya Kowalevsky, Georg Cantor, to name a few—disseminated Weierstrassian rigor to the mathematical centers of Europe. Thus although our modern delta-epsilon definition of derivative cannot be quoted from the *works* of Weierstrass, it is in fact the *work* of Weierstrass [**3**, pp. 284–287]. The rigorous understanding brought to the concept of the derivative by Weierstrass is signaled by his publication in 1872 of an example of an everywhere continuous, nowhere differentiable function. This is a far cry from merely acknowledging that derivatives might not always exist, and the example shows a complete mastery of the concepts of derivative, limit, and existence of limit [**3**, p. 285].

Historical Development Versus Textbook Exposition

The span of time from Fermat to Weierstrass is over two hundred years. How did the concept of derivative develop? Fermat implicitly used it; Newton and Leibniz discovered it; Taylor, Euler, Maclaurin developed it; Lagrange named and characterized it; and only at the end of this long period of development did Cauchy and Weierstrass define it. This is certainly a complete reversal of the usual order of textbook exposition in mathematics, where one starts with a definition, then explores some results, and only then suggests applications.

This point is important for the teacher of mathematics: the historical order of development of the derivative is the reverse of the usual order of textbook exposition. Knowing the history helps us as we teach about derivatives. We should put ourselves where mathematicians were before Fermat, and where our beginning students are now—back on the other side, before we had any concept of derivative, and also before we knew the many uses of derivatives. Seeing the historical origins of a concept helps motivate the concept, which we—along with Newton and Leibniz—want for the problems it helps to solve. Knowing the historical order also helps to motivate the rigorous definition—which we, like Cauchy and Weierstrass, want in order to justify the uses of the derivative, and to show precisely when derivatives exist and when they do not. We need to remember that the rigorous definition is often the end, rather than the beginning, of a subject.

The real historical development of mathematics—the order of discovery—reveals the creative mathematician at work, and it is creation that makes doing mathematics so exciting. The order of exposition, on the other hand, is what gives mathematics its characteristic logical structure and its incomparable deductive certainty. Unfortunately, once the classic exposition has been given, the order of discovery is often forgotten. The task of the historian is to recapture the order of discovery: not as we think it might have been, not as we think it should have been, but as it really was. And this is the purpose of the story we have just told of the derivative from Fermat to Weierstrass.

This article is based on a talk delivered at the Conference on the History of Modern Mathematics, Indiana Region of the Mathematical Association of America, Ball State University, April 1982; earlier versions were presented at the Southern California Section of the M. A. A. and at various mathematics colloquia. I thank the MATHEMATICS MAGAZINE referees for their helpful suggestions.

References

1. Margaret Baron, Origins of the Infinitesimal Calculus, Pergamon, Oxford, 1969.
2. George Berkeley, The Analyst, or a Discourse Addressed to an Infidel Mathematician, 1734. In A. A. Luce and T. R. Jessop, eds., The Works of George Berkeley, Nelson, London, 1951 (some excerpts appear in [**16**, pp. 333–338]).
3. Carl Boyer, History of the Calculus and Its Conceptual Development, Dover, New York, 1959.
4. A.-L. Cauchy, Résumé des leçons données à l'école royale polytechnique sur le calcul infinitésimal, Paris, 1823. In Oeuvres complètes d'Augustin Cauchy, Gauthier-Villars, Paris, 1882-, series 2, vol. 4.
5. Pierre Dugac, Fondements d'analyse, in J. Dieudonné, Abrégé d'histoire des mathématiques, 1700–1900, 2 vols., Hermann, Paris, 1978.
6. Leonhard Euler, Institutiones calculi differentialis, St. Petersburg, 1755. In Operia omnia, Teubner, Leipzig, Berlin, and Zurich, 1911-, series 1, vol. 10.
7. Pierre Fermat, Analysis ad refractiones, 1661. In Oeuvres de Fermat, ed., C. Henry and P. Tannery, 4 vols., Paris, 1891–1912; Supplement, ed. C. de Waard, Paris, 1922, vol. 1, pp. 170–172.
8. ———, Methodus ad disquirendam maximam et minimam et de tangentibus linearum curvarum, Oeuvres, vol. 1, pp. 133–136. Excerpted in English in [**16**, pp. 222–225].

9. Judith V. Grabiner, The Origins of Cauchy's Rigorous Calculus, M. I. T. Press, Cambridge and London, 1981.

10. Morris Kline, Mathematical Thought from Ancient to Modern Times, Oxford, New York, 1972.

11. J.-L. Lagrange, Théorie des fonctions analytiques, Paris, 2nd edition, 1813. In Oeuvres de Lagrange, ed. M. Serret, Gauthier-Villars, Paris, 1867–1892, vol. 9.

12. Michael S. Mahoney, The Mathematical Career of Pierre de Fermat, 1601–1665, Princeton University Press, Princeton, 1973.

13. Isaac Newton, Of Analysis by Equations of an Infinite Number of Terms [1669], in D. T. Whiteside, ed., Mathematical Works of Isaac Newton, Johnson, New York and London, 1964, vol. 1, pp. 3–25.

14. ———, Method of Fluxions [1671], in D. T. Whiteside, ed., Mathematical Works of Isaac Newton, vol. 1, pp. 29–139.

15. ———, Mathematical Principles of Natural Philosophy, tr. A. Motte, ed. F. Cajori, University of California Press, Berkeley, 1934.

16. D. J. Struik, Source Book in Mathematics, 1200–1800, Harvard University Press, Cambridge, MA, 1969.

17. D. T. Whiteside, ed., The Mathematical Papers of Isaac Newton, Cambridge University Press, 1967–1982.

4

The Centrality of Mathematics in the History of Western Thought*

1. Introduction

Since this paper was first given to educators, let me start with a classroom experience. It happened in a course in which my students had read some of Euclid's *Elements of Geometry.* A student, a social science major, said to me, "I never realized mathematics was like this. Why, it's like philosophy!" That is no accident, for philosophy is like mathematics. When I speak of the centrality of mathematics in western thought, it is this student's experience I want to recapture – to reclaim the context of mathematics from the hardware store with the rest of the tools and bring it back to the university. To do this, I will discuss some major developments in the history of ideas in which mathematics has played a central role.

I do not mean that mathematics has by itself caused all these developments; what I do mean is that mathematics, whether causing, suggesting, or reinforcing, has played a key role; it has been there, at center stage. We all know that mathematics has been the language of science for centuries. But what I wish to emphasize is the crucial role of mathematics in shaping views of man and the world held not just by scientists, but by everyone educated in the western tradition.

Given the vastness of that tradition, I will give many examples only briefly, and be able to treat only a few key illustrative examples at any length. Sources for the others may be found in the bibliography. (See also [**26**].)

Since I am arguing for the centrality of mathematics, I will organize the paper around the key features of mathematics which have produced the effects I will discuss. These features are the certainty of mathematics and the applicability of mathematics to the world.

2. Certainty

For over two thousand years, the certainty of mathematics, particularly of Euclidean geometry, has had to be addressed in some way by any theory of knowledge. Why was geometry certain?

---

* ©International Congress of Mathematicians, 1986, Berkeley, California. Reprinted from Math. Mag. 61, 4(Oct. 1988) 220–230.

Was it because of the subject matter of geometry, or because of its method? And what were the implications of that certainty?

Even before Euclid's monumental textbook, the philosopher Plato saw the certainty of Greek geometry—a subject which Plato called "knowledge of that which always is" [**41**, 527b]—as arising from the eternal, unchanging perfection of the objects of mathematics. By contrast, the objects of the physical world were always coming into being or passing away. The physical world changes, and is thus only an approximation to the higher ideal reality. The philosopher, then, to have his soul drawn from the changing to the real, had to study mathematics. Greek geometry fed Plato's idealistic philosophy; he emphasized the study of Forms or Ideas transcending experience: the idea of justice, the ideal state, the idea of the Good. Plato's views were used by philosophers within the Jewish, Christian, and Islamic traditions to deal with how a divine being, or souls, could interact with the material world [**47**, pp. 382–3] [**52**, pp. 17–40] [**34**, p. 305ff] [**23**, p. 46–67]. For example, Plato's account of the creation of the world in his *Timaeus*, where a god makes the physical universe by copying an ideal mathematical model, became assimilated in early Christian thought to the Biblical account of creation [**29**, pp. 21–22]. One finds highly mathematicized cosmologies, influenced by Plato, in the mystical traditions of Islam and Judaism as well. The tradition of Platonic Forms or Ideas crops up also in such unexpected places as the debates in eighteenth- and nineteenth-century biology over the fixity of species. Linnaeus in the eighteenth, and Louis Agassiz in the nineteenth century seem to have thought of species as ideas in the mind of God [**16**, p. 34] [**13**, pp. 36–7], When we use the common terms "certain" and "true" outside of mathematics, we use them in their historical context, which includes the long-held belief in an unchanging reality—a belief stemming historically from Plato, who consistently argued for it using examples from mathematics.

An equally notable philosopher, who lived just before Euclid, namely Aristotle, saw the success of geometry as stemming, not from perfect eternal objects, but instead from its method [*Posterior Analytics*, **I** 10–11; **I** 1–2 (77a5, 71b ff)] [**19**, vol. I, Chapter IX]. The certainty of mathematics for Aristotle rested on the validity of its logical deductions from self-evident assumptions and clearly-stated definitions. Other subjects might come to share that certainty if they could be understood within the same logical form; Aristotle, in his *Posterior Analytics*, advocated reducing all scientific discourse to syllogisms, that is, to logically-deduced explanations from first principles. In this tradition, Archimedes proved the law of the lever, not by experiments with weights, but from deductions *à la* Euclid from postulates like "equal weights balance at equal distances" [**18**, pp. 189–194]. Medieval theologians tried to prove the existence of God in the same way. This tradition culminates in the 1675 work of Spinoza, *Ethics Demonstrated in Geometrical Order*, with such axioms as "That which cannot be conceived through another must be conceived through itself," definitions like "By substance I understand that which is in itself and conceived through itself" (compare Euclid's "A point is that which has no parts"), and such propositions as "God or substance consisting of infinite attributes... necessarily exists," whose proof ends with a QED [**49**, pp. 41–50]. Isaac Newton called his famous three laws "Axioms, or Laws of Motion." His *Principia* has a Euclidean structure, and the law of gravity appears as Book III, Theorems VII and VIII [**37**, pp. 13–14, pp. 414–17]. The Declaration of Independence of the United States is one more example of an argument whose authors tried to inspire faith in its certainty by using the Euclidean form. "We hold these truths to be self-evident... " not that all right angles are equal, but "that all men are created equal." These self-evident truths include that if any government does not obey these postulates, "it is the right of the people to alter or abolish it." The central section begins by saying that they will "prove" King George's government does not obey them. The conclusion is "We, *therefore* . . . declare, that these United Colonies are, and of right ought to be, free and independent states." (My italics) (Jefferson's mathematical education, by the way, was quite impressive by the standards of his time.)

Thus a good part of the historical context of the common term "proof" lies in Euclidean geometry—which was, I remind you, a central part of Western education.

However, the certainty of mathematics is not limited to Euclidean geometry. Between the rise of Islamic culture and the eighteenth century, the paradigm governing mathematical research changed from a geometric one to an algebraic, symbolic one. In algebra even more than in the Euclidean model of reasoning, the method can be considered independently of the subject-matter involved. This view looks at the method of mathematics as finding truths by manipulating symbols. The approach first enters the western world with the introduction of the Hindu-Arabic number system in the twelfth-century translations into Latin of Arabic mathematical works, notably al-Khowarizmi's algebra. The simplified calculations using the Hindu-Arabic numbers were called the "method of al-Khowarizmi" or as Latinized "the method of algorism" or algorithm.

In an even more powerful triumph of the heuristic power of notation, François Viète in 1591 introduced literal symbols into algebra: first, using letters in general to stand for any number in the theory of equations; second, using letters for any number of unknowns to solve word problems [**4**, pp. 59–63, 65]. In the seventeenth century, Leibniz, struck by the heuristic power of arithmetical and algebraic notation, invented such a notation for his new science of finding differentials—an algorithm for manipulating the d and integral symbols, that is, a calculus (a term which meant to him the same thing as "algorithm" to us). Leibniz generalized the idea of heuristic notation in his philosophy [**30**, pp. 12–25]. He envisioned a symbolic language which would embody logical thought just as these earlier symbolic languages enable us to perform algebraic operations correctly and mechanically. He called this language a "universal characteristic," and later commentators, such as Bertrand Russell, see Leibniz as the pioneer of symbolic logic [**46**, p. 170]. Any time a disagreement occurred, said Leibniz, the opponents could sit down and say "Let us calculate," and—mechanically—settle the question [**30**, p. 15]. Leibniz's appreciation of the mechanical element in mathematics when viewed as symbolic manipulation is further evidenced by his invention of a calculating machine. Other seventeenth-century thinkers also stressed the mechanical nature of thought in general: for instance, Thomas Hobbes wrote "Words are wise men's counters, they do but reckon by them" [**21**, Chapter 4, p. 143]. Others tried to introduce heuristically powerful notation in different fields: consider Lavoisier's new chemical notation which he called a "chemical algebra" [**14**, p. 245].

These successes led the great prophet of progress, the Marquis de Condorcet, to write in 1793 that algebra gives "the only really exact and analytical language yet in existence. . . . Though this method is by itself only an instrument pertaining to the science of quantities, it contains within it the principles of a universal instrument, applicable to all combinations of ideas" [**9**, p. 238]. This could make the progress of "every subject embraced by human intelligence . . . as sure as that of mathematics" [**9**, pp. 278–9]. The certainty of symbolic reasoning has led us to the idea of the certainty of progress. Though one might argue that some fields had not progressed one iota beyond antiquity, it was unquestionably true by 1793 that mathematics and the sciences had progressed. To quote Condorcet once more: "the progress of the mathematical and physical sciences reveals an immense horizon . . . a revolution in the destinies of the human race" [**9**, p. 237]. Progress was possible; why not apply the same method to the social and moral spheres as well?

No account of attempts to extend the method of mathematics to other fields would be complete without discussing René Descartes, who in the 1630's combined the two methods we have just discussed—that of geometry and that of algebra—into analytic geometry. Let us look at his own description of how to make such discoveries. Descartes depicted the building-up of the deductive structure of a science—proof—as a later task than analysis or discovery. One first needed to analyze the whole into the correct "elements" from which truths could later be deduced. "The first rule," he wrote in his *Discourse on Method*, "was never to accept anything as true unless I recognized it to be evidently such . . . The second was to divide each of the difficulties which I encountered into as many parts as possible, and as might be required for an easier solution . . . " Then, "the third [rule] was to [start] . . . with the things which were simplest and easiest to understand, gradually and by degrees reaching toward more complex knowledge" [**10**, Part II, p. 12]. Descartes presented his method as the key to his own mathematical and scientific discoveries. Consider, for instance, the opening line of his *Geometry:* "All problems in geometry can easily be reduced to such terms that a knowledge of the lengths of certain straight lines suffices for their construction. Just as arithmetic is composed of only four or five operations . . . , so in geometry" [**11**, Book I, p. 3]. Descartes's influence on subsequent philosophy, from Locke's empiricism to Sartre's existentialism, is well known and will not be reviewed here. But for our purposes it is important to note that the thrust of Descartes's argument is that emulating the method successful in mathematical discovery will lead to successful discoveries in other fields [**10**, Part Five].

Descartes' method of analysis fits nicely with the Greek atomic theory, which had been newly revived in the seventeenth century: all matter is the sum of atoms; analyze the properties of the whole as the sum of these parts [**17**, Chapter VIII, esp. p. 217]. Thus the idea of studying something by "analysis" was doubly popular in seventeenth- and eighteenth-century thought. I would like to trace just one line of influence of this analytic method. Adam Smith in his 1776 *Wealth of Nations* analyzed [**48**, p. 12] the competitive success of economic systems by means of the concept of division of labor. The separate elements, each acting as efficiently as possible, provided for the overall success of the manufacturing process; similarly, each individual in the whole economy, while striving to increase his individual advantages, is "led as if by an Invisible Hand to promote ends which were not part of his original intention" [**48**, p. 27]—that is, the welfare of the whole of society. This Cartesian method of studying a whole system by analyzing it into its elements, then synthesizing the elements to produce the whole, was especially popular in France. For instance, Gaspard François de Prony had the job of calculating, for the French Revolutionary government, a set of logarithmic and trigonometric tables. He, himself, said he did it by applying Adam Smith's ideas about the division of labor. Prony organized a group of people into a hierarchical system to compute these tables. A few mathematicians decided which functions to use; competent technicians then reduced the job of calculating the functions to a set of simple additions and subtractions of pre-assigned numbers; and, finally, a large number of low-level human "calculators" carried out the additions and subtractions. Charles Babbage, the early nineteenth-century pioneer of the digital computer, applied the Smith-Prony analysis and embodied it in a machine [**1**, Chapter XIX]. The way Babbage's ideas developed can be found in a chapter in his *Economy of Machinery* entitled "On the Division of Mental Labour"

[**1**, Chapter XIX]. Babbage was ready to convert Prony's organization into a computing machine because Babbage had long been impressed by the arguments of Leibniz and his followers on the power of notation to make such mathematical calculation mechanical, and Babbage, like Leibniz, accounted for the success of mathematics by "the accurate simplicity of its language" [**22**, p. 26]. Since Babbage's computer was designed to be "programmed" by punched cards, Hollerith's later invention of punched-card census data processing, twentieth-century computing, and other applications of the Cartesian "divide-and-solve" approach, including top-down programming, are also among the offspring of Descartes's mathematically-inspired method.

Whatever view of the *cause* of the certainty of mathematics one adopts, the *fact* of certainty in itself has had consequences. The "fact of mathematical certainty" has been taken to show that there exists *some* sort of knowledge, and thus to refute skepticism. Immanuel Kant in 1783 used such an argument to show that metaphysics is possible [**25**, Preamble, Section IV]. If metaphysics exists, it is independent of experience. Nevertheless, it is not a complex of tautologies. Metaphysics, for Kant, had to be what he called "synthetic," giving knowledge based on premises which is not obtainable simply by analyzing the premises logically. Is there such knowledge? Yes, said Kant, look at geometry. Consider the truth that the sum of the angles of a triangle is two right angles. We do not get this truth by analyzing the concept of triangle—all that gives us, Kant says, is that there are three angles. To gain the knowledge, one must make a construction: draw a line through one vertex parallel to the opposite side. (I now leave the proof as an exercise.) The construction is essential; it takes place in space, which Kant sees as a unique intuition of the intellect. (This example [**24**, II "Method of Transcendentalism," Chapter I, Section I, p. 423] seems to require the space to be Euclidean; I will return to this point later on.) Thus synthetic knowledge independent of experience *is* possible, so metaphysics—skeptics like David Hume to the contrary—is also possible.

This same point—that mathematics is knowledge, so there *is* objective truth—has been made throughout history, from Plato's going beyond Socrates' agnostic critical method, through George Orwell's hero, Winston Smith, attempting to assert, in the face of the totalitarian state's overwhelming power over the human intellect, that two and two are four.

Moreover, since mathematics is certain, perhaps we can, by examining mathematics, find which properties *all* certain knowledge must have. One such application of the "fact of mathematical certainty" was its use to solve what in the sixteenth century was called the problem of the criterion [**43**, Chapter I]. If there is only one system of thought around, people might well accept that one as true—as many Catholics did about the teachings of the Church in the Middle Ages. But then the Reformation developed alternative religious systems, and the Renaissance rediscovered the thought of pagan antiquity. Now the problem of finding the criterion that identified the true system became acute. In the seventeenth and eighteenth centuries, many thinkers looked to mathematics to help find an answer. What was the sign of the certainty of the conclusions of mathematics? The fact that nobody disputed them [**43**, Chapter VII]. Distinguishing mathematics from religion and philosophy, Voltaire wrote, "There are no sects in geometry. One does not speak of a Euclidean, an Archimedean" [**50**, Article "Sect"]. What every reasonable person agrees upon—that is the truth. How can this be applied to religion? Some religions forbid eating beef, some forbid eating pork; therefore, since they disagree, they both are wrong. But, continues Voltaire, all religions agree that one should worship God and be just; that must therefore be true. "There is but one morality," says Voltaire, "as there is but one geometry" [**50**, Miscellany, p. 225].

3. Applicability

Let us turn now from the certainty of mathematics to its applicability. Since applying mathematics to describe the world works so well, thinkers who reflect on the applicability of mathematics find that it affects their views not only about thought, but also about the world. For Plato, the applicability of mathematics occurs because this world is merely an approximation to the higher mathematical reality; even the motions of the planets were inferior to pure mathematical motions [**41**, 529d]. For Aristotle, on the other hand, mathematical objects are just abstracted from the physical world by the intellect. A typical mathematically-based science is optics, in which we study physical objects—rays of light—as though they were mathematical straight lines [*Physics* II, Chapter 2; 194a]. We can thus use all the tools of geometry in that science of optics, but it is the light that is real.

One might think that Plato is a dreamer and Aristotle a hard-headed practical man. But today's engineer steeped in differential equations is the descendant of the dreamer. From Plato—and his predecessors the Pythagoreans who taught that "all is number"—into the Renaissance, many thinkers looked for the mathematical reality beyond the appearances. So did Copernicus, Kepler, and Galileo [**7**, Chapters 3, 5, 6]. The Newtonian world-system that completed the Copernican revolution was embodied in a mathematical model, based on the laws of motion and inverse-square gravitation, and set in Platonically absolute space and time ([**6**]; cf. [**7**, Chapter 7]). The success of Newtonian physics not only strongly reinforced the view that mathematics was the appropriate language of science, but also strongly reinforced the emerging ideas of progress and of truth based on universal agreement.

Another consequence of the Newtonian revolution was Newton's explicit help to theology, strongly buttressing what was called the argument for God's existence from design. The mathematical perfection of the solar system—elliptical orbits nearly circular, planets moving all in the same plane and direction—could not have come about by chance, said Newton, but "from the counsel and dominion of an intelligent and powerful Being" [**37**, General Scholium, p. 544]. "Natural theology," as this doctrine was called, focussed on examples of design and adaptation in nature, inspiring considerable research in natural history, especially on adaptation, research which was to play a role in Darwin's discovery of evolution by natural selection [**14**, pp. 263–266].

Just as the "fact of mathematical certainty" made certainty elsewhere seem achievable, so the "fact of mathematical applicability" in physical science inspired the pioneers of the idea of social science, Auguste Comte and Adolphe Quetelet. Both Comte and Quetelet were students of mathematical physics and astronomy in the early nineteenth century; Comte, while a student at the Ecole polytechnique in Paris, was particularly inspired by Lagrange, Quetelet, and Laplace. Lagrange's great *Analytical Mechanics* was an attempt to reduce all of mechanics to mathematics. Comte went further: if physics was built on mathematics, so was chemistry built on physics, biology on chemistry, psychology on biology, and finally his own new creation, sociology (the term is his) would be built on psychology [**8**, Chapter II]. The natural sciences were no longer (as they had once been) theological or metaphysical; they were what Comte called "positive"—based only on observed connections between things. Social science could now also become positive. Comte was a reformer, hoping for a better society through understanding what he called "social physics." His philosophy of positivism influenced twentieth-century logical positivism, and his ideas on history—"social dynamics"—influenced Feuerbach and Marx [**32**, Chapter 4]. Still, Comte only prophesied but did not create quantitative social science; this was done by Quetelet.

For Quetelet's conception of quantitative social science, the fact of applicability of mathematics was crucial. "We can judge of the perfection to which a science has come," he wrote in 1828, "by the ease with which it can be approached by calculation" (quoted in [**27**, p. 250]). Quetelet noted that Laplace had used probability and statistics in determining planetary orbits; Quetelet was especially impressed by what we call the normal curve of errors. Quetelet found empirically that many human traits—height, for instance—gave rise to a normal curve. From this, he defined the statistical concept, and the term, "average man" (*homme moyen*). Quetelet's work demonstrates that, just as the Platonic view that geometry underlies reality made mathematical physics possible, so having a statistical view of data is what makes social science possible.

Quetelet found also that many social statistics—the number of suicides in Belgium, for instance, or the number of murders—produced roughly the same figures every year. The constancy of these rates over time, he argued, dictated that murder or suicide had constant social causes. Quetelet's discovery of the constancy of crime rates raised an urgent question: whether the individuals are people or particles, do statistical laws say anything about individuals? or are the individuals free?

Laplace, recognizing that one needed probability to do physics, said that this fact did not mean that the laws governing the universe were ultimately statistical. In ignorance of the true causes, Laplace said, people thought that events in the universe depended on chance, but in fact all is determined. To an infinite intelligence which could comprehend all the forces in nature and the "respective situation of the beings who composed it," said Laplace, "nothing would be uncertain" [**28**, Chapter II]. Similarly, Quetelet held that "the social state prepares these crimes, and the criminal is merely the instrument to execute them" [**27**].

Another view was held by James Clerk Maxwell. In his work on the statistical mechanics of gases, Maxwell argued that statistical regularities in the large told you nothing about the behavior of individuals in the small [**33**, Chapter 22, pp. 315–16]. Maxwell seems to have been interested in this point because it allowed for free will. And this argument did not arise from Maxwell's physics; he had read and pondered the work of Quetelet on the application of statistical thinking to society [**44**]. The same sort of dispute about the meaning of probabilistically-stated laws has of course recurred in the twentieth-century philosophical debates over the foundations of quantum mechanics.

Thus discussions of basic philosophical questions—is the universe an accident or a divine design? is there free will or are we all programmed?—owe surprisingly much to the applicability of mathematics.

4. More Than One Geometry?

Given the centrality of mathematics to western thought, what happens when prevailing views of the nature of mathematics change? Other things must change too. Since geometry had been for so long the canonical example both of the certainty and of the applicability of mathematics, the rise of non-Euclidean geometry was to have profound effects.

As is well known, in attempts to prove Euclid's parallel postulate and thus, as Saccheri put it in 1733, remove the single blemish from Euclid, mathematicians deduced a variety of surprising consequences from denying that postulate. Gauss, Bolyai, and Lobachevsky in the early nineteenth century each separately recognized that these consequences were not absurd, but rather were valid results in a consistent, non-Euclidean (Gauss's term) geometry.

Recall that Kant had said that space (by which he meant Euclidean 3-space) was the form of all our perceptions of objects. Hermann von Helmholtz, led in mid-century to geometry by his interest in the psychology of perception, asked whether Kant might be wrong: could we imagine ordering our perceptions in a non-Euclidean space? Yes, Helmholtz said. Consider the world as reflected in a convex mirror. Thus, the question of which geometry describes the world is no longer a matter for intuition—or for self-evident assumptions—but for *experience* [**20**].

What did this view—expressed as well by Bernhard Riemann and W. K. Clifford, among others—do to the received accounts of the relation between mathematics and the world? It detached mathematics from the world. Euclidean and non-Euclidean geometry give the first clear-cut historical example of two mutually contradictory mathematical structures, of which at most one can actually represent the world. This seems to indicate that the choice of mathematical axioms is one of intellectual freedom, not empirical constraint; this view, reinforced by Hamilton's discovery of a non-commutative algebra, suggested that mathematics is a purely formal structure, or as Benjamin Peirce put it, "Mathematics is the science which draws necessary conclusions," [**40**]—*not* the science of number (even symbolic algebra had been just a generalized science of number) or the science of space. Now that the axioms were no longer seen as necessarily deriving from the world, the applicability of mathematics to the world became turned upside down. The world is no longer, as it was for Plato, an imperfect model of the true mathematical reality; instead, mathematics provides a set of different models for one empirical reality. In 1902 the physicist Ludwig Boltzmann expressed a view which had become widely held: that models, whether physical or mathematical, whether geometric or statistical, had become the means by which the sciences "comprehend objects in thought and represent them in language" [**3**]. This view, which implies that the sciences are no longer claiming to speak directly about reality, is now widespread in the social sciences as well as the natural sciences, and has transformed the philosophy of science. As applied to mathematics itself—the formal model of mathematical reasoning—it has resulted in Gödel's demonstration that one can never prove the consistency of mathematics, and the resulting conclusion among some philosophers that there is no certainty anywhere, not even in mathematics [**2**, p. 206].

5. Opposition

The best proof of the centrality of mathematics is that every example of its influence given so far has provoked strong and significant opposition. Attacks on the influence of mathematics have been of three main types. Some people have simply favored one view of mathematics over other views; other people have granted the importance of mathematics but have opposed what they consider its overuse or extension into inappropriate domains; still others have attacked mathematics, and often all of science and reason, as cold, inhuman, or oppressive.

Aristotle's reaction against Platonism is perhaps the first example of opposition to one view of mathematics (eternal objects) while championing another (deductive method). Another example is Newton's attack on Descartes's attempt to use nothing but "self-evident" assumptions to figure out how the universe worked. There are many mathematical systems God could have used to set up the world, said Newton. One could not decide *a priori* which occurs; one must, he says, observe in nature which law actually holds. Though mathematics is the tool one uses to discover the laws, Newton concludes that God set up the world by free choice, not mathematical necessity

[**35**, pp. 7–8] [**36**, p. 47]. This point is crucial to Newton's natural theology: that the presence of order in nature proves that God exists.

Another example of one view based on mathematics attacking another can be found in Malthus's *Essay on Population* of 1798. He accepts the Euclidean deductive model—in fact he begins with two "postulata": man requires food, and the level of human sexuality remains constant [**31**, Chapter **I**]. His consequent analysis of the growth of population and of food supply rests on mathematical models. Nonetheless, one of Malthus's chief targets is the predictions by Condorcet and others of continued human progress modelled on that of mathematics and science. As in Newton's attack on Descartes, Malthus applied one view of mathematics to attack the conclusions others claimed to have drawn from mathematics.

Our second category of attacks—drawing a line that mathematics should not cross—is exemplified by the seventeenth-century philosopher and mathematician Blaise Pascal. Reacting against Cartesian rationalism, Pascal contrasted the "esprit géometrique" (abstract and precise thought) with what he called the "esprit de finesse" (intuition) [**39**, *Pensée* 1], holding that each had its proper sphere, but that mathematics had no business outside its own realm. "The heart has its reasons," wrote Pascal, "which reason does not know" [**39**, *Pensée* 277]. Nor is this contradicted by the fact that Pascal was willing to employ mathematical thinking for theological purposes—recall his "wager" argument to convince a gambling friend to try acting like a good Catholic [**39**, *Pensée* 233]; the point here was to use his friend's own probabilistic reasoning style in order to convince him to go on to a higher level.

Similarly, the mathematical reductionism of men like Lagrange and Comte was opposed by men like Cauchy. Cauchy, whom we know as the man who brought Euclidean rigor to the calculus, opposed both Lagrange's attempt to reduce mechanics to calculus and calculus to formalistic algebra [**15**, pp. 51–54], and opposed the positivists' attempt to reduce the human sciences to an ultimately mathematical form. "Let us assiduously cultivate the mathematical sciences," Cauchy wrote in 1821, but "let us not imagine that one can attack history with formulas, nor give for sanction to morality theorems of algebra or integral calculus" [**5**, p. vii]. Analogously, in our own day, computer scientist Joseph Weizenbaum attacks the modern, computer-influenced view that human beings are nothing but processors of symbolic information, arguing that the computer scientist should "teach the limitations of his tools as well as their power" [**51**, p. 277].

Finally, we have those who are completely opposed to the method of analysis, the mathematization of nature, and the application of mathematical thought to human affairs. Witness the Romantic reaction against the Enlightenment: Goethe's opposition to the Newtonian analysis of white light, or, even more extreme, William Wordsworth in *The Tables Turned:*

Sweet is the lore which Nature brings;

Our meddling intellect

Mis-shapes the beauteous forms of things:—

We murder to dissect.

Again, Walt Whitman, in his poem "When I heard the learn'd astronomer," describes walking out on a lecture on celestial distances, having become "tired and sick," going outside instead to look "up in perfect silence at the stars."

Reacting against statistical thinking on behalf of the dignity of the individual, Charles Dickens in his 1854 novel *Hard Times* satirizes a "modern school" in which a pupil is addressed as "Girl number twenty" [**12**, Book I, Chapter II]; the schoolmaster's son betrays his father, justifying himself by pointing out that in any given population a certain percentage will become traitors, so there is no occasion for surprise or blame [**12**, Book III, Chapter VII]. In a more political point, Dickens through his hero denounces the analytically-based efficiency of industrial division of labor, saying it regards workers as though they were nothing but "figures in a sum" [**12**, Book II, Chapter V].

The Russian novelist Evgeny Zamyatin, in his early-twentieth-century antiutopian novel *We* (a source for Orwell's *1984*), envisions individuals reduced to being numbers, and mathematical tables of organization used as instruments of social control. Though the certainty of mathematics, and thus its authority, has sometimes been an ally of liberalism, as we have seen in the cases of Voltaire and Condorcet, Zamyatin saw how it could also be used as a way of establishing an unchallengeable authority, as philosophers like Plato and Hobbes had tried to use it, and he wanted no part of it.

6. Conclusion

As the battles have raged in the history of Western thought, mathematics has been on the front lines. What does it all (to choose a phrase) add up to?

My point is not that what these thinkers have said about mathematics is right, or is wrong. But this history shows that the nature of mathematics has been—and must be—taken into account by anyone who wants to say anything important about philosophy or about the world. I want, then, to conclude by advocating that we teach mathematics *not* just to teach quantitative reasoning, *not* just as the language of science—though these are very important—but that we teach mathematics to let people know that one cannot fully understand the humanities, the sciences, the world of work, and the world of man without understanding mathematics in its central role in the history of Western thought.

Acknowledgements

I thank the students in my Mathematics 1 classes at Pomona and Pitzer Colleges for stimulating discussions and suggestions on some of the topics covered in this paper, especially David Bricker, Maria Camareña, Marcelo D'Asero, Rachel Lawson, and Jason Gottlieb. I also thank Sandy Grabiner for both his helpful comments and his constant support and encouragement.

References

1. Charles Babbage, *On the Economy of Machinery*, Charles Knight, London, 1832.
2. William Barrett, *Irrational Man: A Study in Existential Philosophy*, Doubleday, New York, 1958. Excerpted in William L. Schaaf, ed., *Our Mathematical Heritage*, Collier, New York, 1963.
3. Ludwig Boltzmann, Model, *Encyclopedia Britannica*, 1902.
4. Carl Boyer, *History of Analytic Geometry*, Scripta Mathematica, New York, 1956.
5. A.-L. Cauchy, *Cours d'analyse*, 1821. Reprinted in A.-L. Cauchy, *Oeuvres*, series 2, vol. 3, Gauthier-Villars, Paris, 182.
6. I. Bernard Cohen, *The Newtonian Revolution*, Cambridge University Press, Cambridge, 1980.
7. ———, *The Birth of a New Physics*, 2d edition. W. W. Norton, New York and London, 1985.

8. Auguste Comte, *Cours de philosophie positive*, vol. I, Bachelier, Paris, 1830.
9. Marquis de Condorcet, *Sketch for a Historical Picture of the Progress of the Human Mind*, 1793, tr, June Barraclough. In Keith Michael Baker, ed., *Condorcet: Selected Writing*. Bobbs-Merrill, Indianapolis, 1976.
10. René Descartes, *Discourse on Method*, 1637, tr. L. J. Lafleur, Liberal Arts Press, New York, 1956.
11. ———, *La géométrie*, 1637, tr., D. E. Smith and Marcia L. Latham, *The Geometry of René Descartes*, Dover, NY, 1954.
12. Charles Dickens, *Hard Times*, 1854. Norton Critical Edition edited by George Ford and Sylvere Monod, W. W. Norton, New York and London, 1966.
13. Neal C. Gillespie, *Charles Darwin and the Problem of Creation*, University of Chicago Press, Chicago and London, 1979.
14. Charles C. Gillispie, *The Edge of Objectivity*, Princeton University Press, Princeton, NJ, 1960.
15. Judith V. Grabiner, *The Origins of Cauchy's Rigorous Calculus*, M.I.T. Press, Cambridge, Mass., 1981.
16. John C. Greene, *Science, Ideology, and World View: Essays in the History of Evolutionary Ideas*, University of California Press, Berkeley, 1981.
17. A. Rupert Hall, *From Galileo to Newton*. 1630–1720. Harper and Row, New York and Evanston, 1963.
18. T. L. Heath (Editor), *The Works of Archimedes with the Method of Archimedes*, New York, n.d.
19. ———, *The Thirteen Books of Euclid's Elements*, Volume I, Cambridge University Press, 1925, reprinted by Dover, New York, 1956.
20. Hermann von Helmholtz, On the origin and significance of geometrical axioms, 1870, reprinted in Hermann von Helmholtz, *Popular Scientific Lectures*, edited by Morris Kline, Dover, NY, 1962.
21. Thomas Hobbes, *Leviathan, or the Matter, Form, and Power of a Commonwealth, Ecclesiastical and Civil*, 1651, in E. Burtt, ed., *The English Philosophers from Bacon to Mill*, Modern Library, New York, 1939.
22. Anthony Hyman, *Charles Babbage: Pioneer of the Computer*. Princeton University Press, Princeton, NJ, 1982.
23. Werner Jaeger, *Early Christianity and Greek Paideia*, Oxford University Press, London, Oxford, and New York, 1961.
24. Immanuel Kant, *Critique of Pure Reason*, 1781, tr. F. Max Müller, Macmillan, New York, 1961.
25. ———, *Prolegomena to Any Future Metaphysics*, 1783, ed. Lewis W. Beck, Liberal Arts Press, New York, 1951.
26. Morris Kline, *Mathematics in Western Culture*, Oxford University Press, New York, 1953.
27. D. Landau and P. Lazarsfeld, Quetelet, *International Encyclopedia of the Social Sciences*, Vol. 13, Macmillan, New York, 1968, pp. 247–257.
28. Pierre Simon Laplace, *A Philosophical Essay on Probabilities*, 1819 tr. F. W. Truscott and F. L. Emory, Dover, NY, 1951.
29. Desmond Lee, 1965 Introduction, in Desmond Lee, ed., *Plato: Timaeus and Critias*, Penguin Books of Great Britain, London, 1971.
30. Leibniz, *Preface to the General Science* and *Towards a Universal Characteristic*; 1677; reprint, *Selections*, Philip P. Wiener, editor, 1951, pp. 12–17, 17–25.
31. Thomas R. Malthus, An Essay on the Principle of Population, as it Affects the Future Improvement of Society: with Remarks on the Speculations of Mr. Godwin, M. Condorcet, and other writers, 1798, in Garrett Hardin, ed., *Population, Evolution and Birth Control*, Freeman, San Francisco, 1964, pp. 4–16.

32. Maurice Mandelbaum, The Search for a Science of Society: From Saint-Simon to Marx and Engels, Chapter 4 in Maurice Mandelbaum, *History, Man, and Reason*, Johns Hopkins University Press, Baltimore and London, 1971, pp. 63–76.

33. James Clerk Maxwell, *Theory of Heat*, Longmans, Green, & Co., London, 1871.

34. Seyyed Hossein Nasr, *Science and Civilization in Islam*, New American Library, New York, Toronto and London, 1968.

35. Isaac Newton, Letter to Henry Oldenburg, July, 1672, in [38].

36. ———, Letter to Richard Bentley, December 10, 1692, in [38].

37. ———, *Sir Isaac Newton's Mathematical Principles of Natural Philosophy and His System of the World*, tr. Andrew Motte, revised and edited by Florian Cajori, University of California Press, Berkeley, 1934.

38. ———, *Newton's Philodophy of Nature: Selections from His Writings*, ed. H. S. Thayer, Hafner, New York, 1951.

39. Blaise Pascal, Pensées, E. P. Dutton, New York, 1958.

40. Benjamin Peirce, Linear Associative Algebra, *Amer. J. Math.*, 4 (1881).

41. Plato, *Republic*, tr. A.D. Lindsay, E. P. Dutton, New York, 1950.

42. Plato, Timaeus, in [29].

43. Richard H. Popkin, *The History of Scepticism from Erasmus to Spinoza*, University of California Press, Berkeley and Los Angeles, 1979.

44. Theodore M. Porter, A Statistical Survey of Gases: Maxwell's Social Physics, *Historical Studies in the Physical Sciences* 12 (1981), 77–116.

45. A. Quetelet, *Instructions populaires sur le calcul des probabilités*, Tarlier, Brussels, 1828.

46. Bertrand Russell, *A Critical Exposition of the Philosophy of Leibniz*, 2d edition, Allen and Unwin, London, 1937.

47. William G. Sinnigen and Arthur E. R. Boak, *A History of Rome to A.D. 565*, sixth edition, Macmillan, New York, 1977.

48. Andrew Skinner, Introduction, Adam Smith, *The Wealth of Nations*, (1776), Penguin Books, London, 1974, pp. 11–97.

49. Benedict de Spinoza, *Ethics*, preceded by *On the Improvement of the Understanding*, ed. James Gutmann, Hafner, New York, 1953.

50. François-Marie Arouet de Voltaire, *The Portable Voltaire*, The Viking Press, New York, 1949, selections from *Dictionnaire Philosophique* (1764), pp. 53–228.

51. Joseph Weizenbaum, *Computer Power and Human Reason: From Judgment to Calculation*, Freeman, San Francisco, 1976.

52. Harry Austryn Wolfson, What is New in Philo?, in Harry A. Wolfson, *From Philo to Spinoza: Two Studies in Religious Philosophy*, Behrman House, New York, 1977.

5

Descartes and Problem-Solving*

Introduction

What does Descartes have to teach us about solving problems? At first glance it seems easy to reply. Descartes says a lot about problem-solving. So we could just quote what he says in the *Discourse on Method* [**12**] and in his *Rules for Direction of the Mind* ([**2**], pp. 9–11). Then we could illustrate these methodological rules from Descartes' major mathematical work, *La Géométrie* [**13**]. After all, Descartes claimed he did his mathematical work by following his "method." And the most influential works in modern mathematics—calculus textbooks—all contain sets of rules for solving word problems, rather like this:

1. Draw a figure.
2. Identify clearly what you are trying to find.
3. Give each quantity, unknown as well as known, a name (e.g., $x, y, \ldots$).
4. Write down all known relations between these quantities symbolically.
5. Apply various techniques to these relations until you have the unknown(s) in equations that you can solve.

The calculus texts generally owe these schemes to George Pólya's *Mathematical Discovery*, especially Chapter 2, "The Cartesian Pattern," and Pólya himself credits them to Descartes' *Rules for Direction of the Mind* ([**32**], pp. 22–23, 26–28, 55–59, 129ff). So I studied those philosophical works as I began to write about Descartes and problem-solving. But the more I re-read Descartes' *Geometry*, the more convinced I became that it is from this work that his real lessons in problem-solving come. One could claim that, just as the history of Western philosophy has been viewed as a series of footnotes to Plato, so the past 350 years of mathematics can be viewed as a series of footnotes to Descartes' *Geometry.*

Now Descartes said in the *Discourse on Method* that it didn't matter how smart you were; if you didn't go about things in the right way—with the right method—you would not discover anything. Descartes' *Geometry* certainly demonstrates a successful problem-solving method in action. Accordingly, this article will bring what historians of mathematics know about Descartes' *Geometry* to bear on the question, what can Descartes teach the mathematics community about

* Reprinted from Math. Mag. 68, 2(April 1995) 83–97.

problem-solving? To answer this question, let us look at the major types of problems addressed in the *Geometry* and at the methods Descartes used to solve them.

A First Look at Descartes' Geometry

We have all heard that Descartes' *Geometry* contains his invention of analytic geometry. So when we look at the work, we may be quite surprised at what is *not* there. We do not see Cartesian coordinates. Nor do we see the analytic geometry of the straight line, or of the circle, or of the conic sections. In fact we do not see *any* new curve plotted from its equation. And what curves did Descartes allow? Not, as we might think, any curve that has an equation; that is secondary. He allowed only curves constructible by some mechanical device that draws them according to specified rules. Finally, we do not find the term "analytic geometry," nor the claim that he had invented a new subject—just a new (and revolutionary) method to deal with old problems.

What we *do* see is a work that is problem-driven throughout. Descartes' *Geometry* has a purpose. It is to solve problems. Some are old, some are new; all are hard. For all the lip service in Descartes' *Discourse on Method* to mathematics as logical deduction from self-evident first principles ([**12**], pp, 12–13, 18–19), the *Geometry* is not like that at all. It discovers; it does not present a finished logical structure. The specific purpose of the book is to answer questions like "What is the locus of a point such that a specified condition is satisfied?" And the answer to these questions must be geometric. Not "it is such-and-such a curve," or even "it has this equation," but "it is this curve, it has this equation, and it can be constructed in this way." Everything else in the *Geometry*—and that does include algebra, theory of equations, classifying curves by degree, etc.—are just means to this geometric end. To solve a problem in geometry, one must be able to construct the curve that is its solution.

The Background of Descartes' Geometry

To appreciate how much Descartes accomplished, we must first look at some achievements of the ancient Greeks. They solved a range of locus problems, some quite complicated. To find their solutions, they too had "methods." Greek mathematics recognized two especially useful problem-solving strategies: *reduction* and *analysis* ([**25**], pp. 23–24).

First, let us describe the method of reduction [in Greek, *apogōgē*]. Given a problem, we observe that we could solve it if only we could solve a second, simpler problem, and so we attack the second one instead. For instance, consider the famous problem of duplicating the cube. In modern notation, the problem is, given a^3, to find x such that $x^3 = 2a^3$. Hippocrates of Chios showed that this problem could be reduced to the problem of finding two mean proportionals between a and $2a$. That is, again in modern notation, if we can find x and y such that:

$$a/x = x/y = y/2a, \tag{1}$$

then, eliminating y, we obtain $x^3 = 2a^3$ as required ([**25**], p. 23). But more geometric knowledge led to a further reduction ([**25**], p. 61). If we consider just the first two terms of (1),

$$a/x = x/y,$$

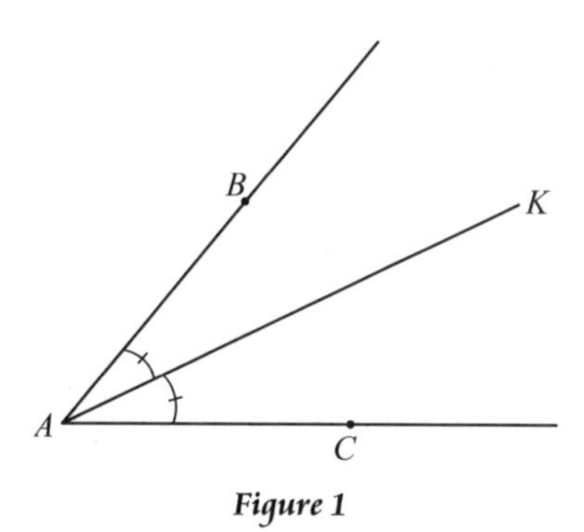

Figure 1

we obtain $x^2 = ay$, which represents a parabola. The equation involving the first and third terms in (1) yields

$$a/x = y/2a$$

or $xy = 2a^2$, which represents a hyperbola. Thus the problem of duplicating the cube is reducible to the problem of finding the intersection of a parabola and a hyperbola. This reduction promoted Greek interest in the conic sections.

The other problem-solving strategy is what the Greeks called "analysis"—literally, "solution backwards" (*àrnápalin lýsin* [**20**], Vol. ii, p. 400; [**25**], p. 9; cp. pp. 354–360). The Greek "analysis" works like this. Suppose we want to learn how to construct an angle bisector, and suppose that we already know how to bisect a line segment. We proceed by first assuming that we have the problem solved. Then, from the assumed existence of that angle bisector, we work backward until we reach something we do know. In Figure 1, take the angle A, and draw AK bisecting it. Then, mark off any length AB on one side of the angle, and an equal length AC on the other side. Connect B and C with the straight line BC, as in Figure 2. Now let M be the intersection of the angle bisector with the line BC. Since angle $BAM =$ angle MAC, $AB = AC$, and $AM = AM$, triangle ABM is congruent to triangle ACM. Thus M bisects BC. But wait. Recall that we already know how to bisect a line segment. Thus, we can find such an M. Now we can construct the angle bisector by reversing the process we just went through. That is, suppose we are given an angle A. To construct the angle bisector, construct $AB = AC$, construct the line BC, bisect it at M, and connect the points A and M. AM bisects the angle. This method—assuming that we have the thing we are looking for and working backwards from that assumption until we

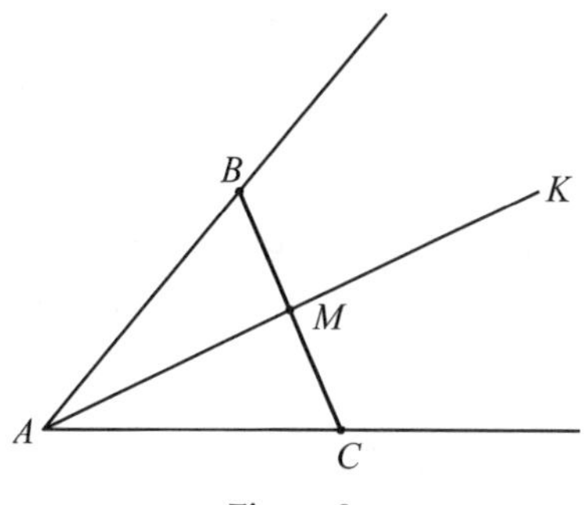

Figure 2

reach something we do know—was well-named "solution backwards." Pappus of Alexandria, in the early fourth century C.E., compiled a "treasury of analysis" in which he gave the classic definition of "analysis" as "solution backwards"; described 33 works, now mostly lost, by Euclid, Apollonius, Aristaeus, and Eratosthenes, which included substantial problems solvable by the method of analysis; and provided some lemmas that illustrate problem-solving by analysis ([**20**], Vol. ii, pp. 399–427).

In our example of bisecting an angle, the mathematical knowledge needed was minimal. But the Greeks knew all sorts of properties of other geometric figures, notably the conic sections, and so had an extensive set of theorems to draw on in using "analysis" to solve problems in geometry ([**6**], pp. 21–39; [**10**], pp. 43–58; [**20**], passim; [**25**]). (The best and fullest account is that of Knorr [**25**].)

Thus we see that Descartes, though he championed these techniques, clearly did not invent the method of analysis and the method of reduction. Descartes' ideas on problem-solving, moreover, have other antecedents besides the Greek mathematical tradition. First, a preoccupation with finding a universal "method" to find truth appears in the work of earlier philosophers, including the thirteenth-century Raymond Lull, whose method was to list all possible truths and select the right one, the sixteenth-century Petrus Ramus, who saw method as the key to effective teaching and to allowing learners to make their own discoveries ([**29**], pp. 148–9), and the seventeenth-century philosopher of science Francis Bacon, whose method to empirically discover natural laws was one of systematic induction and testing [**1**]. All of these seekers for method suggested that intellectual progress, unimpressive earlier in history, could be achieved once the right method for finding truth was employed. Descartes shared this view.

A second, more specific antecedent of Descartes' work was the invention of symbolic algebra as a problem-solving tool, a tool that was explicitly recognized as a kind of "analysis" in the Greek sense by its discoverer, Vieta, in 1591 ([**6**], p. 65; cp pp. 23, 157–173). To say "let $x =$" the unknown, and then calculate with x—square it, add it to itself, etc., *as if it were known*—is a powerful technique when applied to word problems both in and outside of geometry. Vieta recognized that naming the unknown and then treating it as if it were known was an example of what the Greeks called "analysis," so he called algebra "the analytic art." Incidentally, Vieta's use of this term is the origin of the way we use the word "analysis" in mathematics. In the seventeenth and early eighteenth centuries, the term "analysis" was often used interchangeably with the term "algebra," until by the mid-eighteenth century "analysis" became used for the algebra of infinite processes as opposed to that of finite ones [**4**].

Descartes was quite impressed with the power of symbolic algebra. But, although he had all these predecessors, Descartes combined, extended, and then exploited these earlier ideas in an unprecedented way. To see how his new method worked, we need to look at a specific problem.

Descartes' Method in Action

We begin with the first important problem Descartes described solving with his new method ([**13**], pp. 309–314, 324–335). The problem is taken from Pappus, who said in turn that it came from Euclid and Apollonius ([**13**], p. 304). The problem is illustrated in Figure 3 (from [**13**], p. 309).

Given four lines in a plane, and given four angles. Take an arbitrary point C. Consider now the distances (dotted lines) from C to the various given lines, where the distances are measured

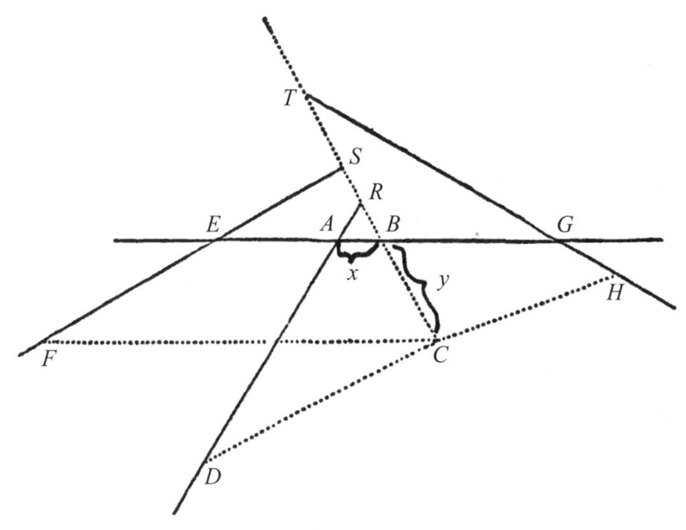

Figure 3

along lines making the given angles with the given lines. (For instance, the distance *CD* makes the given angle *CDA* with the given line *AD*.) A further condition on *C* is that the four distances *CD*, *CF*, *CB*, and *CH* satisfy

$$(CD \cdot CF)/(CB \cdot CH) = \text{a given constant.} \tag{2}$$

The problem is to find the locus of all such points C. For Descartes, that means to discover what curve it is, and then to construct that curve. (At this time, any reader who does not already know the answer is encouraged to conjecture what kind of curve it is—or to imagine constructing even *one* such point C.)

Here is how Descartes attacked this problem. First assume, as we must in order to draw Figure 3, that we already have one point on the curve. We will then work backwards, by the method of analysis. Draw the point *C*, and draw the distances. Label the distance from *C* to the line *EG* as y, and the line segment between that distance and the given line *DA* as x. Given these labels x and y, we use them and look for other relationships that can be derived in terms of them. For instance, independent of the choice of C, the angles in the triangle *ABR* are all known (since angle *CBG* is one of the given angles in the problem, we have angle *ABR* by vertical angles; angle *RAB* is determined by the position of the two given lines that include the segments *DR* and *GE*). Thus the shape of triangle *ABR* is determined, so the side *RB* is a fixed multiple of x. Descartes therefore called that side $(b/z) \cdot x$, where he took b/z to be a known ratio. Thus $CR = y + (b/z) \cdot x$ ([**13**], p. 310). Using his knowledge of geometry in this fashion, Descartes found many more such relationships, and was able to express each of the distances *CD*, *CF*, *CB*, and *CH* as a different linear function of the line segments x and y. For the case where $(CD \cdot CF)/(CB \cdot CH) = 1$, those expressions let him derive an equation between the unknowns x and y and various constants he called m, n, z, o, and p:

$$y = m - (n/z) \cdot x + \sqrt{\{m^2 + ox + (p/m) \cdot x^2\}} \tag{3}$$

([**13**], p. 326). Now perhaps the modern reader can guess what type of curve that equation represents. So could Descartes. From his studies of Greek geometry, Descartes knew quite a lot about the conic sections, so he said, though he did not explain, that if the coefficient of the x^2 term is zero, the points C lie on a parabola; if that coefficient is positive, on a hyperboia; if negative, on an ellipse; etc. The positions, diameters, axes, centers, of these curves can be determined also, and he briefly discussed how to do this ([**13**], p. 329–332).

The reader will have observed that there is no fixed coordinate system here. Descartes labeled as x and y the lengths of line segments that arose in this particular situation. Let us also make a comment about his choice of notation. Vieta had used uppercase vowels for the unknowns, consonants for knowns. Since matters of notation are relatively arbitrary, the fact that we use Descartes' lowercase x and y, rather than Vieta's A and E, testifies to the great influence of Descartes' work on our algebra and geometry. Further, though Descartes himself wrote mm and xx rather than m^2 and x^2 ([**13**], p. 326), he did use raised numbers, exponents, for integer powers greater than two (e.g. [**13**], p. 337, p. 344). Today we follow Descartes here too, using exponential notation for all powers.

The Greeks already knew that the Pappus four-line locus was a conic section. Nonetheless, the way Descartes derived this result is impressive. In line with our overall purpose, let us reflect on the method Descartes used. Why is "let x equal the first unknown" so powerful here? Because the technique of "reduction" was used by Descartes to effectively reduce a problem in geometry to a problem in algebra. Once he had done this, he could use the algorithmic power and generality of algebra to solve a formerly difficult problem with relative ease. It is an old problem-solving method, to reduce a problem to a simpler one, but because the simpler one is algebraic, Descartes had something different in kind from what had been done before. Algebra puts muscles on the problem-solving methods of analysis and reduction.

Beyond the Greeks

To fully exploit the power of algebra—to go beyond the Greeks—Descartes had to make a major break with the past. The earlier symbolic algebra of Vieta was based on the theory of geometric magnitudes inherited from the Greeks. Because of this geometric basis, the product of three magnitudes was spoken of as a volume. This created a problem: What might the product of five magnitudes be? Also, Greek geometry presupposed the Archimedean axiom: Quantities cannot be compared unless some multiple of one can exceed the other, so one cannot add a point to a line, or an area to a solid. How then could one write $x^2 + x$ ([**6**], p. 61, p. 84)? Descartes, like his predecessors, did not envision pure numbers, but only geometrical magnitudes. He too felt constrained to interpret all algebraic operations in geometric terms. But he invented a new geometric interpretation for algebraic equations that freed algebraists from crippling restrictions like being unable to write x^5 or $x^2 + x$. He freed himself, and therefore freed his successors, including us. Here is how he did it.

He took a line that he called "unity," of length one, which could be chosen arbitrarily. This let him interpret the symbol x as the area of a rectangle with one side of length x, the other of length one. He could now write $x^2 + x$ with a clear conscience, since it could be thought of as the sum of two areas. Even more important, he interpreted products as lengths of lines, so that he could interpret any arbitrary power as the length of a line. That is, the product of the line segments a and b for Descartes did not have to be the area ab, but could be another length such that $ab/a = b/1$. And the length ab could be constructed, as in Figure 4 ([**13**], p. 298).

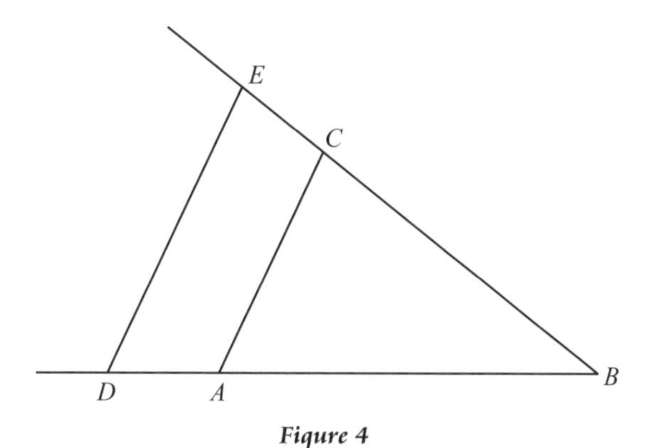

Figure 4

In this example, the product of the lines BD and BC is constructed, given a unit line AB. Let the line segments AB and BD be laid off on the same line originating at B, and let the segment BC be laid off on a line intersecting BD. Extending BC and constructing ED parallel to AC yields the proportion $BE/BD = BC/1$, since $AB = 1$. Thus BE is the required product $BD \cdot BC$. Of course this is an easy construction, but he had to give it explicitly. Descartes' philosophy of geometry did not let him merely assert that there was a length equal to the product of the two lines; he had to construct it. Now there was no problem in writing such expressions as x^5. This was just the length such that $x^5/x^3 = x^2/1$.

By showing that all the basic algebraic operations had geometric counterparts, Descartes could use them later at will. Furthermore, he had made a major advance in writing general algebraic expressions. Because of Descartes' innovations, later mathematicians came to consider algebra as a science of numbers, not geometric magnitudes, even though Descartes himself did not explicitly take this step. Descartes took his notational step in the service of solving geometric problems, in order to legitimize the algebraic manipulations needed to solve these geometric problems. What became a major conceptual breakthrough, then, was in the service of Descartes' problem-solving.

Descartes could now go beyond the Greeks, extending the Pappus four-line problem to five, six, 12, 13, or arbitrarily many lines. With these more elaborate problems, he still followed the same method: Label line segments, work out equations. But when he found the final equation and it was not recognizable as the equation of a conic, what then? To answer this, let me give the simple example he gave, a special case of the five-line problem. He considered four parallel lines separated by a constant distance, with the fifth line perpendicular to the other four ([**13**], pp. 336–337). (See Figure 5.) What, he asked, is the locus of all points C such that

$$CF \cdot CD \cdot CH = CB \cdot CM \cdot AI, \tag{4}$$

where AI is the constant distance between the equally spaced parallel lines and where the distances are all measured at right angles?

Again, Descartes proceeded by analysis. Assuming that he had such a point C, he labelled the appropriate line segments x and y ($x = CM$, $y = CB$), designated the known distance AI as a, and wrote down algebraic counterparts of all known geometric relationships. For this problem they

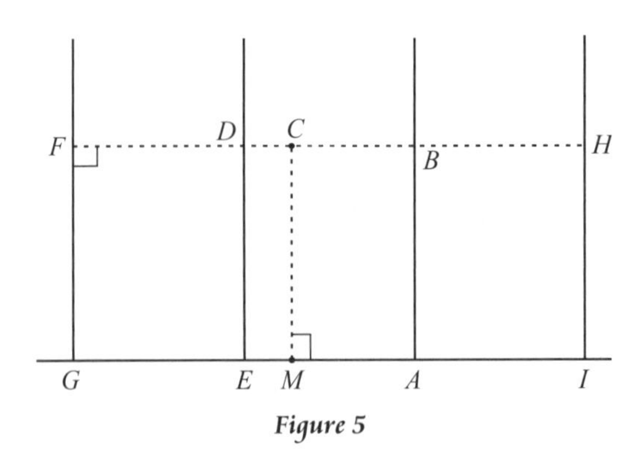

Figure 5

are simple ones. For instance, $CD = a - y$ and $CF = a + (a - y) = 2a - y$. Thus condition (4) becomes

$$(2a - y)(a - y)(y + a) = y \cdot x \cdot a,$$

which, multiplied out, yields the equation

$$y^3 - 2ay^2 - a^2y + 2a^3 = axy \text{ ([13], p. 337).} \tag{5}$$

This is not a conic (it is now often called the cubical parabola of Descartes), so the next question must be, can the curve this represents be constructed? That is, given x, can we find the corresponding value of y and thus construct any point C on the curve? Until these questions are answered affirmatively, Descartes would not consider the five-line problem solved, because, for him, it is a problem in geometry. The algebraic equation was just a means to the end for Descartes; it was not in itself the solution.

So precisely what does "constructible" mean for Descartes? Can the curve represented by that cubic equation be constructed, and, if so, how?

Here another of Descartes' methodological commitments helped him solve this problem: his commitment to generality. The ancients allowed the construction of straight lines and circles, said Descartes, but classified more complex curves as "mechanical, rather than geometrical" ([**13**], p. 315). Presumably this was because instruments were needed to construct them. (For instance, Nicomedes had generated the conchoid by the motion of a linkage of rulers ([**25**], pp. 219–220), and then used the curve in duplicating the cube and trisecting the angle.) But even the ruler and compass are machines, said Descartes, so why should one exclude other instruments ([**13**], p. 315; tr., p. 43)? Descartes decided to add to Euclid's construction postulates that "two or more lines can be moved, one upon the other, determining by their intersections other curves" ([**13**], p. 316). The curves must be generated according to a definite rule. And for Descartes, such a rule, at least in principle, was given by the use of a mechanical device that generated a continuous motion. Exactly what this means is complex—for instance, the machine is not allowed to convert an arc length to a straight line—but Bos has provided an enlightening discussion ([**3**], pp. 304–322, esp. p. 314).

Figure 6 reproduces one of Descartes' curve-constructing devices ([**13**], p. 320). The first curve he generated using it was produced by the intersection of moving straight lines. The straight line KN (extended as necessary) is at a fixed distance KL from a ruler GL. The ruler is attached to the point G, around which it can rotate. The point L can slide along the ruler GL. The segment KL moves up the fixed line AB (extended as needed). As KL moves up, the ruler, which has a

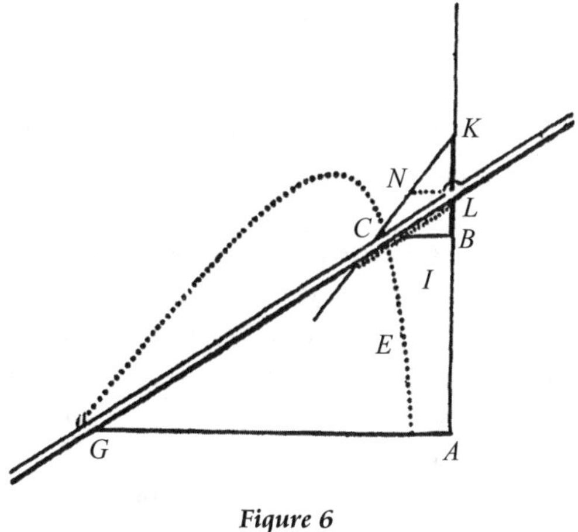

Figure 6

"sleeve" attached to L, rotates about G. Note that KL, KN, and the angle between them are all fixed. Then the point at which the ruler GL intersects the straight line KN extended, namely C, will be a point on the curve generated by this device.

To help the reader understand the operation of this device, I show, in Figure 7, the construction of a second point C' by this device. KL has moved up; KN thus has a new position; the ruler has rotated to a new position. Where the ruler and KN extended now intersect is another point C' on the curve. If one continues moving KL up and down, the points C, C', etc., trace a new curve.

But what kind of curve is it? Descartes solved this problem in his usual way. He labelled the key line segments (he let the unknowns $y = CB$ and $x = BA$, and the knowns $a = AG$, $b = KL$, and $c = NL$), and algebraically represented the geometric relationships between them. He then showed that if KNC is, as it is in our diagram, a straight line, the new curve generated by the points C, C', etc., is a hyperbola ([**13**], p. 322). (In fact AB is one of the hyperbola's asymptotes,

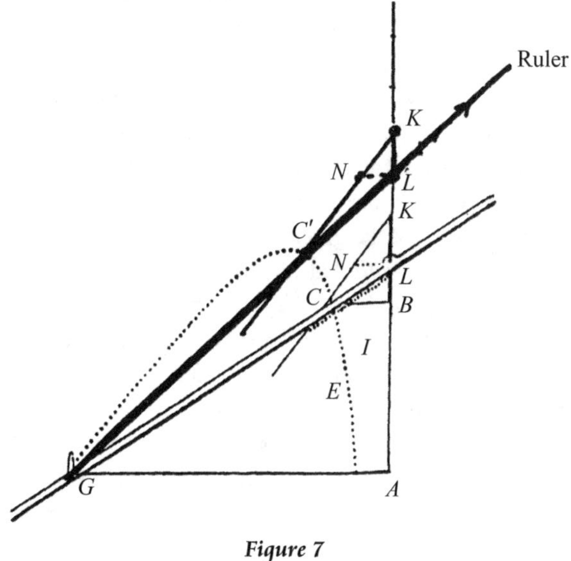

Figure 7

and the other asymptote is parallel to *KN*, as was shown by Jan van Schooten in his Latin edition of Descartes' *Geometry* ([**13**], p. 55n).) If instead of the straight line *KNC*, one uses a parabola whose axis is the straight line *KB*, the new curve constructed by the device can again be identified once its equation is found. In this case, Descartes showed by his usual method that the curve produced was precisely the cubic curve of (5) that he got for the simple five-line problem! ([**13**], p. 322.)

This coincidence must have suggested to Descartes that his construction method could obtain any desired curve. Also, using algebra, Descartes showed that his device would produce curves of successively higher degrees ([**13**], p. 321–323). For instance, when *KN* was a straight line, it produced a curve represented by a quadratic; when *KN* was a parabola, it produced a third-degree curve. Descartes, struck by the generality of these results, said that any algebraic curve could be defined as a Pappus n-line locus ([**13**], p. 324), but here he went too far. (For a proof that this is incorrect, see [**3**], pp. 332–338; incidentally, Newton was the first to try to prove that Descartes was in error on-this point ([**3**], p. 338).) Descartes also seems to have believed that any curve with an algebraic equation could be constructed by one of his devices. And here he was right, as was shown in the nineteenth century by A. B. Kempe ([**22**], cited in [**3**], p. 324). Thus Descartes' methods really did yield results of the generality he sought. We can now understand and appreciate the claim with which Descartes' *Geometry* begins: "Every problem in geometry can easily [!] be reduced to such terms that a knowledge of the lengths of certain straight lines is sufficient for its construction." (See [**13**], p. 297.)

The Power of Descartes' Methods: Tangents and Equations

Descartes held that curves were admissible in geometry only if they could be constructed, but of course he also had equations for them. Thus the study of the curves, and of many of their properties, could be advanced by the study of the corresponding equations. Let us briefly consider one example where Descartes did this.

All properties of geometric curves he had not yet discussed, he said, depend on the angles curves make with other curves ([**13**], pp. 341–342). This problem could be completely solved, he continued, if the normal to a curve at a given point could be found. The reader will recognize that this is an example of the reduction of one problem to another. And how does one find the normal to a curve? Again, by a reduction. It is easy to find the normal to a circle, so we can find the normal to a curve at a point by finding the normal to the circle tangent to the curve at the same point. Thus we must find such a tangent circle. And how did Descartes begin his search for that circle? By yet another reduction, this time to algebra: He sought an algebraic equation for the circle tangent to the given curve at the given point.

He did this by starting with a circle that hit the curve at two points, and then letting the two points get closer and closer together. This required, first, writing an algebraic equation for a circle that hit the curve twice. The equation for the points of intersection of that circle and the original curve would have two solutions. But "the nearer together the points . . . are taken, the less difference there is between the roots; and when the points coincide, the roots are equal" —that is, the equation has only one solution when the points coincide, and thus has only one solution when the intersecting circle becomes the tangent circle ([**13**], pp. 346–7). To find when the two solutions of the algebraic equation became one, Descartes in effect set the discriminant equal to zero, providing another demonstration of the power of algebraic methods to solve geometric problems. Thus, the algebraic equation let him find the tangent circle. Finally, the normal to that circle at the point of tangency gave him the normal to the curve ([**6**], pp. 94–95). Quite a triumph for the method of reduction!

Descartes applied this technique to find normals to several curves. For instance, he did it for the so-called ovals of Descartes ([**13**], pp. 360–2), whose properties, including normals, he used in optics. He also discussed finding the normal to the cubical parabola whose equation is (5) ([**13**], pp. 343–4). Descartes' method was the first treatment of a tangent as the limiting position of a secant to appear in print ([**6**], p. 95). Thus his method of normals was a step in the direction of the calculus, as was Fermat's contemporary, independent, simpler, and more elegant method of tangents ([**6**], pp. 80, 94–5; [**30**], pp. 165–169; [**5**], pp. 166–169, 157–8).

There is one more important class of problems taken up in Descartes' *Geometry*, the solution of algebraic equations. As we have mentioned, classical problems like duplicating the cube required solving equations. So did constructing arbitrary points on the curve that solved a locus problem. Descartes said in fact that "all geometric problems reduce to a single type, namely the question of finding the roots of an equation." (See [**13**], p. 401.) Since this process was so important, if one were given an equation, it would be good to learn as much about the solutions as possible before trying to construct them geometrically.

In the last section of the *Geometry*, Descartes tried to do just this, by developing a great deal of what is now called the theory of equations. One example will suffice to illustrate his approach:

$$(x-2)(x-3)(x-4)(x+5)=0. \tag{6}$$

Using this numerical example and multiplying it out, he obtained

$$x^4-4x^3-19x^2+106x-120=0. \tag{7}$$

Descartes pointed out that one can see from the way the polynomial in (7) is generated from (6) that it has three positive roots and one negative one, and that the number of positive roots is given by the number of changes of sign of the coefficients (this is the principal case of what is now called Descartes' Rule of Signs). Also, a polynomial with several roots is divisible by x minus any root, and it can have as many distinct roots as its degree ([**13**], pp. 372–4). Descartes was not the first to have pointed out these things, but his presentation was systematic and influential, and the context made clear the importance of the results. The algebra was not an end in itself; it was all done to solve geometric problems.

The last major class of problems addressed in the *Geometry* was constructing the roots of equations of degree higher than two. Going beyond the Greek example of a cubic solved by intersecting conics, Descartes solved fifth- and sixth-degree equations. Why? They come up, he said, in geometry, if one tries to divide an angle into five equal parts ([**13**], pp. 412), or if one tries to solve the Pappus 12-line problem ([**13**], p. 324). To illustrate his solution method, he solved a sixth-degree equation with six positive roots by using intersecting cubic curves. The curve he used was not $y=x^3$, which we might think of as simple, but the cubics he had defined as the intersections of moving conic sections and lines. In Figure 8, the diagram for one such solution is shown ([**13**], p. 404). The cubic curve, a portion of which is shown as *NCQ*, intersects the circle *QNC* at the points that solve the sixth-degree equation. The cubic curve involved in this construction, generated by the motion of the parabola *CDF*, is the cubic curve (5) once again.

Descartes said that he could construct the solution to every problem in geometry. We can now see why he thought that!

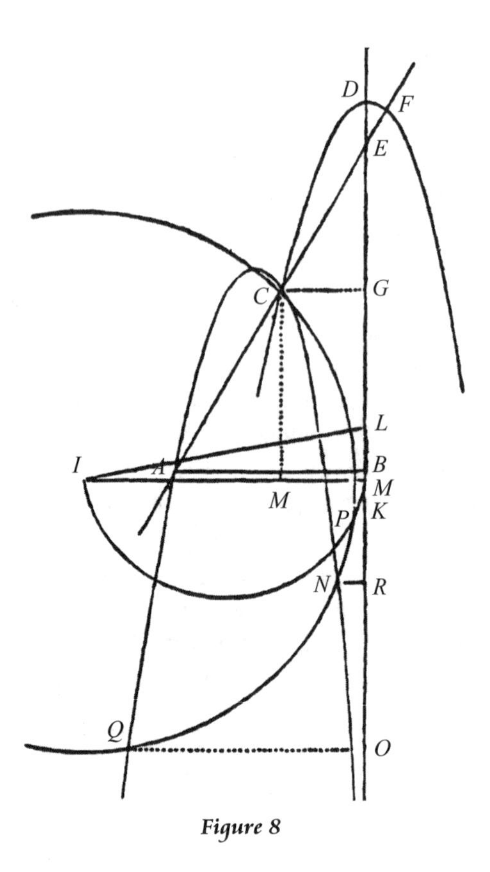

Figure 8

Conclusion

Now that we have seen Descartes in action, let us assess his influence on problem-solving. First, consider the mathematics that we now call "analysis." Descartes' *Geometry* solved hard problems by novel methods. There was, as an additional aid for his successors, the simultaneous and analogous work of Fermat; though Fermat's work on analytic geometry, tangents, and areas was not printed until the 1670s, it was circulated among mathematicians in the 1630s and 1640s and exerted great influence. *Geometry* itself attracted many followers. Continental mathematicians, especially Frans van Schooten and Florimond Debeaune, wrote commentaries and added explanations for Descartes' often cryptic statements. They also extended Descartes' methods to construct other loci. The second edition of Schooten's commentary on Descartes' *Geometry* (with a Latin translation) was published in 1659–1661 together with several other influential works based on Descartes. One was Jan de Witt's *Elements of Curves*, which systematized analytic geometry, including a discussion of constructing conic sections from their equations ([**6**], p. 115–116); another was Hendrik van Heuraet's work on finding arc lengths. Schooten's collection helped inspire both John Wallis and Isaac Newton. Wallis "seized upon the methods and aims of Cartesian geometry" ([**6**], p. 109) and went even further in replacing geometric concepts by algebraic or arithmetic ones. Many mid-seventeenth-century mathematicians, including Wallis, James Gregory, and Christopher Wren, influenced both by Descartes and by Fermat, used algebraic methods to make further progress on the problem of tangents, and—as Descartes had suggested, but did not do—to find areas. Men like van Heuraet, William Neil, and Wren also found arc lengths for some curves this way ([**5**], p. 162), which Descartes, who couldn't do it, had said couldn't be done ([**13**], p. 340). Wallis also extended the algebraic approach of Descartes to infinitesimals. In the 1660s, Isaac Newton carefully studied Schooten's edition of Descartes, using it (together with the work of men like Barrow, Wallis, and Gregory) as a key starting point in his invention of the calculus ([**35**], pp. 106–111, 128–130). In 1674, less than two years before his own invention of the calculus, Gottfried Wilhelm Leibniz worked his way through Descartes' *Geometry*; he was especially interested in the algebraic ideas ([**21**], p. 143). He later even examined some of Descartes' unpublished manuscripts ([**21**], p. 182–183).

Some scholars have credited Descartes with bringing about a revolution in analysis ([**7**], pp. 157–159, 506; [**3**], p. 304; for dissenting views, see [**31**], pp. 110–111; [**21**], pp. 202–210; [**18**], p. 55; [**19**], p. 164). But at the very least we may say of the *Geometry* what Thomas Kuhn once said about Copernicus' *On the Revolution of the Celestial Orbs* ([**26**], p. 134); though it may not have been revolutionary, it was "a revolution-making text." The problem-solving methods introduced in Descartes' *Geometry* and developed in the commentaries on it were clearly seminal throughout the seventeenth century, influencing both Newton and Leibniz, whether or not Descartes was the first inventor of these techniques. And such influence continued through the eighteenth century and beyond ([**17**], pp. 156–158, 505–507).

Incidentally, the systematic approach to analytic geometry we all learned in school is not in either Descartes or Fermat (though Fermat, unlike Descartes, did plot elementary curves from their equations), but dates from various eighteenth-century textbooks, especially those from the hands of Euler, Monge, Lagrange, and Lacroix ([**16**], pp. 192–224). Descartes, though, was not a textbook writer, but a problem-solver. The essence of his influence was in his new approach and his self-consciousness about method. These highlight his achievement as a problem-solver.

Second, then, let us look at his influence on problem-solving in general. The problem-solving methods we teach our students are the direct descendants of Descartes' methods. This is not because he passed them down to us in a set of rules (although he did). Nor is it because his methods work for the problems in elementary textbooks (although they do). It is because his methods solved many outstanding problems of his day. Descartes saw himself as a problem-solver because he had a method. He saw himself also as a teacher of problem-solving. One can see this even in the way he left hard questions as exercises to the reader, as he put it at the end of the *Geometry*, "to leave for others the pleasures of discovery." (See [**13**], p. 413.) His *Geometry* teaches us how to solve problems because it contains a set of solved problems whose successful solutions validate his methods. We may not care about the Pappus four-line problem, but we certainly prize the problem-solving power of a generalized algebra. Descartes' methods have come to us indirectly—who reads the *Geometry* nowadays?—but they have come to us because they are embedded in the work of his successors: In algebraic notation and equation theory, in analytic geometry, in calculus, in Lagrange's view that algebra is the study of general systems of operations, and in the more abstract and general subjects built upon these achievements. Because of his influence on later mathematicians, Descartes' methods are embedded also in the way we teach mathematics, in the standard collections of problems and solutions. In fact, for routine problems, the task of applying Descartes' analytic methods is, as he intended, fairly mechanical. Some of the *Rules for Direction of the Mind* explicitly parallel the method of the *Geometry*,

([**2**], pp. 177–178) and Pólya is thus right to have made such rules explicit for modern students. Inventing new mathematical methods—say, like analytic geometry— is, however, not a routine task. Even here, for Descartes, "method" is crucial.

Third, then, for those of us who want to invent great and new things like analytic geometry, to teachers and students of mathematics, Descartes has something else he wants us to learn, and that is his emphasis on method in general. Here he, together with his great contemporary Sir Francis Bacon, has inspired many. For instance, Leibniz saw his differential calculus as a problem-solving method, explicitly comparing it with analytic geometry, saying "From [my differential calculus] flow all the admirable theorems and problems of this kind with such ease that there is no more need to teach and retain them than for him who knows our present algebra to memorize many theorems of ordinary geometry" ([**27**], excerpted in [**34**], p. 281). Or, in our century, there is Pólya's sophisticated emphasis on teaching about method. Let me put Descartes' lesson this way: Raise problem-solving techniques to consciousness. Reflect on the methods that are successful and on their strengths and weaknesses. Then apply them systematically in attacking new problems. That is how Descartes himself invented analytic geometry, as he said in the *Discourse on Method:* "I took the best traits of geometrical analysis and algebra, and corrected the faults of one by the other." (See [**12**], p. 13, 20.)

Fourth and last, let us briefly consider a key point in Descartes' philosophy: that the methods of mathematics could solve the problems of science. Here, Descartes the philosopher learned from Descartes the mathematician that method was important, that the right method could solve previously intractable problems. He used the ideas of reduction and analysis in his philosophy of science. For instance, he argued that all macroscopic phenomena could be explained by analyzing nature into its component parts, bits of matter in motion. (See [**14**], pp. 409–414) and ([**36**], pp. 32–38). Descartes came to believe that the most powerful methods were both general and mathematical. His *Principles of Philosophy* (1644) attempted to deduce all the laws of nature from self-evident first principles; his principles XXXVII and XXXIX are equivalent to Newton's First Law of Motion (1687) ([**8**], pp. 182–183).) In fact, Descartes went so far as to state that everything that could be known could be found by a method modelled on that of mathematics. He wrote,

> Those long chains of reasoning, so simple and easy, which enabled the geometers to reach the most difficult demonstrations, had made me wonder whether all things knowable to man might not fall into a similar logical sequence. If so, we need only refrain from accepting as true that which is not true, and carefully follow the order necessary to deduce each one from the others, and there cannot be any propositions so abstruse that we cannot prove them, not so recondite that we cannot discover them ([**12**], pp. 12–13; 19).

Descartes' vision is clearly echoed by what Leibniz wrote in 1677 about his own search for a general symbolic method of finding truth: "If we could find characters or signs appropriate for expressing all our thoughts as definitely and as exactly as arithmetic expresses numbers or geometric analysis expresses lines, we could in all subjects in so far as they are amenable to reasoning accomplish what is done in Arithmetic and Geometry." (See [**28**], p. 15.) Again, consider the prediction of the great prophet of progress of the Enlightenment, the Marquis de Condorcet, that Descartes' methods could solve all problems. Although the "method" of algebra "is by itself only an instrument pertaining to the science of quantities," Condorcet wrote, "it contains within it the principles of a universal instrument, applicable to all combinations of ideas." This could make the progress of "every subject embraced by human intelligence... as sure as that of mathematics." (See **[9]**, pp. 238, 278–279; quoted in **[17]**, p. 222.)

Descartes has been attacked as a methodological imperialist and a reductionist, and lauded as an intellectual liberator and one of the founders of modern thought (e.g., **[11]**, **[18]**, **[24]**, **[33]**). For good or ill, the power of Descartes' vision has shaped Western thought since the seventeenth century, and his mathematical work helped inspire his philosophy. But whatever our assessment of Descartes the philosopher may be, his importance for the mathematician is clear. The history of the past 350 years of mathematics can fruitfully be viewed as the story of the triumph of Descartes' methods of problem-solving.

Acknowledgement

I thank Professor Tatiana Deretsky for suggesting this topic to me, and two MATHEMATICS MAGAZINE referees for helpful suggestions.

References

1. Bacon, Francis, *Novum Organum* [1620], Often reprinted, e.g., in E. A. Burtt, ed., *The English Philosophers from Bacon to Mill*, Modern Library, New York, 1913, pp. 24–123.
2. Beck, L. J., *The Method of Descartes: A Study of the Regulae*, Clarendon, Oxford, 1952. [Note: Descartes' *Rules for Direction of the Mind* (*Regulae*) were written about 1628, and published posthumously in 1701].
3. Bos, H. J. M., "On the representation of curves in Descartes' *Géométrie*," *Arch. Hist. Ex. Sci.* 1981, 295–338.
4. Boyer, Carl B., "Analysis: Notes on the evolution of a subject and a name," *Math Teacher* XLVII (1954), 450–62.
5. ———, *The Concepts of the Calculus*, Columbia, New York, 1939.
6. ———, *History of Analytic Geometry*, Scripta Mathematica, New York, 1956.
7. Cohen, I. B., *Revolution in Science*, Harvard, Cambridge, MA, 1985.
8. ———, *The Newtonian Revolution, with Illustrations of the Transformation of Scientific Ideas*, Cambridge University Press, Cambridge, MA, 1980.
9. Condorcet, Marquis de, *Sketch for a Historical Picture of the Progress of the Human Mind*, 1793, tr. J. Barraclough, in Keith Baker, ed., *Condorcet: Selected Writings*, Bobbs-Merrill, New York, 1976.
10. Coolidge, J. L., *A History of Geometrical Methods,* Oxford University Press, Oxford, 1940.
11. Davis, P. J., and Hersh, Reuben, *Descartes' Dream*, Harcourt, Brace, Jovanovich, San Diego and New York, 1986.
12. Descartes, René, *Discourse on the Method of Rightly Conducting the Reason to Seek the Truth in the Sciences*, 1637, tr. L. J. Lafleur, Bobbs-Merrill, New York, 1956. The first set of page numbers in each citation in the present paper are from this translation; the second set are from the edition of Ch. Adam et P. Tannery, eds., *Oeuvres de Descartes*, Paris, 1879–1913, Vol. VI.
13. Descartes, René, *The Geometry*, tr. from the French and Latin by D. E. Smith and M. L. Latham, Dover Reprint, New York, 1954. Contains a facsimile reprint of the original 1637 French edition. In the present paper, page references from Decartes, which appear in the Dover reprint, are given from the French edition, while citations from the Smith-Latham commentary are identified by the page numbers from the Dover reprint itself.

14. Dijksterhuis, E. J., *The Mechanisation of the World-Picture*, Tr. C. Dikshoorn, Oxford University Press, Oxford, 1961.

15. Gaukroger, Stephen, ed., *Descartes: Philosophy, Mathematics, and Physics*, Barnes and Noble, New York, 1980.

16. Gillies, Donald, ed., *Revolutions in Mathematics*, Oxford University Press, Oxford, 1992.

17. Grabiner, Judith V., "The centrality of mathematics in the history of Western thought," this MAGAZINE 61 (1988), pp. 220–230.

18. Grosholz, Emily, *Cartesian Method and the Problem of Reduction*, Clarendon Press, Oxford, 1991.

19. ———, "Descartes' Unification of Algebra and Geometry," in [**15**, pp. 156–168].

20. Heath, Thomas L., *Greek Mathematics*, 2 vols., Clarendon Press, Oxford, 1921.

21. Hofmann, J. E., *Leibniz in Paris*, 1672–1676: *His Growth to Mathematical Maturity*, Tr. A. Prag and D. T. Whiteside, Cambridge University Press, Cambridge, 1974.

22. Kempe, A. B., "On a general method of describing plane curves of the nth degree by linkwork," *Proc. Lond. Math. Soc.* 7 (1876), 213–16.

23. Klein, Jacob, *Greek Mathematical Thought and the Origin of Algebra*, The MIT Press, Cambridge, MA, 1968.

24. Kline, Morris, *Mathematics in Western Culture*, Oxford University Press, Oxford, 1964.

25. Knorr, Wilbur, *The Ancient Tradition of Geometric Problems*, Birkhäuser, Boston, 1986.

26. Kuhn, Thomas S., *The Copernican Revolution*, Harvard, Cambridge, MA, 1957.

27. Leibniz, G. W., "De geometria recondita et analysi indivisibilium atque infinitorum," *Acta Eruditorum* 5 (1686). Excerpted in [**34**, pp. 281–282].

28. ———, "Preface to the General Science," in [**37**, pp. 12–17].

29. Mahoney, Michael S., "The Beginnings of Algebraic Thought in the Seventeenth Century," in [**15**, pp. 141–155].

30. ———, *The Mathematical Career of Pierre de Fermat, 1601–65*, Princeton University Press, Princeton, NJ, 1973.

31. Mancosu, Paolo, "Descartes' *Géométrie* and revolutions in mathematics," [**16**, pp. 83–116].

32. Pólya, George, *Mathematical Discovery: On Understanding, Learning, and Teaching Problem Solving*, John Wiley & Sons, Inc., New York, 1981.

33. Russell, Bertrand, *A History of Western Philosophy*, Simon and Schuster, New York, 1945.

34. Struik, Dirk, J., *A Source Book in Mathematics, 1200–1800*, Harvard, Cambridge, MA, 1969.

35. Westfall, Richard S., *Never At Rest: A Biography of Isaac Newton*, Cambridge University Press, Cambridge, 1980.

36. ———, *The Construction of Modern Science*, John Wiley & Sons, Inc., New York, 1971.

37. Wiener, P., ed. *Leibniz: Selections*, Scribner's New York, 1951.

6

The Calculus as Algebra, the Calculus as Geometry: Lagrange, Maclaurin, and their Legacy*

Prelude: The Ways Mathematicians Think

Given a regular hexagon and a point in its plane: draw a straight line through the given point that divides the given hexagon into two parts of equal area.[1] Please stop for a few moments, solve the problem, and think about the way you solved it.

Did you draw an actual diagram? Did you draw a mental diagram? Were you able to solve the problem without drawing one at all? If you had a diagram, do you find it hard to understand how others could proceed without one? Were you motivated at all by analytic considerations? If you were not, do you understand how others were? Did you use kinematic ideas, such as imagining the line to be moving or rotating about a point? If not, can you understand how others might have? In solving the problem for the hexagon, were you consciously motivated by the analogy to the circle? Now that the analogy has been mentioned, do you think you were unconsciously so motivated? If not, do you understand how others were? Did you have any other method not already mentioned? (For instance, some people, often chemists, say they thought of folding the plane.) Finally, did you try to prove your solution? Did you nevertheless "know" that you were right?

The diversity of problem-solving approaches and of levels of conviction, even for a simple problem like this one, makes clear that there are many kinds of successful mathematical thinking. Distinctions like that between the diagram-drawers and the others, between those who use a geometric analogy and those who do not, between those who use kinematic ideas and those who do not, and between those who cannot imagine doing the problem analytically and those who naturally proceed that way, have been analyzed by Jacques Hadamard in his *Psychology*

---

* Reprinted from *Vita Mathematica: Historical Research and Integration with Teaching*, edited by Ronald Calinger, MAA, 1996.

[1] G. Pólya, *How to Solve It: A New Aspect of Mathematical Method*, 2d edition (Princeton: Princeton University Press, 1971) p. 234.

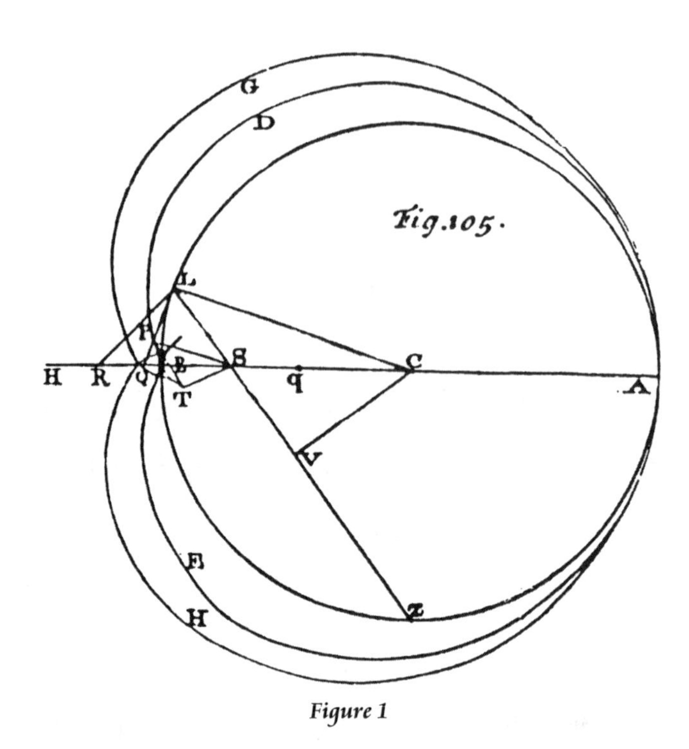

Figure 1

of Invention in the Mathematical Field.[2] One might, he said, distinguish between "geometric" and "algebraic" approaches to mathematics, for instance. Henri Poincaré did not find this distinction sufficient, so he divided mathematicians between what he called "intuitionalists" and "logicians."[3] For instance, Bernhard Riemann was intuitive, Karl Weierstrass logical. Hadamard explained Poincaré's distinction by citing the cases of Hermite (whom Hadamard saw as intuitive) and Poincaré himself (whom he saw as logical), saying, "Reading one of [Poincaré's] great discoveries, I should fancy . . . that, however magnificent, one ought to have found it long before, while . . . memoirs of Hermite . . . arouse in me the idea, 'What magnificent results! How could he dream of such a thing?'"[4]

Hadamard explained the different types of mathematical thinking according to how deeply the unconscious is tapped in the process, by whether the thought is broadly directed in contrast to narrowly directed, and according to what kinds of mental pictures or other concrete representations are used.[5] Hadamard's general discussion can be nicely illustrated on a small scale by observing the diversity of responses obtained when the Pólya hexagon problem is given to a group of mathematicians. Reflecting on Hadamard's discussion on the grander scale serves to set the scene for the contrast I want to draw between the work of two very different eighteenth-century

[2] Jacques Hadamard, *The Psychology of Invention in the Mathematical Field* (Princeton: Princeton University Press, 1945; Dover reprint, n.d.), p. 108n.

[3] Hadamard, op. cit., pp. 108–109.

[4] Hadamard, op. cit., p. 108n.

[5] Hadamard, op. cit., pp. 112–115.

Figure 2 This is not a misprint.

mathematicians, Colin Maclaurin and Joseph-Louis Lagrange. Each contributed to the development of the calculus in the eighteenth century. But Maclaurin viewed the calculus as geometry, while Lagrange saw it as algebra. It will not be surprising if different readers of Maclaurin and Lagrange respond "of course!" or "how could he dream of such a thing?" to different arguments. Such differing responses illustrate our main point: there are many modes of creative mathematical thought.

Introduction to the Geometric and Algebraic Approaches

Figure 1 is a diagram from Maclaurin's *Treatise of Fluxions* of 1742.[6] Figure 2 is a diagram from Joseph-Louis Lagrange's *Théorie des fonctions analytiques* of 1797. This contrast underlines the distinction between the way these two men—Joseph-Louis Lagrange (1736–1813) and Colin Maclaurin (1698–1746)—thought about the calculus and its applications. For Maclaurin, the calculus was at heart geometric; for Lagrange, the calculus was algebraic. Maclaurin's great *Treatise of Fluxions* (1742) has over 350 diagrams; Lagrange's masterwork on the calculus, the *Théorie des fonctions analytiques*, search as one will, contains none—just pages of text and formulas.[7]

To see how contemporaries thought about these approaches, let us look at two illustrations from the early nineteenth century. First, here is an example of the Edinburgh tradition to which Maclaurin belonged, from the Scottish philosopher William Hamilton:

> The process in the algebraic method is like running a railroad through a tunneled mountain; in the geometrical like crossing the mountain on foot. The former carries us by a short and easy transit to our destined point, but in miasma, darkness and torpidity, whereas the latter allows us to reach it only after time and trouble, but feasting us at each turn with glances of the earth and of the heavens, while we . . . gather new strength at every effort we put forth."[8]

---

[6] All diagrams in this paper are taken from Colin Maclaurin, *A Treatise of Fluxions in Two Books* (Edinburgh: Ruddimans, 1742).

[7] Joseph-Louis Lagrange, *Théorie des fonctions analytiques* (Paris: Imprimérie de la République, An V [1797]; cp. the second edition (Paris: Courcier, 1813), reprinted in *Oeuvres de Lagrange*, pub. M. J.-A Serret, 14 volumes (Paris: Gauthier-Villars, 1867–1892), reprinted (Hildesheim and New York: Georg Oms Verlag, 1973), volume 9.

[8] George Elder Davie, *The Democratic Intellect: Scotland and her Universities in the Nineteenth Century* (Edinburgh: The University Press, 1966), p. 127.

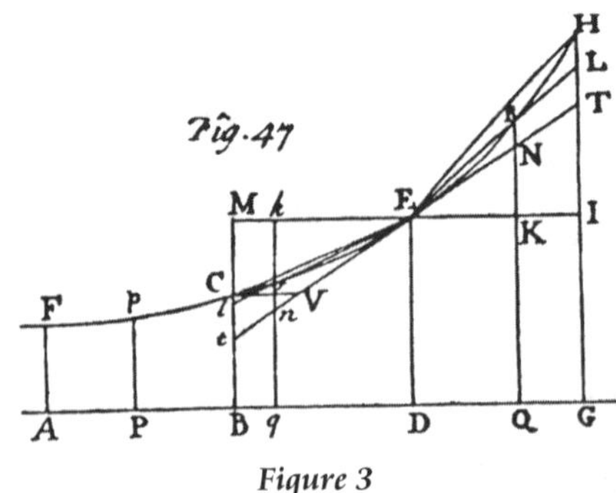

Figure 3

In contrast, followers of Lagrange saw mathematical truth arising from algebra, because of what Charles Babbage called "the accurate simplicity of its language."[9] Lagrange preferred algebra, not only for calculus, but even for mechanics. He wrote in 1797 that mechanics is just an analytic geometry with four coordinates, x, y, z, and t.[10] As is well known, Lagrange boasted in the preface to his *Analytical Mechanics* that "no diagrams are to be found in this work—only algebraic operations."[11] Thus, we avoid being misled by geometric intuition or by ideas about speeds and motion, and follow only the logic of pure, abstract systems of operations. The development of the calculus in the century after Isaac Newton and Gottfried Wilhelm Leibniz is the history of these two approaches: the calculus as geometry and the calculus as algebra, as exemplified in the present paper by Maclaurin and Lagrange.[12]

Let us now look at examples of the actual mathematical work of Maclaurin and Lagrange. The calculus is, of course, one subject, and Maclaurin and Lagrange had much in common. Nonetheless, as we shall see, their different approaches conditioned the problems they solved and how they solved them. The fact that successful mathematicians—even at the same time and in the same field—think in such different ways has implications for our work with students and for how we think about mathematical research.

Maclaurin

First, we will consider Maclaurin, whose calculus is Newtonian. He viewed the derivative—which he called the fluxion—as a mathematically-idealized velocity. Maclaurin wrote,

> The velocity by which a quantity flows . . . is called its *Fluxion* which is therefore always measured by the increment or decrement that would be generated in a given time by this motion, if it was continued uniformly from that term without acceleration or retardation.[13]

---

[9] Preface to *Memoirs of the Analytical Society* (Cambridge: J. Smith, 1813). Attributed to Babbage by Anthony Hyman, *Charles Babbage: Pioneer of the Computer* (Princeton, Princeton University Press, 1982), p. 26.

[10] Lagrange, *Fonctions analytiques*, in *Oeuvres*, vol. 9, p. 337.

[11] Lagrange, *Mécanique analytique*, 2d edition, in Oeuvres, vols. 11–12.

[12] Of course others shared these approaches; in particular, Lagrange's idea of the calculus as algebra owes much to Euler, especially his *Introductio in analysin infinitorum* of 1748. See Judith V. Grabiner, *The Origins of Cauchy's Rigorous Calculus*, (Cambridge, Mass.: The MIT Press, 1981), pp. 51–52.

[13] Maclaurin, *Fluxions*, p. 57.

In Maclaurin's diagram reproduced above, for instance, the fluxion of the curve at E is measured by the line TI. Given this background, we will review a specific example of Maclaurin's approach to calculus: his theory of maxima and minima.

Long before the invention of calculus, it was well known that at the highest point of a smooth curve, the tangent is horizontal. The Greeks had treated not only tangents, but also the concavity and convexity of curves, and seventeenth-century mathematicians had gone beyond the Greeks to study many more curves, including those with points of inflexion and many branches. Maclaurin was interested in geometry; he knew about many kinds of curves; because of this, he wanted to give a systematic and complete theory of maxima, minima, concavity, and convexity for a wide variety of curves. He saw calculus as a means of doing so. Here are some of the curves that motivate and illustrate his theory (Figure 4).

Maclaurin defined the maximum not by inequalities, but by words. The words defined it in terms of velocities, but he immediately related his definition to graphs. Graphs are almost always intended to go with his verbal statements. For instance, he said,

> When for some time the variable quantity first increases till a certain assignable term, and then decreases . . . its magnitude at that term is considered as a maximum. . . . In problems of this kind . . . the variable quantity is represented by an ordinate of a curve. . . . The ordinate from a point of the curve is a maximum . . . when it is greater . . . than the ordinates which may be drawn from the parts of either branch of the curve adjoining to that point. . . .[14]

Minimum is analogously defined. And what of concavity? Here Maclaurin began by following the ancient Greeks, saying that when a curve "has its concavity turned one way, the tangent at any point of it is on the convex side."[15] But he continued by using the language of velocities. He gave a lemma which said that, if a curve is convex toward the base [what we call concave up],

> the ordinate increases with a motion that is continually accelerated, and decreases with a motion that is continually retarded.[16]

This Lemma about acceleration and retardation is made plausible immediately by a diagram, where we look at successive values of the ordinate and see that the tangent is always under the curve.[17] (Look again at Figure 3.)

Similar considerations apply when the curve is concave down. Thus, for instance, for extrema of smooth curves:

> The arch being supposed to have its concavity turned one way, and the tangent at E being supposed parallel to the base, if the arch meet the base we may conclude that DE is a maximum" (or in the other case a minimum).[18]

---

[14] Maclaurin, op. cit., pp. 214–215. Maclaurin's text here refers to three diagrams, Figure 81, 82, and 83. Figure 81 shows two smooth curves and is our Figure 5; 82 and 83 show curves with various types of cusps.

[15] Maclaurin, op. cit., p. 179; he refers immediately to his Figure 47—our Figure 3.

[16] Maclaurin, op. cit., p. 179.

[17] Maclaurin, op. cit., Figure 47, following p. 190, our Figure 3.

[18] Maclaurin, op. cit., Figure 81, our Figure 5; see p. 217.

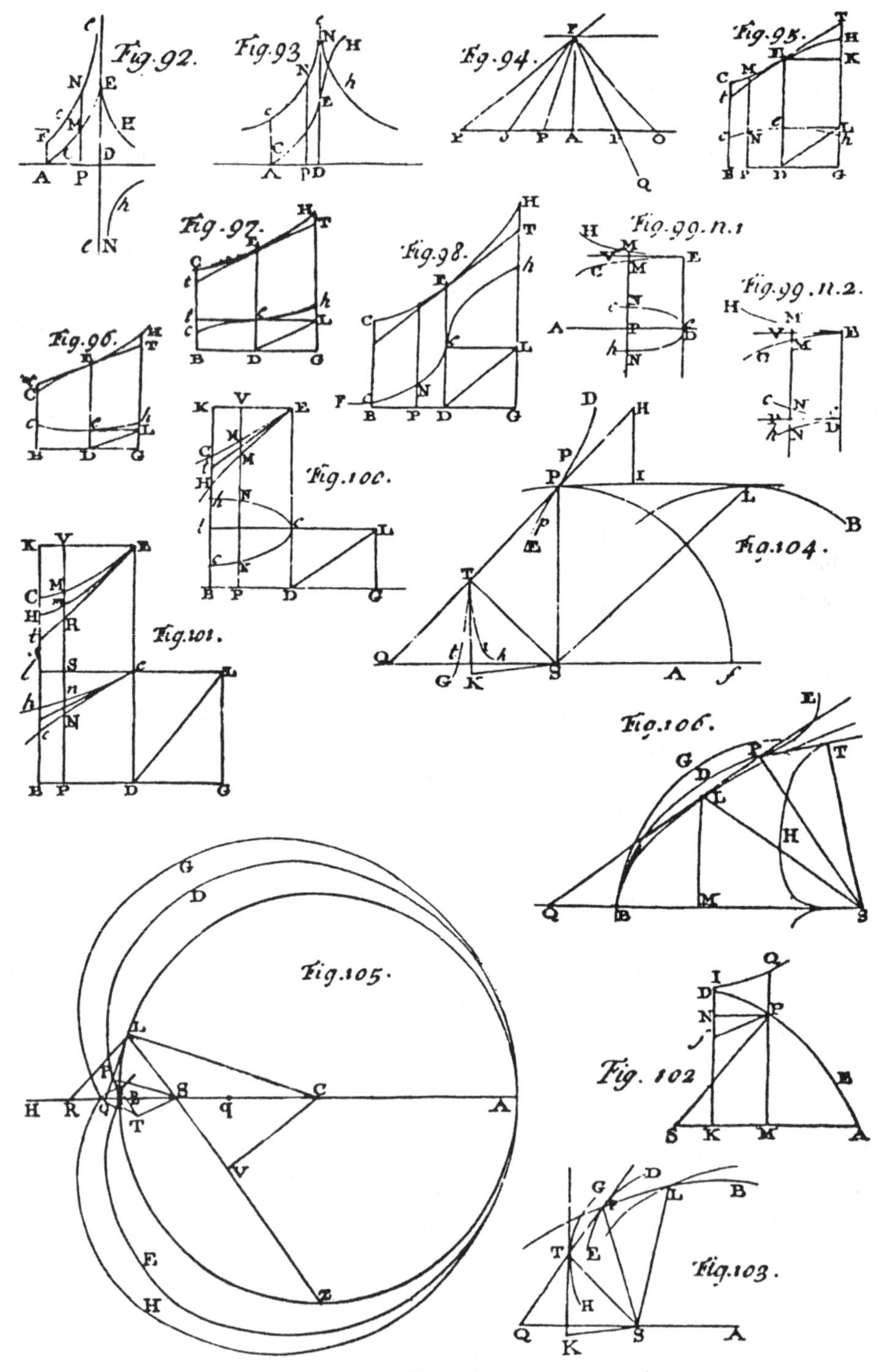
Fig. 92.
Fig. 93
Fig. 94.
Fig. 95.
Fig. 96.
Fig. 97.
Fig. 98.
Fig. 99. n. 1
Fig. 99. n. 2.
Fig. 100.
Fig. 101.
Fig. 102
Fig. 103.
Fig. 104.
Fig. 105.
Fig. 106.

Figure 4

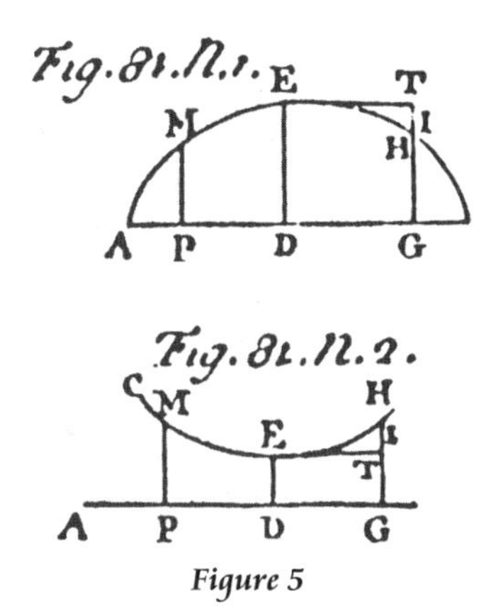

Figure 5

Thus he *defined* the maximum and minimum in terms of velocities, speaking of "increasing" or "decreasing" variables, but he *characterized* the difference in terms of *geometric* notions of concavity and convexity, illustrated by diagrams. The properties seem to flow from the underlying geometric intuition.

Using these properties, as one would expect, Maclaurin characterized the maximum, minimum, and point of inflexion of a smooth curve in terms of the values of the first and second fluxions. But he extended this method further to discover a new and general result. By a complicated geometric argument, drawing a succession of curves which we would call the graphs of $f(x)$, $f'(x)$, $f''(x)$, etc., and repeatedly using convexity and concavity, Maclaurin showed that there will be a maximum or minimum when $0, 2, 4, \ldots$ fluxions vanish. The sign of the first fluxion different from zero determines on which side of the tangent the curve lies, because the sign of this fluxion determines whether the ordinate is being accelerated or decelerated. Similarly, there will be a point of inflexion when $1, 3, 5, \ldots$ fluxions vanish.[19]

Later in his *Treatise*, Maclaurin derived the same result by using Taylor series—a method which may be more familiar to us. Yet even this series derivation was geometrically motivated. In my exposition of that derivation, I shall use Lagrange's notation for derivatives to make the mathematics easier to follow.[20] Suppose we have a maximum; now let the first derivative $f'(x) = 0$. If h is small enough, then the diagrams we have looked at, and the lemma about acceleration and retardation quoted above, make clear that $f(x+h)$ and $f(x-h)$ will exceed $f(x)$ when $f''(x)$ is positive, and will be smaller than $f(x)$ when $f''(x)$ is negative. We see this from the geometry of the situation.

Maclaurin then used Taylor's series to express the same result. Maclaurin introduced Taylor's series in this context by giving the diagram in Figure 7 as well as the series. We give the series here in Lagrange's notation: Suppose

$$f(x+h) = f(x) + hf'(x) + h^2/2!f''(x) + h^3/3!f'''(x) + \cdots$$

---

[19] Maclaurin, op. cit., pp. 226–227.

[20] Maclaurin used E for the ordinate of the curve (that is, the value of the function) at the given point. He used x for the increment we have called h, and he designated the fluxions of E by the appropriate number of dots over the E, divided by the fluxion of x to the appropriate power. The coefficient of the x^2 term, for instance, is

$$\ddot{E}/2(\dot{x})^2.$$

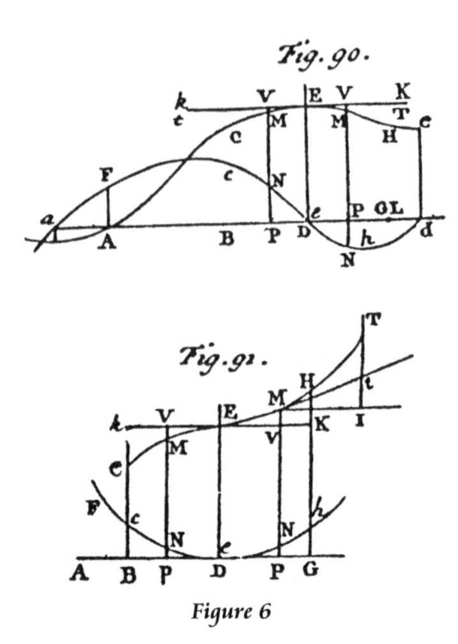

Figure 6

where, in his diagram, AF is $f(x)$, PM is $f(x+h)$, and $AP = h$. Similarly, taking Ap in the opposite direction:

$$f(x-h) = f(x) - hf'(x) + h^2/2f''(x) - h^3/3!f'''(x) + \cdots$$

Now, the geometric result that $f''(x) < 0$ at the maximum is confirmed by looking at the Taylor series; we see that, when $f'(x) = 0$ and when $AP = Ap = h$ is sufficiently small, the h^2 term (that is, the nonzero term with the smallest power of h), which here contains $f''(x)$, dominates the rest and thus $f''(x)$ must be positive for a minimum, negative for a maximum. But thus translating the geometric inequalities into series now took Maclaurin beyond what is visible from the geometry, and again gave him the general result that there is a maximum or minimum when the first n derivatives for any odd n vanish; and then, if the next derivative is positive we get a minimum, otherwise a maximum. But, if n is even and the first n derivatives vanish, we get neither a minimum nor a maximum. Thus Maclaurin was led by geometric considerations of convexity and concavity and inequalities to formulate a series result which leads once again to the complete solution of the general problem of identifying maxima, minima, and points of inflection

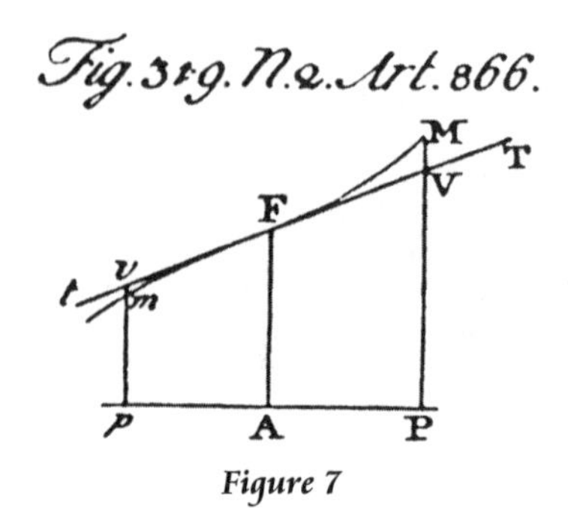

Figure 7

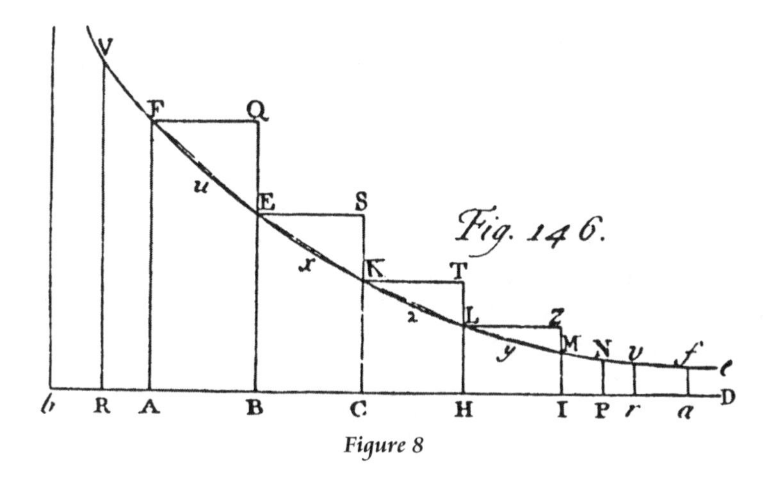

Figure 8

by computing higher-order derivatives.[21] The motivation for this set of results was far from being the calculus of polynomials, or even of functions defined by infinite power series. Maclaurin developed his theory of maxima, minima, points of inflexion, convexity and concavity, and orders of contact between curves because he wanted to study curves of all types, including those which cross over themselves, have many branches, or have cusps.[22] Thus this analytic theory had a geometric motivation as well as a geometric mode of discovery. But the analytic result—the Taylor-series version—was highly influential, being adopted to treat the same problems, without any diagrams at all, by Euler and then by Lagrange.[23]

Finding relationships between functions represented by curves and their series expansions, motivated by inequalities derived from geometric diagrams, is characteristic of Maclaurin's *Treatise of Fluxions*. For instance, consider briefly what is now called the Euler-Maclaurin series, discovered independently and almost simultaneously by Euler and by Maclaurin (who published it first).[24] Suppose we have the curve $F(x)$, and a point a. Following the modern notation used by Herman Goldstine, we have:

$$\sum_{h=0}^{\infty} F(a+h) = \int_a^{\infty} F(x)\,dx + 1/2F(a) + 1/12F'(a) - 1/720F'''(a) + 1/30240F^{(v)}(a) + \cdots$$

How did Maclaurin get it? Characteristically, he started from a geometric diagram (Figure 8).[25] The left-hand side of Goldstine's version of the Euler-Maclaurin series is the sum of the successive

[21] Maclaurin, op. cit., pp. 694–696; compare section 261.

[22] Maclaurin, op. cit., pp. 217–218. Look again at our Figure 4, above.

[23] Leonhard Euler, *Institutiones calculi differentialis*, 1755, sections 253–255. In L. Euler, *Opera Omnia*, (Leipzig, Berlin, Zurich: Teubner, 1911–), Series 1, Vol XI. For Lagrange, see below.

[24] James Stirling, congratulating Euler in a letter of 16 April, 1738, on Euler's publication (in *Comm Petrop* (22), 1738) of that formula, told him that Maclaurin had already made it public in the first part of the *Treatise of Fluxions*, printed and circulating in Britain in 1737. See Charles Tweedie, *James Stirling* (1922), p. 178, and A. P. Juškevič and R. Taton, eds., *Leonhard Euleri Commercium Epistolicum* (Basel, Birkhäuser, 1980). On this early publication, see also Maclaurin, *Fluxions*, p. iii and p. 691n. Compare Herman H. Goldstine, *A History of Numerical Analysis from the 16th through the 19th Century* (New York, Heidelberg, Berlin: Springer, 1977), pp. 84–86.

[25] For the derivation, see Maclaurin, *Fluxions*, p. 292ff; for the diagram, see Figure 146, following p. 310.

ordinates AF, BE, CK, HT, etc. Maclaurin saw his result as showing that this sum is equal to the limit approached by the area $APNF$, plus $1/2$ of AF, plus $1/12$ of the fluxion of AF, etc. Maclaurin obtained the coefficients, which are now known as Bernoulli numbers, from representing the various quantities involved by Taylor series, integrating them term by term, equating like powers, and performing various substitutions. Still, the motivation was geometric.

Maclaurin applied his result in two ways: using various series to approximate the value of the area (what we call approximating the integral); and using the area's value to find the sums of various series. For instance, he used it to sum the powers of arithmetic progressions, to derive Stirling's formula for factorials, to obtain the Newton-Cotes numerical integration formula, and to get Simpson's rule as a special case.[26] In these applications we notice again the link between analytic results and their geometric motivation.

In fact Maclaurin based his understanding of the limit of the sum of a series on a geometric analogy that goes back to the ancient Greeks: the idea that the area under a curve is a limit of rectilinear figures. On the basis of this geometric model he gave the first clear definition of the sum of infinite series:

> There are progressions... which may be continued at pleasure, and yet the sum of the terms be always less than a certain finite number. If the difference betwixt their sum and this number decrease in such a manner, that by continuing the progression it may become less than any fraction how small soever it be assigned, this number is the *limit of the sum of the progression*, and is what is understood by the value of the progression when it is supposed to be continued infinitely.[27]

The geometric motivation for this notion of the limit of the sum of a series is made clear in what immediately follows: "These limits are analogous to the limits of [geometric] figures which we have been considering, and they serve to illustrate each other mutually."[28]

Another geometrically-motivated topic for Maclaurin was elliptic integrals. Some integrals (Maclaurin used the Newtonian term "fluent" where we use "integral"), he observed, can be evaluated by being reduced to algebraic functions. Others cannot, but some of them can be reduced to finding circular arcs or logarithms. By analogy Maclaurin suggested that perhaps a class of integrals can be reduced to finding the length of an elliptic or hyperbolic arc.[29] His general approach was to use clever geometric transformations to reduce the integral that represented the length of a hyperbolic or elliptic arc to a "nice" form, and then reduce some previously intractable integral to the same form. For instance, he showed that the integral $\int dx/\sqrt{x\sqrt{(1+x^2)}}$ can be reduced to finding the length of an elliptic arc. In 1746 Jean d'Alembert used Maclaurin's work as a stepping-stone to his own analytic treatment, which in turn helped produce the influential work of Leonhard Euler and Adrien-Marie Legendre.[30] Thus Maclaurin initiated a very important type of investigation, again motivated by geometry, and foresaw its generality.

In physics Maclaurin created the subject of the gravitational attraction of ellipsoids and applied it to the problem of the shape of the earth. For instance, in 1740 he showed that an oblate

---

[26] Maclaurin, op. ct., pp. 676–693.

[27] Maclaurin, op. cit., p. 289.

[28] Ibid.

[29] Maclaurin, op. cit., p. 652ff.

[30] See Alfred Enneper, *Elliptische Functionen: Theorie und Geschichte*, 2d. ed. (Halle: Nebert, 1896), pp. 526ff.

spheroid is a possible figure of equilibrium under Newtonian mutual gravitation.[31] He devised some of the key methods used in studying the equilibrium of fluids,[32] including the methods of balancing columns and level surfaces.[33] His approach to the subject was geometric,[34] though his successors often began by putting his results in analytic form. Alexis-Claude Clairaut's classic *La Figure de la Terre* (1743) explicitly and repeatedly credited Maclaurin's work.[35] Lagrange began his own memoir on the attraction of ellipsoids by praising Maclaurin's prize paper on the subject as a masterwork of geometry, comparing the beauty and ingenuity of Maclaurin's work to that of Archimedes—before translating Maclaurin's geometric work into the language of the calculus.[36]

In every case that we have considered, Maclaurin contributed not to antiquarian geometry but to mainstream mathematics, and he did it in the same way. He started by defining a geometric problem, and used geometric insight to advance it. He also used the kinematic notion of fluxion as an idealized velocity, algorithmic tools of calculus, and nontrivial results from elsewhere in the calculus. In some cases—as when he used the Euler-Maclaurin formula to sum series—the final result is not geometric; in others (as for points of inflexion) it is. But always, he applied and advanced the calculus via his kinematic and geometric understanding of it.

Lagrange

Now, let us turn to calculus as practiced by Lagrange, beginning with the way he conceptualized the derivative. The contrast with Maclaurin is striking.

Lagrange wanted calculus to be "pure analysis," without appeals to intuitions about motion or geometry. Since he thought that algebra alone was rigorous, he believed that the way to make calculus rigorous was to banish the ideas of infinitesimals, geometry, and velocity, substituting instead "the algebraic analysis of finite quantities."[37] Lagrange also believed that expanding functions into infinite power series was an entirely algebraic process—as it appears to be, say, when one divides $1/(1-x)$ to obtain $1 + x + x^2 + \cdots$. He seems to have based this opinion on Leonhard Euler's *Introductio in analysin infinitorum* of 1748, in which Euler successfully derived a wealth of power-series expansions by what appeared to Lagrange to be algebraic means.[38]

---

[31] See Isaac Todhunter, *A History of the Mathematical Theories of Attraction and the Figure of the Earth, from the Time of Newton to That of Laplace* (London: Macmillan, 1873), p. 374; C. Truesdell, "Rational Fluid Mechanics, 1687–1765," introduction to Euler *Opera*, Ser. 2, vol. 12, p. xix.

[32] For "Maclaurin ellipsoids" in twentieth-century classical dynamics, see S. Chandrasekhar, *Ellipsoidal Figures of Equilibrium* (New Haven: Yale, 1969), pp. 77–100.

[33] Todhunter, *Attraction*, pp. 136–137.

[34] See Todhunter, *Attraction*, pp. 145, 175, 409, 474. Compare Truesdell, "Rational Fluid Mechanics," pp. ix–cxxv, especially xix.

[35] A.-C. Clairaut, *Théorie de la figure de la terre* (Paris: Duraud, 1743).

[36] J.-L. Lagrange, "Sur l'attraction des sphéroädes elliptiques," *Mémoires de l'académie de Berlin*, 1773, 121–148. Reprinted in *Oeuvres de Lagrange*, vol. III, p 619ff.

[37] The full title of Lagrange's *Théorie des fonctions analytiques* is translated as "Theory of analytic functions, detached from any consideration of infinitely small or evanescent quantities, of limits or of fluxions, and reduced to the algebraic analysis of finite quantities." Amy Dahan Dalmedico gives a clear statement of Lagrange's program of using algebra "to obtain a priori the most general and uniform procedures and demonstrations independent of any specific geometric representation" in "L'ideal analytique de Lagrange," pp. 185–187 (esp. p. 185) in her "La méthode critique du <<mathématicien-philosophe>>," in Jean Dhombres, *L'école normale de l'an III: Leçons de mathématiques*, Paris, 1992, pp. 171–192.

[38] On this point, see Grabiner, *Origins of Cauchy's Rigorous Calculus*, p. 51.

Because Lagrange thought that generating power-series expansions was purely algebraic, he thought it legitimate to define the derivative of a function as the coefficient of the linear term in the power-series expansion of the function, as follows. Note first that Lagrange's definition of function is algebraic: a function is an "expression de calcul" into which the variable enters in any way.[39] Suppose that

$$f(x+h) = f(x) + ph + qh^2 + rh^3 + \cdots,$$

where p, q, r, are all functions of x.[40] Then he defined $f'(x)$ as the coefficient of h, the function $p(x)$.[41] Analogously, he defined the second derivative as the coefficient of the linear term in the expansion of $f'(x+h)$, and so on recursively. The algebraic manipulation of power series then gave Lagrange the Taylor series

$$f(x+h) = f(x) + hf'(x) + h^2/2!f''(x) + h^3/3!f'''(x) + \cdots.$$

Lagrange invented this notation for derivatives to emphasize that the derivative $f'(x)$, like $f(x)$, is a function of x. The notation also reminds us that the function $f'(x)$ is *derived* from $f(x)$.[42] (Lagrange's term "fonction derivée" is the origin of the term derivative.)

From the Taylor series, assuming implicitly that all the derivatives are bounded, Lagrange said that one can always find h small enough so that the h^n term (for any n) dominates the rest of the series. Of this result, which he called a "fundamental principle," he declared, "it is because of this... that the calculus is the most fruitful, especially in its application to the problems of geometry and mechanics."[43] (The reader will recall that Maclaurin used this result to derive the theory of extrema from the Taylor series.) As an example of Lagrange's own application of it, let us turn to his *Leçons sur le calcul des fonctions*, where he considered the result for the case $n = 1$. In this case, letting hH representing all the terms with derivatives of higher order than 1, he obtains:

$$f(x+h) = f(x) + hf'(x) + hH,$$

where H goes to zero with h.[44] And Lagrange meant something very precise when he said that H went to zero with h:

> Given any quantity D, h can be chosen sufficiently small so that $f(x+h) - f(x)$ is included between $h(f'(x) \pm D)$.[45]

---

[39] Lagrange, *Fonctions analytiques*, in *Oeuvres* 9, p. 15.

[40] The notation is his except that I use h for the increment instead of his i.

[41] Lagrange, *Fonctions analytiques*, in Oeuvres 9, p. 32.

[42] For a full account of Lagrange's philosophy of the calculus, see Judith V. Grabiner, *The Calculus as Algebra: J.-L. Lagrange, 1736–1813* (Boston: Garland, 1990).

[43] Lagrange, *Fonctions analytiques*, Oeuvres 9, vol. 9, p. 29. Lagrange probably knew the result from Euler as well as from Maclaurin. For Euler, see *Institutiones calculi differentialis*, sections 253–254. Lagrange's knowledge of Euler's work is well known; see Grabiner, *Origins of Cauchy's Rigorous Calculus*, for example, pp. 118–120; for Lagrange's admiration for Maclaurin's work on maxima and minima, see Maria Teresa Borgato and Luigi Pepe, "Lagrange a Torino (1750–1759) e le sue lezioni inedite nelle R. Scuole di Artiglieria," *Bollettino di Storia delle Scienze Matematiche* 1987, 7: 3–180, p. 154.

[44] J.-L. Lagrange, *Leçons sur le calcul des fonctions*, new ed. (Paris: Courcier, 1806). In *Oeuvres*, vol. 10, pp. 86–87. (Compare *Fonctions analytiques*, in *Oeuvres* 9, p. 77). The h and H notations are mine, in place of his i and I, but otherwise the presentation is Lagrange's.

[45] Lagrange, *Calcul des fonctions, Oeuvres* 10, p. 87.

Because of its importance, I have called this characterization of $f'(x)$ the *Lagrange property of the derivative*.[46] If we rewrite the Lagrange property with absolute value signs and modern notation, we get, for h sufficiently small,

$$\left|[f(x+h)-f(x)]/h - f'(x)\right| < D.$$

This is exactly Cauchy's delta-epsilon characterization of the derivative, to within an alphabetical isomorphism. And Lagrange was Cauchy's source.[47]

What did Lagrange do with this property of the derivative? First, he used it to prove that a function with a positive derivative on an interval is increasing on that interval. The proof is too complicated to give here.[48] For our present purposes it will suffice to say that it is entirely algebraic and proceeds by manipulating inequalities in a delta-epsilon way (though Lagrange did not distinguish between convergence and uniform convergence and assumed that, given D, his choice of h will work for any value of x). Still, it was an influential, pioneering effort.

The key point for us about Lagrange's lemma is this: the result is obvious to anybody who visualizes curves. Only someone who wanted to make calculus purely analytic and eliminate all geometric intuition would imagine that one needs to prove that a function with a positive derivative on an interval is increasing on that interval. Thus an algebraic view of the calculus was necessary for Lagrange even to think of this as a theorem, let alone prove it.

Once Lagrange had that result, he used it together with the intermediate-value property for continuous functions (which he had also tried to prove),[49] to give an algebraic proof for the mean-value theorem for derivatives, which he gave in this form:

$$f(x)+hf'(p) < f(x+h) < f(x)+hf'(q).$$

where $f'(p)$ is the minimum, $f'(q)$ the maximum, value of the derivative $f'(x)$ on the interval he was considering. The idea of his proof is this: If the minimum of $f'(x)$ is $f'(p)$ and its maximum is $f'(q)$, then the functions

$$g'(h) = f'(x+h) - f'(p)$$

and

$$k'(h) = f'(q) - f'(x+h)$$

are both positive. Integrating with respect to the initial condition $g(0) = k(0) = 0$, he obtained the new functions

$$g(h) = f(x+h) - f(x) - hf'(p)$$

---

[46] Grabiner, *Origins of Cauchy's Rigorous Calculus*, pp. 118–121.

[47] A.-L. Cauchy, *Résumé des leçons données . . . l'école royale polytechnique sue le calcul infinitésimal* (Paris: Imprimérie royale, 1823); in *Oeuvres complètes d'Augustin Cauchy* (Paris: Gauthier-Villars, 1882–), Series 2, vol. 4, pp. 44–45. For an English translation, see Grabiner, *Origins of Cauchy's Rigorous Calculus*, pp. 168–170, and for Cauchy's debt to Lagrange's theory of the derivative, see *Origins*, chapter 5.

[48] Lagrange, *Calcul des fonctions*, in *Oeuvres* 10, p. 86ff; compare *Fonctions analytiques, Oeuvres* 9, pp. 78–80. The *Calcul des fonctions* version is described in detail in Grabiner, *Origins of Cauchy's Rigorous Calculus*, pp. 123–126.

[49] J.-L. Lagrange, *Traité de la résolution des équations numérique de tous les degrés*, 2d ed. (Paris: Courcier, 1808); in *Oeuvres*, vol. 8, pp. 19–20 and 134. This proof is not very successful logically; it uncharacteristically used the idea of motion, and was criticized on that account by Bolzano. See Grabiner, *Origins of Cauchy's Rigorous Calculus*, p. 73.

and

$$k(h) = hf'(q) - f(x+h) + f(x).$$

Since the lemma just proved shows that the functions g and k are increasing, $g(h)$ and $k(h)$ must be greater than 0, and thus

$$f(x) + hf'(p) < f(x+h) < f(x) + hf'(q).$$

The intermediate-value property for continuous functions then ensures that there is an X between p and q such that $f(x+h) = f(x) + hf'(X)$.[50] Reasoning in an analogous fashion but using higher-order derivatives, Lagrange found that

$$f(x+h) = f(x) + hf'(x) + h^2/2f''(x) + \cdots + h^n/n!f^{(n)}(X),$$

the Taylor series with Lagrange remainder.[51]

Lagrange applied his remainders to solve a variety of problems. In particular he used the Taylor series with Lagrange remainder to prove his version of the Fundamental Theorem of Calculus, that the derivative of the area under the curve is the function itself. That is, he considered the function $y = f(x)$, and the function $F(x)$ that defines the area under $y = f(x)$ up to x. Then he showed that $F'(x) = f(x)$.

Lagrange's proof begins by observing (without a diagram!) that, for a monotonic function $f(x)$ where the area under $f(x)$ up to x is called $F(x)$,

$$hf(x) < F(x+h) - F(x) < hf(x+h).^{52} \qquad (*)$$

Using his mean-value theorem and the Taylor series with second-order remainder, Lagrange obtained:

$$f(x+h) = f(x) + hf'(x+j) \quad \text{for some } j \text{ between 0 and } h$$
$$F(x+h) = F(x) + hF'(x) + h^2/2F''(x+J) \quad \text{for some } J \text{ between 0 and } h.$$

The inequality $(*)$ then yields

$$h\big[F'(x) - f(x)\big] + h^2/2F''(x+J) < h^2(f'(x+j)). \qquad (**)$$

If one chooses h sufficiently small (and he explicitly calculated how small), the inequality $(**)$ cannot be true unless the $h[F'(x) - f(x)]$ term vanishes, so one must conclude that $F'(x) = f(x)$.

Lagrange also applied the Taylor series with Lagrange remainder to obtain the results Maclaurin got about extrema, and also to treat orders of contact between curves—again, without diagrams.[53] In addition, Lagrange applied his remainders to mechanics. For instance, he

---

[50] Lagrange, *Calcul des fonctions, Oeuvres* 10, p. 91ff; compare *Fonctions analytiques, Oeuvres* 9, pp. 80–81. A modern analyst would note that Lagrange, in these pioneering arguments, did not consistently distinguish between "less than," "less than or equal to," and "bounded away from."

[51] Lagrange, *Calcul des fonctions, Oeuvres* 10, pp. 91–95; *Fonctions analytiques, Oeuvres*, vol. 9, pp. 80–85.

[52] Lagrange, *Fonctions analytiques, Oeuvres* 9, pp. 238–239. In what follows, I have for clarity used j and J where Lagrange used the same letter for both, though he made clear in the text that the two quantities are not the same.

[53] On extrema, see Lagrange, *Fonctions analytiques, Oeuvres* 9, pp. 233–237. On orders of contact, see *Fonctions analytiques, Oeuvres* 9, e.g. pp. 189, 198.

considered motion along a line such that $x = f(t)$. Between the time t and $t + \phi$, the distance traversed is

$$f(t + \phi) - f(t) = \phi f'(t) + \phi^2/2 f''(t) + \phi^3/3! f'''(t + L\phi)$$

where L is between 0 and 1.[54] Lagrange then pointed out that the distance produced by the motion can be decomposed, via the right-hand side of the equation, into terms that represent the results of different partial motions, where the first term arises from a uniform motion, the second from a uniformly accelerated motion, and the third term represents all the other motions. For ϕ sufficiently small, he said that the motion composed of the first two terms gets very close to the actual motion.[55]

One more application of series to mechanics is Lagrange's explanation of an error that Nikolaus Bernoulli claimed to have found in Newton's *Principia* (Book II, Prop. X) on motion in a resisting medium; Lagrange said in effect that Newton had not considered terms of sufficiently high order.[56] Characteristically, Lagrange began his treatment of this subject by saying, "To discover the source of the error, we are going to reduce Newton's solution to analysis."[57] Thus, even in doing something which we think of as geometry or physics, Lagrange's method is self-consciously analytic. That Lagrange saw calculus as algebra—sometimes the algebra of inequalities, sometimes formal manipulations with power series—was the key to his success.

Why the Difference?

Contrasting the approaches to the calculus of Maclaurin and Lagrange raises a general historical question. Why were Maclaurin and Lagrange so different? The source of this difference is not just one of temperament, but arose from differences in education and cultural traditions. Maclaurin's mathematical philosophy, as Erik Sageng has shown, was shaped in part by the views of Sir Francis Bacon.[58] Maclaurin was a an initiator of the Scottish Enlightenment, whose ideas are linked with those of British philosophers of the empirical—and Newtonian—traditions. For instance, among Maclaurin's contemporaries in Edinburgh was David Hume, and among his successors was Adam Smith. In mathematics itself, Maclaurin's mentor was the classical geometer Robert Simson. Maclaurin's dissertation at Glasgow was a defense of Newton's ideas on physics, and Maclaurin's later mathematics was in the Newtonian tradition. Finally, Maclaurin taught at the University of Edinburgh, which, especially in the eighteenth and early nineteenth centuries, valued what we call general education, where students pursued classics and philosophy together with the exact sciences. Mathematics there, said Professor John Leslie, was "a branch of liberal education, [not] a mechanical knack."[59]

It was different for Lagrange, who was schooled in the major works of Continental mathematics and philosophy, notably those of René Descartes and Gottfried Wilhelm Leibniz. The

---

[54] I use ϕ and L for Lagrange's theta and lambda. Lagrange, *Fonctions analytiques*, p. 341.

[55] Lagrange, op. cit., pp. 341–342.

[56] Lagrange, op. cit., pp. 365–376.

[57] Lagrange, op. cit., p. 368.

[58] Erik Lars Sageng, *Colin Maclaurin and the Foundations of the Method of Fluxions*, unpublished Ph. D. Dissertation, Princeton University, 1989, chapter II.

[59] Davie, *Democratic Intellect*, p. 108.

prevailing Continental view of the philosophy of mathematics, arising from the influence of Descartes and Leibniz, was clearly expressed by the Enlightenment *philosophe*, the Marquis de Condorcet, when he said that algebra is "the *only* really exact and analytical language in existence."[60] Lagrange taught calculus, not as part of a liberal education in an institution like the University of Edinburgh, but first at the military school in Turin, and later at the *École polytechnique* in Revolutionary Paris, where most of his students were being educated to be engineers. Lagrange found his major mathematical influence in the works of Euler and was inspired by Euler's unparalleled algorithmic power. Calculus, of course, was still one subject. But Maclaurin and Lagrange learned it from different traditions, thought about it and taught it in different settings, and came to see its basic problems through different eyes.

Influence

Histories of the calculus sometimes deplore the influence of the approaches of both Maclaurin and Lagrange. Maclaurin, they say, looked backward toward ancient Greek geometry and away from modern analysis.[61] Lagrange did harm too, by overformalizing the subject.[62] But their immediate successors thought otherwise. Eighteenth- and early nineteenth-century mathematicians found much value in Mclaurin's work, even when they did not share his geometrical approach. As we have seen, Lagrange himself praised Maclaurin's treatment of extrema, and compared Maclaurin's work on the attraction of ellipsoids to that of Archimedes (before translating it into analysis).[63] Lagrange's contemporary Silvestre-François Lacroix, in his classic textbook on the calculus, praised and used Maclaurin's work on series.[64] The German mathematician Carl G. J. Jacobi worked out the remainder for the Euler-Maclaurin formula; Jacobi called the result simply the Maclaurin summation formula and cited it directly from the *Treatise of Fluxions*.[65]

As for Lagrange, his primary influence was on Augustin-Louis Cauchy, Bernhard Bolzano, and Karl Weierstrass. Cauchy used the Lagrange property of the derivative, which for Lagrange was a byproduct of his power-series definition of derivative, as the defining property. Once he had done so, all the proofs Lagrange had based on that property became legitimate.[66] Bolzano enthusiastically adopted Lagrange's philosophy of pure analysis; in fact, that is a major point of Bolzano's famous 1817 paper which gave a "Rein analytischer Beweis" (purely analytic proof) of the intermediate-value theorem for continuous functions.[67] Also, Bolzano's *Functionenlehre*

---

[60] Condorcet, *Sketch for a Historical Picture of the Progress of the Human Mind* (1793), tr. June Barraclough, in Keith Michael Baker, ed., *Condorcet: Selected Writings* (Indianapolis: Bobbs-Merrill, 1976), p. 238. Italics added.

[61] See, e.g., Morris Kline, *Mathematical Thought from Ancient to Modern Times* (New York: Oxford, 1972), p. 429; F. Cajori, *A History of the Conceptions of Limits and Fluxions in Great Britain from Newton to Woodhouse* (Chicago and London: Open Court, 1919), p. 187.

[62] See, e.g., E. T. Bell, *The Development of Mathematics* (New York: McGraw-Hill, 1945), pp. 289–290; N. Bourbaki, *Elements d'histoire des mathématiques* (Paris: Hermann, 1960), pp. 217–218.

[63] J.-L. Lagrange, "Sur l'attraction des sphéroïdes elliptiques," *Mémoires de l'académie de Berlin*, 1773, pp. 121–148; in *Oeuvres*, vol. 3, p. 619.

[64] S.-F. Lacroix, *Traité du calcul différentiel et du calcul intégral*, 3 vols., 2d. ed. (Paris: Courcier, 1810–1819), vol. 1, p. xxvii.

[65] C. G. J. Jacobi, "De usu legitimo formulae summatoriae Maclaurinianae," *Crelles Journal* 18 (1834), 263–272.

[66] This point is documented at length in Grabiner, *Origins of Cauchy's Rigorous Calculus*, esp. chapter 5.

[67] B. Bolzano, *Rein analytischer Beweis der Lehrsatzes, dass zwischen zwey Werthen, die ein entgegengesetzes Resultat gewaehren, wenigstens eine relle Würzel liegt* (Prag, 1817).

(whose title is just a German rendering of Lagrange's *Théorie des fonctions*) bases many results on Lagrangian Taylor-series expansions.[68] Weierstrass adopted Lagrange's term "analytic functions" from the real-variable case to the complex, defined analytic functions as those with convergent power-series expansions, and then made use of some of Lagrange's techniques.

For the Newtonians, including Maclaurin, geometry played a key role in conceptualizing the calculus; Lagrangians later emphasized and exploited the algorithmic and abstracting power of algebra. But neither Maclaurin's nor Lagrange's understanding of calculus met the nineteenth-century standards of someone like Cauchy. Cauchy agreed with Lagrange that geometric intuition could not serve as a basis for proofs in analysis, but criticized Lagrange's overconfidence in what Cauchy called the "generality of algebra." This generality was used to permit jumps from results about the finite to the infinite, or from the real to the complex. Cauchy said instead that algebraic results hold only for particular values of the variables they contain and that infinite series can be used only when their convergence has been established.[69] Cauchy's was a more cautiously employed algebra, an algebra of inequalities rather than an algebra of power series, and it was constrained by the tradition of proof inherited from geometry. Thus Cauchy turned the calculus—the extensive body of results and techniques he inherited from his predecessors—from an overly algebraic subject toward the Euclidean model, where proofs rest on definitions and on explicit and intuitively reasonable assumptions. Cauchy thus synthesized the two ways of thought traced here.

Conclusion

Today we say that neither Maclaurin nor Lagrange was totally correct in his understanding of calculus. Nevertheless, each emphasized one way of thinking about it, and as a result each made distinct—and essential—contributions to the subject. The story I have told, like the Pólya experiment with which we began, makes clear that the question, "How do mathematicians think?" has many answers. As we teach, we need to be aware of the different ways our students think; we need to provide a range of examples and explanations to suit their different cognitive styles.

But the way mathematicians think is not just a matter of individual inclination. Rather, it is a product of education and culture. Further, the very multiplicity of ways that mathematicians think has great value. Partisans of individual points of view make discoveries that others could not see. It is thus worth recognizing and encouraging a wealth of approaches. We may wish to keep this in mind as we work to recruit non-traditional students into mathematical research, so that both we and they will value what their backgrounds have taught them. The story of calculus as geometry and calculus as algebra suggests that progress in mathematics is made by those who sharpen their thinking by exercising the courage of their sometimes idiosyncratic convictions.

---

[68] For Lagrange's influence on Bolzano, see Grabiner, *Origins of Cauchy's Rigorous Calculus*, for example, p. 45, 52–53, 74, 95, 192n; compare Judith V. Grabiner, "Cauchy and Bolzano: Tradition and Transformation in the History of Mathematics," in E. Mendelsohn, ed., *Transformation and Tradition in the Sciences* (Cambridge: Cambridge University Press, 1984), pp. 105–124.

[69] A.-L. Cauchy, *Cours d'analyse de l'école royale polytechnique*, 1821. In *Oeuvres*, Ser. 2, vol. 3, pp. ii–iii.

7

Was Newton's Calculus a Dead End? The Continental Influence of Maclaurin's Treatise of Fluxions*

1. Introduction

Eighteenth-century Scotland was an internationally-recognized center of knowledge, "a modern Athens in the eyes of an enlightened world." [**74**, p. 40] [**81**] The importance of science, of the city of Edinburgh, and of the universities in the Scottish Enlightenment has often been recounted. Yet a key figure, Colin Maclaurin (1698–1746), has not been highly rated. It has become a commonplace not only that Maclaurin did little to advance the calculus, but that he did much to retard mathematics in Britain—although he had (fortunately) no influence on the Continent. Standard histories have viewed Maclaurin's major mathematical work, the two-volume *Treatise of Fluxions* of 1742, as an unread monument to ancient geometry and as a roadblock to progress in analysis. Nowadays, few people read the *Treatise of Fluxions.* Much of the literature on the history of the calculus in the eighteenth and nineteenth centuries implies that few people read it in 1742 either, and that it marked the end—the dead end—of the Newtonian tradition in calculus. [**9**, p. 235], [**49**, p. 429], [**10**, p. 187], [**11**, pp. 228–9], [**43**, pp. 246–7], [**42**, p. 78], [**64**, p. 144]

But can this all be true? Could nobody on the Continent have cared to read the major work of the leading mathematician in eighteenth-century Scotland? Or, if the work was read, could it truly have been "of little use for the researcher" [**42**, p. 78] and have had "no influence on the development of mathematics"? [**64**, p. 144]

We will show that Maclaurin's *Treatise of Fluxions* did develop important ideas and techniques and that it did influence the mainstream of mathematics. The Newtonian tradition in calculus did not come to an end in Maclaurin's Britain. Instead, Maclaurin's *Treatise* served to transmit Newtonian ideas in calculus, improved and expanded, to the Continent. We will look at what these ideas were, what Maclaurin did with them, and what happened to this work afterwards. Then, we will ask what by then should be an interesting question: why has Maclaurin's role been so consistently underrated? These questions will involve general matters of history and historical

---

* Reprinted from Amer. Math. Monthly 104, 5(May 1997) 393–410.

writing as well as the development of mathematics, and will illustrate the inseparability of the external and internal approaches in understanding the history of science.

2. The Standard Picture

Let us begin by reviewing the standard story about Maclaurin and his *Treatise of Fluxions.* The calculus was invented independently by Newton and Leibniz in the late seventeenth century. Newton and Leibniz developed general concepts—differential and integral for Leibniz, fluxion and fluent for Newton—and devised notation that made it easy to use these concepts. Also, they found and proved what we now call the Fundamental Theorem of Calculus, which related the two main concepts. Last but not least, they successfully applied their ideas and techniques to a wide range of important problems. [**9**, p. 299] It was not until the nineteenth century, however, that the basic concepts were given a rigorous foundation.

In 1734 George Berkeley, later Bishop of Cloyne, attacked the logical validity of the calculus as part of his general assault on Newtonianism. [**12**, p. 213] Berkeley's criticisms of the rigor of the calculus were witty, unkind, and—with respect to the mathematical practices he was criticizing—essentially correct. [**6**, v. 4, pp. 65–102] [**38**, pp. 33–34] [**82**, pp. 332–338] Maclaurin's *Treatise* was supposedly intended to refute Berkeley by showing that Newton's calculus was rigorous because it could be reduced to the methods of Greek geometry. [**10**, pp. 181–2, 187] [**9**, pp. 233, 235] Maclaurin himself said in this preface that he began the book to answer Berkeley's attack, [**63**, p. i] and also to rebut Berkeley's accusation that mathematicians were hostile to religion. [**78**, p. 50]

The majority of Maclaurin's treatise is contained in its first Book, which is called "The Elements of the Method of Fluxions, Demonstrated after the Manner of the Ancient Geometricians." That title certainly sounds as though it looks backward to the Greeks, not forward to modern analysis. And the text is full of words—lots of words. So much time is spent on preliminaries that it is not until page 162 that he can show that the fluxion of **ay** is **a** times the fluxion of **y**. Florian Cajori, whose writings have helped spread the standard story, compared Maclaurin to the German poet Klopstock who, Cajori said, was praised by all, read by none. [**10**, p. 188] While British mathematicians, bogged down with geometric baggage, studied and revered the work and notation of Newton and argued with Berkeley over foundations, Continental mathematicians went onward and upward analytically with the calculus of Leibniz. The powerful analytic results and techniques in eighteenth-century Continental mathematics were all that mathematicians like Cauchy, Riemann, and Weierstrass needed for their nineteenth-century analysis with its even greater power, together with its improved rigor and generality. [**9**, ch. 7] [**49**, p. 948] This story became so well known that it was cited by the literary critic Matthew Arnold, who wrote, "The man of genius [Newton] was continued by . . . completely powerless and obscure followers. . . . The man of intelligence [Leibniz] was continued by successors like Bernoulli, Euler, Lagrange, and Laplace—the greatest names in modern mathematics." [**1**, p. 54; cited by [**61**, p. 15]]

Now since I myself have contributed to the standard story, especially in delineating the links among Euler, Lagrange, and Cauchy, [**38**, chs. 3–6] I have a good deal of sympathy for it, but I now think that it must be modified. Maclaurin's *Treatise of Fluxions* is an important link between the calculus of Newton and Continental analysis, and Maclaurin contributed to key developments in the mathematics of his contemporaries. Let us examine the evidence for this statement.

3. The Nature of Maclaurin's Treatise of Fluxions

Why—the standard story notwithstanding—might Maclaurin's *Treatise of Fluxions* have been able to transmit Newtonian calculus, improved and expanded, to the Continent? First, because the *Treatise of Fluxions* is not just one "Book," but two. While Book I is largely, though not entirely, geometric, Book II has a different agenda. Its title is "On the *Computations* in the Method of Fluxions." [my italics] Maclaurin began Book II by championing the power of symbolic notation in mathematics. [**63**, pp. 575–576] He explained, as Leibniz before him and Lagrange after him would agree, that the usefulness of symbolic notation arises from its generality. So, Maclaurin continued, it is important to demonstrate the rules of fluxions once again, this time from a more algebraic point of view. Maclaurin's appreciation of the algorithmic power of algebraic and calculus notation expresses a common eighteenth-century theme, one developed further by Euler and Lagrange in their pursuit of pure analysis detached from any kind of geometric intuition. To be sure, Maclaurin, unlike Euler and Lagrange, did not wish to detach the calculus from geometry. Nonetheless, Maclaurin's second Book in fact, as well as in rhetoric, has an algorithmic character, and most of its results may be read independently of their geometric underpinnings, even if Maclaurin did not so intend. (In his Preface to Book I, he even urged readers to look at Book II before the harder parts of Book I.) [**63**, p iii] The *Treatise of Fluxions*, then, was not foreign to the Continental point of view, and may have been written in part with a Continental audience in mind.

Nor was this algebraic character a secret open only to the reader of English. There was a French translation in 1749 by the Jesuit R. P. Pézénas, including an extensive table of contents. [**62**] Lagrange, among others, seems to have used this French edition (since he cited it by the French title [**58**, p. 17] though he cited other English works in English [**58**, p. 18]). Pézénas' translation, moreover, was neither isolated nor idiosyncratic, but part of the activity of a network of Jesuits interested in mathematics and mathematical physics, especially work in English, with Maclaurin one of the authors of interest to them. [**84**, pp. 33, 221, 278, 517, 655] For instance, Pézénas himself translated other English works, including those by Desaguliers, Gardiner's logarithmic tables, and Seth Ward's *Young Mathematician's Guide* [**83**, pp. 571–2] Thus there was a well-worn path connecting English-language work with interested Continental readers. Furthermore, the two-fold character of the *Treaties of Fluxions* was noted, with special praise for Book II's treatment of series, by Silvestre-François Lacroix in the historical introduction to the second edition of his highly influential three-volume calculus textbook. [**52**, p. xxvii] Unfortunately, though, recognition of the two-fold character has been absent from the literature almost completely from Lacroix's time until the recent work by Sageng and Guicciardini. [**42**] [**78**] We shall address the reasons for this neglect in due course.

4. The Social Context: The Scottish Enlightenment

Another reason for doubting the standard picture comes from the social context of Maclaurin's career. Eighteenth-century Scotland, Maclaurin's home, was anything but an intellectual backwater. It was full of first-rate thinkers who energetically pursued science and philosophy and whose work was known and respected throughout Europe. One would expect Scotland's leading mathematician to share these connections and this international renown, and he did.

Although Scotland had been deprived of its independent national government by the Act of Union of 1707, it still retained, besides its independent legal system and its prevailing religion, its own educational system. The strength and energy of Scottish higher education in Maclaurin's time is owed in large part to the Scottish ruling classes, landowners and merchants alike, who saw science, mathematics, and philosophy as keys to what they called the "improvement" of their yet underdeveloped nation. [**65**, p. 254] [**80**, pp. 7–8, 10–11] [**17**, pp. 127, 132–3] Eighteenth-century Scotland, with one-tenth the population of England, had four major universities to England's two. [**80**, p. 116] Maclaurin, when he wrote the *Treatise of Fluxions*, was Professor of Mathematics at the University of Edinburgh. Edinburgh was about to become the heart of the Scottish Enlightenment, and Maclaurin until his death in 1746 was a leading figure in that city's cultural life.

Mathematics played a major role in the Scottish university curriculum. This was in part for engineers; Scottish military engineers were highly in demand even on the Continent. [**17**, p. 125] Maclaurin himself was actively interested in the applications of mathematics, and just before his untimely death had planned to write a book on the subject. [**36**] [**68**, p. xix] In addition, mathematics and Newtonian physics were part of the course of study for prospective clergyman. [**80**, p. 20] The influential "Moderate" party in the Church of Scotland appreciated the Newtonian reconciliation of science and religion. [**16**, pp. 53, 57]

Maclaurin's position in Edinburgh's cultural life was not just that of a technically competent mathematician. For instance, he was part of the Rankenian society, which met at Ranken's Tavern in Edinburgh to discuss such things as the philosophy of Bishop Berkeley; the society introduced Berkeley's philosophy to the Scottish university curriculum. [**24**, p. 222] [**17**, p. 133] [**65**, p. 197] Maclaurin and his physician friend Alexander Monro were the founders and moving spirits of the Edinburgh Philosophical Society. [**65**, p. 198] With Newton's encouragement, Maclaurin had become the chief spokesman in Scotland for the new Newtonian physics. His posthumously published book, *An Account of Sir Isaac Newton's Philosophical Discoveries*, was based on material Maclaurin used in his classes at Edinburgh, and the book was of great interest to philosophers. [**24**, p. 137] That book became well known on the Continent. It was translated into French almost as soon as it appeared, by Louis-Anne Lavirotte in 1749, and the first part appeared in Italian in Venice in 1762.

Anoher branch of Scottish science, namely medicine, also had many links with the Continent and was highly regarded there. Medical students went back and forth between Scotland, Holland, and France. [**17**, p. 135] [**80**, p. 7]

The best-known figures of eighteenth-century Scotland had major interactions with, and influence upon, Continental science and philosophy. [**39**] [**81**] Let it suffice to mention the names of four: the philosopher David Hume, who was a student at Edinburgh in Maclaurin's time; the geologist James Hutton, who attended and admired Maclaurin's lectures; [**34**, pp. 577–8] and, a bit after Maclaurin's time but still subject to his influence on Scottish higher education, the chemist Joseph Black and the economic and political philosopher Adam Smith. Maclaurin himself had twice won prizes from the Académie des Sciences in Paris, once in 1724 for a memoir on percussion, and then in 1740 (dividing the prize with Daniel Bernoulli, P. Antoine Cavalleri, and Leonhard Euler) for a memoir on the tides. [**79**, p. 611] [**39**, pp. 400–401]

Scotland in the eighteenth century nurtured first-rate intellectual work on mathematics, philosophy, science, medicine, and engineering, and did it all as part of a general European culture. [**39**, p. 412] [**81**, passim] The *Treatise of Fluxions* was the major mathematical work of a Scottish mathematician of considerable reputation on the Continent, a major work philosophically attuned to the enormously influential Newtonian physics and the Continentally popular algebraic symbolism. Such a work would certainly be of interest to Continental thinkers. Social considerations may not suffice to determine mathematical ideas, but they certainly affect the mathematician's ability to make a living, to get research support, and to promote contact and communication with other mathematicians and scientists at home and abroad. And so it was with Maclaurin.

5. Maclaurin's Continental Reputation

An even better reason for not accepting the traditional view of Maclaurin is that his work demonstrably *was* read in the eighteenth century, and was read by the big names of Continental mathematics. He had a Continental acquaintance through travel and correspondence. Even before the *Treatise of Fluxions*, his reputation had been enhanced by his Académie prizes and by his books on geometry. He was thus a respected member of an international network of mathematicians with interests in a wide range of subjects, and the publication of the *Treatise of Fluxions* was eagerly anticipated on the Continent.

The *Treatise of Fluxions* of 1742 was Maclaurin's major work on analysis, incorporating and somewhat dwarfing what he had done earlier. It contains an exposition of the calculus, with old results explained and many new results introduced and proved. Maclaurin seems to have included almost everything he had done in analysis and its applications to Newtonian physics. In particular, the findings of his Paris prize paper on the tides were included and expanded. His other papers, the posthumous and relatively elementary *Algebra*, and his works on geometry as such—though highly regarded—do not concern us here, but his Continental reputation was enhanced by these as well.

Let us turn now to some specific evidence for the Continental reputation of Maclaurin's major work. In 1741, Euler wrote to Clairaut that, though he had not yet seen the Paris prize papers on the tides, "from Mr. Maclaurin I expect only excellent ideas." [**47**, p. 87] Euler added that he had heard from England (presumably from his correspondent James Stirling) that Maclaurin was bringing out a book on "differential calculus," and asked Clairaut to keep him posted about this. In turn, Clairaut asked Maclaurin later in 1741 about his plans for the book, [**66**, p. 348] which Clairaut wanted to see before publishing his own work on the shape of the earth. [**47**, p. 110] Euler did get the *Treatise of Fluxions*, and read enough of it quickly to praise it in a letter to Goldbach in 1743. [**48**, p. 179] Jean d'Alembert, in his *Traité de dynamique* of 1743, [**22**, sec. 37, n.] praised the rigor brought to calculus by the *Treatise of Fluxions.* D'Alembert's most recent biographer, Thomas Hankins, argues that Maclaurin's *Treatise*, appearing at this time, helped persuade d'Alembert that gravity could best be described as a continuous acceleration rather than a series of infinitesimal leaps. [**44**, p. 167] D'Alembert's general approach to the foundations of the calculus in terms of limits clearly was influenced by Newton's and Maclaurin's championing of limits over infinitesimals, in particular by Maclaurin's clear description of limits in one of the parts of his *Treatise of Fluxions* that explicitly responds to Berkeley's objections (and which incidentally may be the first explicit description of the tangent as the limit of secant lines; see Section 7). [**44**, p. 23] [**63**, pp. 422–3] Lagrange in his *Analytical Mechanics* [**55**, p. 243] said that Maclaurin, in the *Treatise of Fluxions*, was the first to treat Newton's laws of motion in the language of the calculus in a coordinate system fixed in space. Though C. Truesdell [**80**, pp. 250–3] has shown that Lagrange was wrong because Johann Bernoulli and Euler were ahead of Maclaurin on this, the fact that Lagrange believed this is one more piece of evidence for the Continental reputation of Maclaurin as mathematician and physicist.

6. Maclaurin's Mathematics and Its Importance

The previous points show that Maclaurin could have been influential, but not that he was. Five examples will reveal both the nature of Maclaurin's techniques and the scope of his influence: a special case of the Fundamental Theorem of Calculus; Maclaurin's treatment of maxima and minima for functions of one variable; the attraction of spheroids; what is now called the Euler-Maclaurin summation formula; and elliptic integrals.

a. Key Methods in the Calculus

Two methods were central to the study of real-variable calculus in the eighteenth and nineteenth centuries. One of these is studying real-valued functions by means of power-series representations. This tradition is normally thought first to flower with Euler; it is then most closely associated with Lagrange, and, later for complex variables, with Weierstrass. The second such method is that of basing the foundations of the calculus on the algebra of inequalities—what we now call delta-epsilon proof techniques—and using algebraic inequalities to prove the major results of the calculus; this tradition is most closely associated with the work of Cauchy in the 1820's. I have traced these traditions back to Lagrange and Euler in my work on the origins of Cauchy's calculus. [**38**, chs. 3–6] It is surprising, at least if one accepts the standard picture of the history of the calculus, that both of these methods—studying functions by power series, basing foundations on inequalities—were materially advanced by Maclaurin in the *Treatise of Fluxions*. It is especially striking that the importance of Maclaurin's work on series—work based, it is well to remember, on Newton's use of infinite series—was recognized and praised in 1810 by Lacroix, who also linked it with the series-based calculus of Lagrange. [**52**, p. xxxiii]

Maclaurin skillfully used algebraic inequalities in his proof of a special case of the Fundamental Theorem of Calculus. He showed, for a particular function, that if one takes the fluxion of the area under the curve whose equation is $y = f(x)$, one gets the function $f(x)$. In his proof, Maclaurin adapted the intuition underlying Newton's argument for this fact in *De Analysi* [**69**]—that the rate of change of the area under a curve is measured by the height of the curve—but Maclaurin's proof is more rigorous. Although Maclaurin's argument proceeds algebraically, the concepts involved resemble those of the Greek "method of exhaustion" (more precisely termed by Dijksterhuis "indirect passage to the limit"). [**26**, p. 130] A key step in this Greek work is first to assume that two equal areas or expressions for areas are *unequal*, and then to argue to a contradiction by using inequalities that hold among various rectilinear areas. Newton in the *Principia* had based proofs of new results about areas and curves on methods akin to those of the Greeks. Maclaurin carried this much further. It was Maclaurin's "conservative" allegiance to Archimedean *geometric* methods that led him to buttress the *kinematic* intuition of Newton's calculus with *algebraic* inequality proofs.

What Maclaurin proved in the example under discussion is that, if the area under a curve up to x is given by x^n, the ordinate of the curve must be $y = nx^{n-1}$, which is known to be the fluxion of x^n. [**63**, pp. 752–754] Maclaurin's diagram for this is much like the one Newton gave in the *De Analysi*. [**69**, pp. 3–4] Maclaurin began by saying that, since x and y increase together,

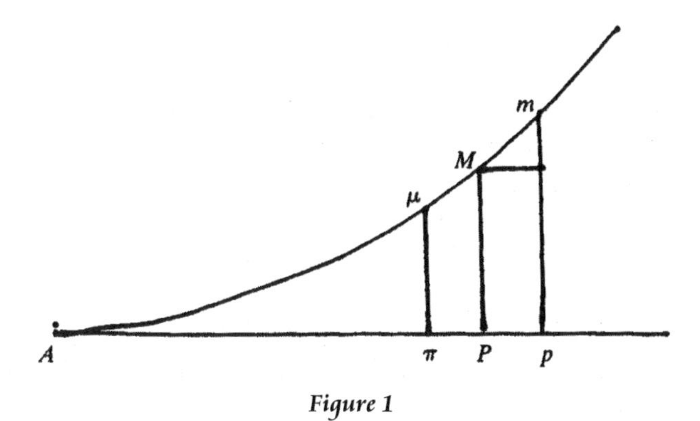

Figure 1

the following inequality holds between the areas shown:

$$x^n - (x-h)^n < yh < (x+h)^n - x^n. \tag{1}$$

(Maclaurin gave this inequality verbally; I have supplied the "<" signs; also, I use "h" for the increment where Maclaurin used "o".) Now Maclaurin recalled an algebraic identity he had proved earlier: [**63**, p. 583; inequality notation added]

$$\text{If } E < F, \text{ then } nF^{n-1}(E-F) < E^n - F^n < nE^{n-1}(E-F). \tag{2}$$

(It may strike the modern reader that, since nx^{n-1} is the derivative of x^n, this second inequality is a special case of the mean-value theorem for derivatives. I shall return to this point later.)

Now, letting $x-h$ play the role of F and x play the role of E, $E-F$ is h and the first inequality in (2) yields

$$n(x-h)^{n-1}h < x^n - (x-h)^n$$

Similarly, if $F = x$ and $E = x+h$, then $E - F = h$ and the second inequality in (2) becomes

$$(x+h)^n - x^n < n(x+h)^n h.$$

Combining these with inequality (1) about the areas, Maclaurin obtained

$$n(x-h)^{n-1}h < yh < n(x+h)^{n-1}h.$$

Dividing by h produces

$$n(x-h)^{n-1} < y < n(x+h)^{n-1}. \tag{3}$$

Recall that, given that the area was x^n, Maclaurin was seeking an expression for y, the fluxion of that area. A modern reader, having reached the inequality (3), might stop, perhaps saying "let h go to zero, so that y becomes nx^{n-1}," or perhaps justifying the conclusion by appealing to the delta-epsilon characterization of limit. What Maclaurin did instead was what Archimedes might have done, a double reductio ad absurdum. But what Archimedes might have done geometrically and verbally, Maclaurin did algebraically. He assumed first that y is *not* equal to nx^{n-1}. Then, he said, it must be equal to $nx^{n-1}+r$ for some r. First, he considered the case when this r was positive. This will lead to a contradiction if h is chosen so that $y = n(x+h)^{n-1}$, since, he observed, inequality (3) will be violated when $h = (x^{n-1}+r/n)^{1/(n-1)}$. Similarly, he calculated the h that produces a contradiction when r is assumed to be negative. Thus there can be no such r, and $y = nx^{n-1}$. [**63**, p. 753]

Maclaurin introduced this proof by saying something surprising for a *Treatise of Fluxions*: that the use of the inequalities makes the demonstration of the value of y "independent of the notion of a fluxion." [**63**, p. 752] (Of course one would need the notion of fluxion to interpret y as the fluxion of the area function x^n, but the proof itself is algebraic.) This proof was presumably part of his agenda in writing the more algebraic Book II of the *Treatise* for an audience on the Continent, where fluxions were suspect as involving the idea of motion. Later Lagrange, in seeking his purely algebraic foundation for the calculus, explicitly said he wanted to free the calculus from fluxions and what he called the "foreign idea" of motion. It is thus striking that Lagrange's *Théorie des fonctions analytiques* (1797) gives a more general version of the kind of argument Maclaurin had given, applying to *any* increasing function that satisfies the geometric inequality expressed in (1). In place of the algebraic inequality (2), Lagrange used the mean-value theorem. [**58**, pp. 238–9] [**38**, pp. 156–158] The similarity of the two arguments does not prove influence, of course, but it certainly demonstrates that Maclaurin's work, which we know Lagrange read (e.g., [**58**, p. 17]), uses the algebra of inequalities in a way consistent with that used by Lagrange and his successors.

Maclaurin's argument exemplifies the way his *Treatise* reconciles the old and the new. The double *reductio ad absurdum* reflects his Archimedean agenda. Treating the area as generated by a moving vertical line, and then searching for the relationship between the area and its fluxion, are Newtonian. Maclaurin did not have a general proof of the Fundamental Theorem in this argument, but relied on an inequality based on the specific properties of a specific function. Nonetheless, he had the precise bounding inequalities for the area function used later by Lagrange, and he used an algebraic inequality proof in a manner that would not disgrace a nineteenth-century analyst.

Inequality-based arguments in the calculus as used by Lagrange and Cauchy owe a lot to the eighteenth-century study of algebraic approximations, and it once seemed to me that this was their origin. But the algebra of inequalities as used in Continental analysis, especially in d'Alembert's pioneering treatment of the tangent as the limit of secants in the article "Différentiel" in the *Encyclopédie*, [**19**] must owe something also to Maclaurin's translation of Archimedean geometry into algebraic dress to justify results in calculus. Throughout the eighteenth century, practitioners of the limit tradition on the Continent use inequalities; a clear line of influence connects Maclaurin's admirer d'Alembert, Simon L'Huilier (who was a foreign member of the Royal Society), the textbook treatment of limits by Lacroix, and, finally, Cauchy. [**38**, pp. 80–87]

Now let us turn to some of Maclaurin's work on series. There is, of course, the Maclaurin series, that is, the Taylor series expanded around zero. This result Maclaurin himself credited to Taylor, and it was known earlier to Newton and Gregory. It was called the Maclaurin series by John F. W. Herschel, Charles Babbage, and George Peacock in 1816 [**51**, pp. 620–21] and by Cauchy in 1823. [**14**, p. 257] Since it was obvious that Maclaurin had not invented it, the attribution shows appreciation by these later mathematicians for the way Maclaurin used the series to study functions. A key application is Maclaurin's characterization of maxima, minima, and points of inflection of an infinitely differentiable function by means of its successive derivatives. When the first derivative at a point is zero, there is a maximum if the second derivative is negative there, a minimum if it is positive. If the second derivative is also zero, one looks at higher derivatives to tell whether the point is a maximum, minimum, or point of inflection. These results can be proved by looking at the Taylor series of the function near the point in question, and arguing on the basis of the inequalities expressed in the definition of maximum and minimum. For instance (in modern [Lagrangian] notation), if $f(x)$ is a maximum, then

$$f(x) > f(x+h) = f(x) + hf'(x) + h^2/2!f''(x) + \cdots, \text{ and} \tag{4}$$
$$f(x) > f(x-h) = f(x) - hf'(x) + h^2/2!f''(x) - \cdots$$

if h is small. If the derivatives are bounded, and if h is taken sufficiently small so that the term in h dominates the rest, the inequalities (4) can both hold only if $f'(x) = 0$. If $f'(x) = 0$, then the h^2 term dominates, and the inequalities (4) hold only if $f''(x)$ is negative. And so on.

I have traced Cauchy's use of this technique back to Lagrange, and from Lagrange back to Euler. [**38**, pp. 117–118] [**37**, pp. 157–159] [**58**, pp. 235–6] [**29**, Secs. 253–254] But this technique is explicitly worked out in Maclaurin's *Treatise of Fluxions.* Indeed, it appears twice: once in geometric dress in Book I, Chapter IX, and then more algebraically in Book II. [**63**, pp. 694–696] Euler, in the version he gave in his 1755 textbook, [**20**] does not refer to Maclaurin on this point, but then he makes few references in that book at all. Still we might suspect, especially knowing that Stirling told Euler in a letter of 16 April 1738 [**91**] that Maclaurin had some interesting results on series, that Euler would have been particularly interested in looking at Maclaurin's applications of the Taylor series. Certainly Lacroix's praise for Maclaurin's work on series must have taken this set of results into account. [**52**, p. xxvii] Even more important, Lagrange, in unpublished lectures on the calculus from Turin in the 1750's, after giving a very elementary treatment of maxima and minima, referred to volume II of Maclaurin's *Treatise of Fluxions* as the chief source for more information on the subject. [7, p. 154] Since Lagrange did not mention Euler in this connection at all, Lagrange could well have not even have seen the *Institutiones calculi differentialis* of 1755 when he made this reference. This Taylor-series approach to maxima and minima (with the Lagrange remainder supplied for the Taylor series) plays a major role in the work of Lagrange, and later in the work of Cauchy. It is because Maclaurin thought of maxima and minima, and of convexity and concavity, in Archimedean geometrical terms that he was led to look at the relevant inequalities, just as the geometry of Archimedes helped Maclaurin formulate some of the inequalities he used to prove his special case of the Fundamental Theorem of Calculus.

6. Ellipsoids

We now turn to work in applied mathematics that constitutes one of Maclaurin's great claims to fame: the gravitational attraction of ellipsoids and the related problem of the shape of the earth. Maclaurin is still often regarded as the creator of the subject of attraction of ellipsoids. [**85**, pp. 175, 374] In the eighteenth century, the topic attracted serious work from d'Alembert, A.-C. Clairaut, Euler, Laplace, Lagrange, Legendre, Poisson, and Gauss. In the twentieth century, Subramanyan Chandrasekhar (later Nobel laureate in physics) devoted an entire chapter of his classic *Ellipsoidal Figures of Equilibrium* to the study of Maclaurin spheroids (figures that arise when homogeneous bodies rotate with uniform angular velocity), the conditions of stability of these spheroids and their harmonic modes of oscillation, and their status as limiting cases of more general figures of equilibrium. Such spheroids are part of the modern study of classical dynamics in the work of scientists like Chandrasekhar, Laurence Rossner, Carl Rosenkilde, and Norman Lebovitz. [**15**, pp. 77–100] Already in 1740 Maclaurin had given a "rigorously exact, geometrical theory" of homogeneous ellipsoids subject to inverse-square gravitational forces, and had shown that an oblate spheroid is a possible figure of equilibrium under Newtonian mutual gravitation, a result with obvious relevance for the shape of the earth. [**39**, p. 172] [**86**, p. xix] [**85**, p. 374]

Of particular importance was Maclaurin's decisive influence on Clairaut. Maclaurin and Clairaut corresponded extensively, and Clairaut's seminal 1743 book *La Figure de la Terre* [**18**] frequently, explicitly, and substantively cites his debts to Maclaurin's work. [**39**, pp. 590–597] A key result, that the attractions of two confocal ellipsoids at a point external to both are proportional to their masses and are in the same direction, was attributed to Maclaurin by d'Alembert, an attribution repeated by Laplace, Lagrange, and Legendre, then by Gauss, who went back to Maclaurin's original paper, and finally by Lord Kelvin, who called it "Maclaurin's splendid theorem." [**15**, p. 38] [**85**, pp. 145, 409] Lagrange began his own memoir on the attraction of ellipsoids by praising Maclaurin's treatment in the prize paper of 1740 as a masterwork of geometry, comparing the beauty and ingenuity of Maclaurin's work with that of Archimedes, [**57**, p. 619] though Lagrange, typically, then treated the problem analytically. Maclaurin's eighteenth- and nineteenth-century successors also credit him with some of the key methods used in studying the equilibrium of fluids, such as the method of balancing columns. [**39**, p. 597] Maclaurin's work on the attraction of ellipsoids shows how his geometric insights fruitfully influenced a subject that later became an analytic one.

c. The Euler-Maclaurin Formula

The Euler-Maclaurin formula expresses the value of definite integrals by means of infinite series whose coefficients involve what are now called the Bernoulli numbers. The formula shows how to use integrals to find the partial sums of series. Maclaurin's version, in modern notation, is:

$$\sum_{h=0}^{\infty} F(a+h) = \int_0^a F(x)\,dx + 1/2F(a) + 1/2F'(a)$$
$$-1/720F'''(a) + 1/30240F^{(v)}(a) - \cdots$$

[**35**, pp. 84–86]

James Stirling in 1738, congratulating Euler on his publication of that formula, told Euler that Maclaurin had already made it public in the first part of the *Treatise of Fluxions*, which was printed and circulating in Great Britain in 1737. [**47**, p. 88n] [**91**, p. 178] (On this early publication, see also [**63**, pp. iii, 691n]). P. L. Griffiths has argued that this simultaneous discovery rests on De Moivre's work on summing reciprocals, which also involves the so-called Bernoulli numbers. [**40**] [**41**, pp. 16–17] [**25**, p. 19] In any case, Euler and Maclaurin derived the Euler-Maclaurin formula in essentially the same way, from a similar geometric diagram and then by integrating various Taylor series and performing appropriate substitutions to find the coefficients. [**31**] [**32**] [**33**] Maclaurin's approach is no more Archimedean or geometric than Euler's; they are similar and independent. [**63**, pp. 289–293, 672–675] [**35**, pp. 84–93] [**67**] In subsequent work, Euler went on to extend and apply the formula further to many other series, especially in his *Introductio in analysin infinitorum* of 1748 and *Institutiones calculi differentialis* of 1755. [**35**, p. 127] But Maclaurin, like Euler, had applied the formula to solve many problems. [**63**, pp. 676–693] For instance, Maclaurin used it to sum powers of arithmetic progressions and to derive Stirling's formula for factorials. He also derived what is now called the Newton-Cotes numerical integration formula, and obtained what is now called Simpson's rule as a special case. It is possible that his work helped stimulate Euler's later, fuller investigations of these important ideas.

In 1772, Lagrange generalized the Euler-Maclaurin formula, which he obtained as a consequence of his new calculus of operators. [**53**] [**35**, pp. 169, 261] In 1834, Jacobi provided the formula with its remainder term, [**46**, pp. 263, 265] in the same paper in which he first introduced what are now called the Bernoulli polynomials. Jacobi, who called the result simply the Maclaurin summation formula, cited it directly from the *Treatise of Fluxions.* [**46**, p. 263] Later, Karl Pearson used the formula as an important tool in his statistical work, especially in analyzing frequency curves. [**72**, pp. 217, 262]

The Euler-Maclaurin formula, then, is an important result in the mainstream of mathematics, with many applications, for which Maclaurin, both in the eighteenth century and later on, has rightly shared the credit.

d. Elliptic Integrals

Some integrals (Maclaurin used the Newtonian term "fluents"), are algebraic functions, Maclaurin observed. Others are not, but some of these can be reduced to finding circular arcs, others to finding logarithms. By analogy, Maclaurin suggested, perhaps a large class of integrals could be studied by being reduced to finding the length of an elliptical or hyperbolic arc. [**63**, p. 652] By means of clever geometric transformations, Maclaurin was able to reduce the integral that represented the length of a hyperbolic arc to a 'nice' form. Then, by algebraic manipulation, he could reduce some previously intractable integrals to that same form. His work was translated into analysis by d'Alembert and then generalized by Euler. [**13**, p. 846] [**23**] [**27**, p. 526] [**28**, p. 258] In 1764, Euler found a much more elegant, general, and analytic version of this approach, and worked out many more examples, but cited the work of Maclaurin and d'Alembert as the source of his investigation. A.-M. Legendre, the key figure in the eighteenth-century history of elliptic integrals, credited Euler with seeing that, by the aid of a good notation, arcs of ellipses and other transcendental curves could be as generally used in integration as circular and logarithmic arcs. [**45**, p. 139] Legendre was, of course, right that "elliptic integrals" encompass a wide range of examples; this was exactly Maclaurin's point. Thus, although his successors accomplished more, Maclaurin helped initiate a very important investigation and was the first to appreciate its generality. Maclaurin's geometric insight, applied to a problem in analysis, again brought him to a discovery.

7. Other Examples of Maclaurin's Mathematical Influence

The foregoing examples provide evidence of direct influence of the *Treatise of Fluxions* on Continental mathematics. There is much more. For instance, Lacroix, in his treatment of integrals by the method of partial fractions, called it "the method of Maclaurin, followed by Euler." [**52**, Vol. II, p. 10] [**63**, pp. 634–644] Of interest too is Maclaurin's clear understanding of the use of limits in founding the calculus, especially in the light of his likely influence on d'Alembert's treatment of the foundations of the calculus by means of limits in the *Encyclopédie*, which in turn influenced the subsequent use of limits by L'Huilier, Lacroix, and Cauchy, [**38**, chapter 3] (and on Lagrange's acceptance of the limit approach in his early work in the 1750's). [**7**] Although the largest part of Maclaurin's reply to Berkeley was the extensive proof of results in calculus using

Greek methods, he was willing to explain important concepts using limits also. In particular, Maclaurin wrote, "As the tangent of an arch [arc] is the right line that limits the position of all the secants that can pass through the point of contact . . . though strictly speaking it be no secant; so a ratio may limit the variable ratios of the increments, though it cannot be said to be the ratio of any real increments." [**63**, p. 423] Maclaurin's statement answers Berkeley's chief objection—that the increment in a function's value is first treated as non-zero, then as zero, when one calculates the limit of the ratio of increments or finds the tangent to a curve. Maclaurin's statement is in the tradition of Newton's *Principia* (Book I, Scholium to Lemma XI), but is in a form much closer to the later work of d'Alembert on secants and tangents. [**20**] Maclaurin pointed out that most of the propositions of the calculus that he could prove by means of geometry "may be *briefly* demonstrated by this method [of limits]." [**63**, p. 87, my italics]

In addition, Maclaurin had considerable influence in Britain, on mathematicians like John Landen (whose work on series was praised by Lagrange), Robert Woodhouse (who sparked the new British interest in Continental work about 1800), and on Edward Waring and Thomas Simpson, whose names are attached to results well known today. [**42**] Going beyond the calculus, Maclaurin's purely geometric treatises were read and used by French geometers of the stature of Chasles and Poncelet. [**90**, p. 145] Thus, though Maclaurin may not have been the towering figure Euler was, he was clearly a significant and respected mathematician, and the *Treatise of Fluxions* was far more than an unread tome whose weight served solely to crush Bishop Berkeley.

8. Why a Treatise of Fluxions?

The *Treatise of Fluxions* was not really intended as a reply to Berkeley. Maclaurin could have refuted Berkeley with a pamphlet. It was not a student handbook either; this work is far from elementary. Nor was it merely written to glory in Greek geometry. Maclaurin wrote several works on geometry per se. But he was no antiquarian. Instead, the *Treatise of Fluxions* was the major outlet for Maclaurin's solution of significant research problems in the field we now call analysis. Geometry, as the examples I gave illustrate, was for Maclaurin a source of motivation, of insight, and of problem-solving power, as well as being his model of rigor.

For Maclaurin, rigor was not an end in itself, or a goal pursued for purely philosophical reasons. It was motivated by his research goals in analysis. For instance, Maclaurin developed his theory of maxima, minima, points of inflection, convexity and concavity, orders of contact, etc., because he wanted to study curves of all types, including those that cross over themselves, loop around and are tangent to themselves, and so on. He needed a sophisticated theory to characterize the special points of such curves. Again, in problems as different as studying the attraction of ellipsoids and evaluating integrals approximately, he needed to use infinite series and know how close he was to their sum. Thus, rigor, to Maclaurin, was not merely a tool to defend Newton's calculus against Berkeley—though it was that—nor just a response to the needs of a professor to present his students a finished subject—though it may have been that as well. In many examples, Maclaurin's rigor serves the needs of his research.

Moreover, the *Treatise of Fluxions* contains a wealth of applications of fluxions, from standard physical problems such as curves of quickest descent to mathematical problems like the summation of power series—in the context of which, incidentally, Maclaurin gave what may be the earliest clear definition of the sum of an infinite series: "There are progressions of fractions which may be continued at pleasure, and yet the sum of the terms be always less than a certain finite number. If the difference betwixt their sum and this number decrease in such a manner, that by continuing the progression it may become less than any fraction how small soever that can be assigned, this number is the *limit of the sum of the progression*, and is what is understood by the value of the progression when it is supposed to be continued indefinitely." [**63**, p. 289] Thus, though eighteenth-century Continental mathematicians did not care passionately about foundations, [**38**, pp. 18–24] they could still appreciate the *Treatise of Fluxions* because they could mine it for results and techniques.

9. Why the Traditional View?

If the reader is convinced by now that the traditional view is wrong, that Maclaurin's Treatise did not mark the end of the Newtonian tradition, and that not all of modern analysis stems solely from the work of Leibniz and his school, the question arises, how did that traditional view come to be, and why it has been so persistent?

Perhaps the traditional view could be explained as follows. Consider the approach to mathematics associated with Descartes: symbolic power, not debates over foundations; problem-solving power, not axioms or long proofs. The Cartesian approach to mathematics is clearly reflected in the work and in the rhetoric of Leibniz, Johann Bernoulli, Euler, Lagrange—especially in the historical prefaces to his influential works—and even Cauchy. These men, the giants of their time, are linked in a continuous chain of teachers, close colleagues, and students. Some topics, like partial differential equations and the calculus of variations, were developed mostly on the Continent. Moreover, the Newton-Leibniz controversy helped drive English and Continental mathematicians apart. Thus the Continental tradition can be viewed as self-contained, and the outsider sees no need for eighteenth-century Continental mathematicians to struggle through 750 pages of a *Treatise of Fluxions*, which is at best in the Newtonian notation and at worst in the language of Greek geometry. Lagrange's well-known boast that his *Analytical Mechanics* [**55**] had (and needed) no diagrams, thus opposing analysis to geometry at the latter's expense, reinforced these tendencies and enshrined them in historical discourse. But the explanation we have just given does not suffice to explain the strength, and persistence into the twentieth century, of the standard interpretation. The traditional view of Maclaurin's lack of importance has been reinforced by some other historiographical tendencies that deserve our critical attention.

The traditional picture of Maclaurin's *Treatise of Fluxions* radically separates his work on foundations, which it regards as geometric, sterile, and antiquarian, from his important individual results, which often are mentioned in histories of mathematics but are treated in isolation from the purpose of the *Treatise*, in isolation from one another, and in isolation from Maclaurin's overall approach to mathematics. Strangely, both externalist and internalist historians, each for different reasons, have reinforced this picture.

For instance, in the English-speaking world, viewing the *Treatise* as only about Maclaurin's foundation for the calculus, and thus as a dead end, has been perpetuated by the "decline of science in England" school of the history of eighteenth-century science, stemming from such early nineteenth-century figures as John Playfair, and, especially, Charles Babbage. [**77**] [**2**] [**4**] Babbage felt strongly about this because he was a founder of the Cambridge Analytical Society, which fought to introduce Continental analysis into Cambridge in the early nineteenth century. This group had an incentive to exaggerate the superiority of Continental mathematics and downgrade the British, as is exemplified by their oft-quoted remark that the principles of "pure d-ism" should replace what they called the "dot-age" of the University. [**5**, ch. 7] [**10**, p. 274] The pun, playing on the Leibnizian and Newtonian notation in calculus, may be found in [**2**, p. 26]. These views continued to be used in the attempt by Babbage and others to reform the Royal Society and to increase public support for British science.

It is both amusing and symptomatic of the misunderstanding of Maclaurin's influence that Lacroix's one-volume treatise on the calculus of 1802, [**50**] translated into English by the Cambridge Analytical Society with added notes on the method of series of Lagrange, [**51**] was treated by them, and has been considered since, as a purely "Continental" work. But Lacroix's short treatise was based on the concept of limit, which was Newtonian, elaborated by Maclaurin, adapted by d'Alembert and L'Huilier, and finally systematized by Lacroix. [**38**, pp. 81–86] Moreover, the translators' notes by Babbage, Herschel, and Peacock supplement the text by studying functions by their Taylor series, thus using the approach that Lacroix himself, in his multi-volume treatise of 1810, had attributed to Maclaurin. This is, of course, not to deny the overwhelming importance of the contributions of Euler and Lagrange, both to the mathematics taught by the Analytical Society and to that included by Lacroix in his 1802 book, nor to deny the Analytical Society's emphasis on a more abstract and formal concept of function. But all the same, Babbage, Herschel, and Peacock were teaching some of Maclaurin's ideas without realizing this.

In any case, the views expressed by Babbage and others have strongly influenced Cambridge-oriented writers like W. W. Rouse Ball, who said that the history of eighteenth-century English mathematics "leads nowhere." [**5**, p. 98] H. W. Turnbull, though he wrote sympathetically about Maclaurin's mathematics on one occasion, [**88**] blamed Maclaurin on another occasion for the decline: "When Maclaurin produced a great geometrical work on fluxions, the scale was so heavily loaded that it diverted England from Continental habits of thought. During the remainder of the century, British mathematics were relatively undistinguished." [**89**, p. 115]

Historians of Scottish thought, working from their central concerns, have also unintentionally contributed to the standard picture. George Elder Davie, arguing from social context to a judgment of Maclaurin's mathematics, held that the Scots, unlike the English, had an anti-specialist intellectual tradition, based in philosophy, and emphasizing "cultural and liberal values." Wishing to place Maclaurin in this context, Davie stressed what he called Maclaurin's "mathematical Hellenism," [**24**, p. 112] and was thus led to circumscribe the achievement of the *Treatise of Fluxions* as having based the calculus "on the Euclidean foundations provided by [Robert] Simson," [**24**, p. 111] who had made the study of the writings of the classical Greek geometers the "national norm" in Scotland. The "Maclaurin is a geometer" interpretation among Scottish historians has been further reinforced by a debate in 1838 over who would fill the Edinburgh chair in mathematics. Phillip Kelland, a candidate from Cambridge, was seen as the champion of Continental analysis, while the partisans of Duncan Gregory argued for a more geometrical approach. Wishing to enlist the entire Scottish geometric tradition on the side of Gregory, Sir William Hamilton wrote, "The great Scottish mathematicians, . . . even Maclaurin, were decidedly averse from the application of the mechanical procedures of algebra." [**24**, p. 155] Though Kelland eventually won the chair, the dispute helped spread the view that Maclaurin had been hostile to analysis. More recently, Richard Olson has characterized Scottish mathematics after Maclaurin as having been conditioned by Scottish common-sense philosophy to be geometric in the extreme. [**70**, pp. 4, 15] [**71**, p. 29] But in emphasizing Maclaurin's influence on this development, Olson, like Davie, has overstated the degree to which Maclaurin's approach was geometric.

By contrast, consider internalist historians. The treatment of Maclaurin's results as isolated reflects what Herbert Butterfield called the Whig approach to history, viewing the development of eighteenth-century mathematics as a linear progression toward what we value today, the collection of results and techniques which make up classical analysis. Thus, mathematicians writing about the history of this period, from Moritz Cantor in the nineteenth century to Hermann Goldstine and Morris Kline in the twentieth, tell us what Maclaurin did with specific results, some named after him, for which they have mined the *Treatise of Fluxions.* [**13**, pp. 655–63] [**35**, pp. 126ff, 167–8] [**49**, pp. 522–3, 452, 442] They either neglect the apparently fruitless work on foundations, or, viewing it as geometric, see it as a step backward. It is of course true that many Continental mathematicians used Maclaurin's results without accepting the geometrical and Newtonian insights that Maclaurin used to produce them. But without those points of view, Maclaurin would not have produced those results.

Both externalist and internalist historians, then, have treated Maclaurin's work in the same way: as a throwback to the Greeks, with a few good results that happen to be in there somewhat like currants in a scone. Further, the fact that Maclaurin's book, especially its first hundred pages, is very hard to read, especially for readers schooled in modern analysis, has encouraged historians who focus on foundations to read only the introductory parts. The fact that there is so much material has encouraged those interested in results to look only at the sections of interest to them. And the fact that the first volume is so overwhelmingly geometric serves to reinforce the traditional picture once again whenever anybody opens the *Treatise.* The recent Ph.D. dissertation by Erik Sageng [**78**] is the first example of a modern scholarly study of Maclaurin's *Treatise* in any depth. The standard picture has not yet been seriously challenged in print.

10. Some Final Reflections

Maclaurin's work had Continental influence, but with an important exception—his geometric foundation for the calculus. Mastering this is a major effort, and I know of no evidence that any eighteenth-century Continental mathematician actually did so. Lagrange perhaps came the closest. In the introduction to his *Théorie des fonctions analytiques*, Lagrange could say only, Maclaurin did a good job basing calculus on Greek geometry, so it can be done, but it is very hard. [**58**, p. 17] In an unpublished draft of this introduction, Lagrange said more pointedly: "I appeal to the evidence of all those with the courage to read the learned treatise of Maclaurin and with enough knowledge to understand it: have they, finally, had their doubts cleared up and their spirit satisfied?" [**73**, p. 30]

Something else may have blunted people's views of the mathematical quality of Maclaurin's *Treatise.* The way the book is constructed partly reflects the Scottish intellectual milieu. The Enlightenment in Britain, compared with that on the Continent, was marked less by violent contrast and breaks with the past than by a spirit of bridging and evolution. [**75**, pp. 7–8, 15] Similarly, Scottish reformers operated less by revolution than by the refurbishment of existing institutions. [**16**, p. 8] These trends are consistent with the two-fold character of the *Treatise of Fluxions*: a synthesis of the old and the new, of geometry and algebra, of foundations and of new results, a refurbishment of Newtonian fluxions to deal with more modern problems. This contrasts with the explicitly revolutionary philosophy of mathematics of Descartes and Leibniz, and thus with the spirit of the *mathématicien* of the eighteenth century on the Continent.

Of course Scotland was not unmarked by the conflicts of the century. During the Jacobite rebellion in 1745, Maclaurin took a major role in fortifying Edinburgh against the forces of Bonnie Prince Charlie. When the city was surrendered to the rebels, Maclaurin fled to York. Before his return, he became ill, and apparently never really recovered. He briefly resumed teaching, but died in 1746 at the relatively young age of forty-eight. Nonetheless, the Newtonian tradition in the calculus was not a dead end. Maclaurin in his lifetime, and his *Treatise of Fluxions* throughout the century, transmitted an expanded and improved Newtonian calculus to Continental analysts. And Maclaurin's geometric insight helped him advance analytic subjects.

We conclude with the words of an eighteenth-century Continental mathematician whose achievements owe much to Maclaurin's work. [**39**, pp. 172, 412–425, 590–597] The quotation [**66**, p. 350] illustrates Maclaurin's role in transmitting the Newtonian tradition to the Continent, the respect in which he was held, and the eighteenth-century social context essential to understanding the fate of his work. In 1741, Alexis-Claude Clairaut wrote to Colin Maclaurin, "If Edinburgh is, as you say, one of the farthest corners of the world, you are bringing it closer by the number of beautiful discoveries you have made."

Acknowledgment

I thank the Department of History and Philosophy of Science of the University of Leeds, England, for its hospitality while I was doing much of this research, and the Mathematics Department of the University of Edinburgh, where I finished it. I also thank Professor G. N. Cantor for material as well as intellectual assistance, and Professors J. R. R. Christie and M. J. S. Hodge for stimulating and valuable conversations.

References

1. Arnold, Matthew, The Literary Importance of Academies, in Matthew Arnold, *Essays in Criticism*, Macmillan, London, 1865, 42–79. Cited in [**61**].
2. Babbage, Charles, *Passages from the Life of a Philosopher*, Longman, London, 1864.
3. [Babbage, Charles], "Preface" to *Memoirs of the Analytical Society* Cambridge, J. Smith, 1813. Attributed to Babbage by Anthony Hyman, *Charles Babbage: Pioneer of the Computer*, Princeton University Press, Princeton, 1982.
4. Babbage, Charles, *Reflections on the Decline of Science in England, and Some of its Causes*, B. Fellowes, London, 1830.
5. Ball, W. W. Rouse, *A History of the Study of Mathematics at Cambridge*, Cambridge University Press, Cambridge, 1889.
6. Berkeley, George, *The Analyst, or a Discourse Addressed to an Infidel Mathematician*, in A. A. Luce and T. R. Jessop, eds., *The Works of George Berkeley*, vol. 4, T. Nelson, London, 1951, 65–102.
7. Borgato, Maria Teresa, and Luigi Pepe, Lagrange a Torino (1750–1759) e le sue lezioni inedite nelle R. Scuole di Artiglieria, *Bollettino di Storia delle Scienze Matematiche*, 1987, 7: 3–180.
8. Bourbaki, Nicolas, *Elements d'histoire des mathématiques*, Paris Hermann, Paris, 1960.
9. Boyer, Carl, *The History of the Calculus and Its Conceptual Development*, Dover, New York, 1959.
10. Cajori, Florian, *A History of the Conceptions of Limits and Fluxions in Great Britain from Newton to Woodhouse*, Open Court, Chicago and London, 1919.
11. Cajori, Florian, *A History of Mathematics*, 2d. ed., Macmillan, New York, 1922.

12. Cantor, G. N., Anti-Newton, in J. Fauvel et al., eds., *Let Newton Be!*, Oxford University Press, Oxford, 1988, pp. 203–222.

13. Cantor, Moritz, *Vorlesüngen über Geschichte der Mathematik*, vol. 3, Teubner, Leipzig, 1898.

14. Cauchy, A.-L., *Résumé des leçons données à l'école royale polytechnique sur le calcul infinitésimal*, in *Oeuvres complètes*, Ser. 2, vol. 4, Gauthier-Villars, Paris, 1899.

15. Chandrasekhar, S., *Ellipsoidal Figures of Equilibrium*, Yale, New Haven, 1969.

16. Chitnis, Anand, *The Scottish Enlightenment: A Social History*, Croom Helm, London, 1976.

17. Christie, John R. R., The Origins and Development of the Scottish Scientific Community, 1680–1760, *History of Science* 12 (1974), 122–141.

18. Clairaut, A.-C., *Théorie de la figure de la terre*, Duraud, Paris, 1743.

19. d'Alembert, Jean, "Différentiel," in [**21**].

20. d'Alembert, Jean, and de la Chapelle, "Limite," in [**21**].

21. d'Alembert, Jean, et al., eds., *Dictionnaire encyclopédique des mathématiques*, Hotel de Thou, Paris, 1789, which collects the mathematical articles from the Diderot-d'Alembert *Encyclopédie.*

22. d'Alembert, Jean, *Traité de dynamique,* David l'Aîné, Paris, 1743.

23. d'Alembert, Jean, Récherches sur le calcul intégral, *Histoire de l'Académie de Berlin* (1746), 182–224.

24. Davie, George Elder, *The Democratic Intellect: Scotland and her Universities in the Nineteenth Century*, The University Press, Edinburgh, 1966.

25. De Moivre, Abraham, *Miscellanea analytica*, Tonson and Watts, London, 1730.

26. Dijksterhuis, E. J., *Archimedes*, 2d. ed., Tr. C. Dikshoorn, Princeton University Press, Princeton, 1987.

27. Enneper, Alfred, *Elliptische Functionen: Theorie und Geschichte*, 2d. ed., Nebert, Halle, 1896.

28. Euler, Leonhard, De reductione formularum integralium ad rectificationem ellipsis ac hyperbolae, *Nov. Comm. Petrop.* 10 (1764), 30–50, in L. Euler, *Opera Omnia*, Teubner, Leipzig, Berlin, Zurich, 1911–, Ser. I, vol. 20, 256–301.

29. Euler, Leonhard, *Institutiones calculi differentialis*, 1755, sections 253–255. In *Opera*, Ser. I, Vol. XI.

30. Euler, Leonhard, *Introductio in analysin infinitorum*, Lausanne, 1748, in *Opera*, Ser. I, vols. 8–9.

31. Euler, Leonhard, Inventio summae cuiusque seriei ex data termino generali, *Comm. Petrop.* 8 (1741), 9–22; in *Opera,* Ser. I, vol. 14, 108–123.

32. Euler, Leonhard, Methodus generalis summandi progressiones, *Comentarii Acad. Imper. Petrop.* 6 (1738), 68–97; in *Opera*, Ser. II, vol. 22.

33. Euler, Leonhard, Methodus universalis serierum convergentium summas quam proxime inveniendi, *Comm. Petrop.* 8 (1741), 3–9, in *Opera*, Ser. I, vol. 14, 101–107.

34. Eyles, V. A., Hutton, *Dictionary of Scientific Biography*, vol. 6, 577–589.

35. Goldstine, Herman, *A History of Numerical Analysis from the 16th through the 19th Century*, Springer-Verlag, New York, Heidelberg, Berlin, 1977.

36. Grabiner, Judith V., A Mathematician Among the Molasses Barrels: MacLaurin's Unpublished Memoir on Volumes, *Proceedings of the Edinburgh Mathematical Society* 39 (1996), 193–240.

37. Grabiner, Judith V., *The Calculus as Algebra: J.-L. Lagrange, 1736–1813*, Garland Publishing, Boston, 1990.

38. Grabiner, Judith V., *The Origins of Cauchy's Rigorous Calculus*, M.I.T. Press, Cambridge, Mass., 1981.

39. Greenberg, John L., *The Problem of the Earth's Shape from Newton to Clairaut,* Cambridge University Press, Cambridge, 1995.

40. Griffiths, P. L., Private communication.

41. Griffiths, P. L., The British Influence on Euler's Early Mathematical Discoveries, preprint.

42. Guicciardini, Niccolo, *The Development of Newtonian Calculus in Britain, 1700–1800*, Cambridge University Press, Cambridge, 1989.

43. Hall, A. Rupert, *Philosophers at War: The Quarrel between Newton and Leibniz*, Cambridge University Press, Cambridge, 1980.

44. Hankins, Thomas, *Jean d'Alembert: Science and the Enlightenment*, Clarendon Press, Oxford, 1970.

45. Itard, Jean, Legendre, *Dictionary of Scientific Biography*, vol. 8, 135–143.

46. Jacobi, C. G. J., De usu legitimo formulae summatoriae Maclaurinianae, *Journ. f. reine u. angew. Math.* 18 (1834), 263–272. Also in *Gesammelte Werke*, vol. 6, 1891, pp. 64–75.

47. Juskevič, A. P., and R. Taton, eds., *Leonhard Euleri Commercium Epistolicum*, Birkhauser, Basel, 1980. In Leonhard Euler, *Opera*, Ser. 4, vol. 5.

48. Juskevič, A. P., and Winter, E., eds. *Leonhard Euler und Christian Goldbach: Briefwechsel, 1729–1764*, Akademie-Verlag, Berlin, 1965.

49. Kline, Morris, *Mathematical Thought from Ancient to Modern Times*, Oxford, New York, 1972.

50. Lacroix, S. F., *Traité élémentaire de calcul différentiel et de calcul intégral*, Duprat, Paris, 1802. Translated as [**51**].

51. Lacroix, S.-F., *An Elementary Treatise on the Differential and Integral Calculus*, translated, with an Appendix and Notes, by C. Babbage, J. F. W. Herschel, and G. Peacock, J. Deighton and Sons, Cambridge, 1816.

52. Lacroix, S.-F., *Traité du calcul différentiel et du calcul intégral*, 3 vols., 2d. ed., Courcier, Paris, 1810–1819. Vol. I, 1810.

53. Lagrange, J.-L., Sur une nouvelle espèce de calcul rélatif a la différentiation et à l'intégration des quantités variables, *Nouvelles Memoires de l'académie... de Berlin*, 1772, 185–221, in *Oeuvres*, vol. 3, 439–476.

54. Lagrange, J.-L., *Leçons sur le calcul des fonctions*, new ed., Courcier, Paris, 1806. In *Oeuvres*, vol. 10.

55. Lagrange, J.-L., *Mécanique analytique*, 2d. ed., 2 vols., Courcier, Paris, 1811–1815, in *Oeuvres*, vols. 11–12.

56. Lagrange, J.-L., Note sur la métaphysique du calcul infinitésimal, *Miscellanea Taurinensia* 2 (1760–61), 17–18; in *Oeuvres*, vol. 7, 597–599.

57. Lagrange, J.-L., Sur l'attraction des sphéroïdes elliptiques, *Mémoires de l'académie de Berlin*, 1773, 121–148. Reprinted in *Oeuvres de Lagrange*, vol. III, 619ff.

58. Lagrange, Joseph-Louis, *Théorie des fonctions analytiques*, Imprimérie de la République, Paris, An V [1797]; compare the second edition, Courcier, Paris, 1813, reprinted in *Oeuvres de Lagrange*, pub. M. J.-A. Serret, 14 volumes, Gauthier-Villars, Paris, 1867–1892, reprinted again, Georg Oms Verlag, Hildesheim and New York, 1973, vol. 9.

59. Legendre, Adrien-Marie, Mémoires sur les intégrations par arcs d'ellipse et sur la comparaison de ces arcs, *Mémoires de l'Academie des sciences*, 1786, 616, 644–673.

60. Legendre, Adrien-Marie, *Traité des fonctions elliptiques et des intégrales eulériennes, avec des tables pour en faciliter le calcul numérique*, 3 vols., Paris, 1825–1828.

61. Loria, Gino, The Achievements of Great Britain in the Realm of Mathematics, *Mathematical Gazette* 8 (1915), 12–19.

62. Maclaurin, Colin, *Traité de fluxions*, Traduit de l'anglois par le R. P. Pézénas, 2 vols., Jombert, Paris, 1749.

63. Maclaurin, Colin, *A Treatise of Fluxions in Two Books*, Ruddimans, Edinburgh, 1742.

64. Mahoney, Michael, Review of [**42**], *Science* 250 (1990), 144.

65. McElroy, Davis, *Scotland's Age of Improvement*, Washington State University Press, Pullman, 1969.

66. Mills, Stella, *The Collected Letters of Colin Maclaurin*, Shiva Publishing, Nantwich, 1982.

67. Mills, Stella, The Independent Derivations by Leonhard Euler and Colin Maclaurin of the Euler-MacLaurin Summation Formula, *Archive for History of Exact Science* 33 (1985), 1–13.

68. Murdoch, Patrick, An Account of the Life and Writings of the Author, in Colin Maclaurin, *Account of Sir Isaac Newton's Philosophical Discoveries*, For the Author's Children, London, 1748, i–xx; reprinted, Johnson Reprint Corp., New York, 1968.

69. Newton, Isaac, *Of Analysis by Equations of an Infinite Number of Terms*, J. Stewart, London, 1745, in D. T. Whiteside, ed., *Mathematical Works of Isaac Newton*, vol. I, Johnson Reprint, New York and London, 1964, 3–25.

70. Olson, Richard, *Scottish Philosophy and British Physics, 1750–1880*, Princeton University Press, Princeton, 1975.

71. Olson, Richard, Scottish Philosophy and Mathematics, 1750–1830, *Journal of the History of Ideas* 32 (1971), 29–44.

72. Pearson, Karl, *The History of Statistics in the Seventeenth and Eighteenth Centuries* [written 1921–1933]. Ed. E. S. Pearson, Charles Griffin & Co., London and High Wycombe, 1976.

73. Pepe, Luigi, Tre 'prime edizioni' ed un' introduzione inedita della Fonctions analytiques di Lagrange, *Boll. Stor. Sci. Mat.* 6 (1986), 17–44.

74. Phillipson, Nicholas, The Scottish Enlightenment, in [**76**], pp. 19–40.

75. Porter, Roy, The Enlightenment in England, in [**76**], pp. 1–18.

76. Porter, Roy, and Mikulas Teich, eds., *The Enlightenment in National Context*, Cambridge University Press, Cambridge, 1981.

77. Playfair, John, Traité de Mechanique Celeste, *Edinburgh Review* 22 (1808), 249–84.

78. Sageng, Erik Lars, *Colin Maclaurin and the Foundations of the Method of Fluxions*, unpublished Ph.D. Dissertation, Princeton University, 1989.

79. Scott, J. F., Maclaurin, *Dictionary of Scientific Biography*, vol. 8, 609–612.

80. Shapin, Stephen, and Arnold Thackray, Prosopography as a Research Tool in History of Science: The British Scientific Community, 1700–1900, *History of Science* 12 (1974), 95–121.

81. Stewart, M. A., ed., *Studies in the Philosophy of the Scottish Enlightenment*, Clarendon Press, Oxford, 1990.

82. Struik, D. J., *A Source Book in Mathematics, 1200–1800*, Harvard University Press, Cambridge, Mass., 1969.

83. Taton, Juliette, Pézénas, *Dictionary of Scientific Biography*, vol. 10, Scribner's, New York, 1974, 571–2.

84. Taton, René, ed., *Enseignement et diffusion des sciences en France au XVIIIe siècle*, Hermann, Paris, 1964.

85. Todhunter, Isaac, *A History of the Mathematical Theories of Attraction and the Figure of the Earth, from the Time of Newton to That of Laplace*, Macmillan, London, 1873.

86. Truesdell, C., Rational Fluid Mechanics, 1687–1765, introduction to Euler *Opera*, Ser. 2, vol. 12.

87. Truesdell, C., The Rational Mechanics of Flexible or Elastic Bodies, 1638–1788, in Euler, *Opera*, Ser. 2, vol. 11.

88. Turnbull, H. W., *Bicentenary of the Death of Colin Maclaurin (1698–1746)*, The University Press, Aberdeen, 1951.

89. Turnbull, H. W., *The Great Mathematicians*, Methuen, London, 1929.

90. Tweedie, Charles, A Study of the Life and Writings of Colin Maclaurin, *Mathematical Gazette* 8 (1915), 132–151.

91. Tweedie, Charles, *James Stirling*, Clarendon Press, Oxford, 1922.

8

Newton, Maclaurin, and the Authority of Mathematics*

1. Introduction: Maclaurin, the Scottish Enlightenment, and the "Newtonian Style"

Sir Isaac Newton revolutionized physics and astronomy in his book *Mathematical Principles of Natural Philosophy* [**27**]. This book of 1687, better known by its abbreviated Latin title as the *Principia*, contains Newton's three laws of motion, the law of universal gravitation, and the basis of all of classical mechanics. As one approaches this great work, a key question is: How did Newton do all of this? An equally important question is: Can Newton's methods work on any area of inquiry? Newton's contemporaries hoped that the answer to the second question was yes: that his methods would be universally effective, whether the area was science, society, or religion. What are the limits of the Newtonian method? In 1687, nobody knew. But his followers wanted to find out, and they tried to find out by applying these methods to every conceivable area of thought.

In Great Britain, Newton's most successful follower was Colin Maclaurin (1698–1746). Maclaurin was the most significant Scottish mathematician and physicist of the eighteenth century, and was highly influential both in Britain and on the Continent. He was one of the key figures in what is called the Scottish Enlightenment, the eighteenth-century intellectual movement that includes philosophers like Francis Hutcheson and David Hume, scientists like James Hutton and Joseph Black, and social philosophers like Adam Smith [**3**], [**31**]. And I have come to think that Newton's method—what has been called "the Newtonian style"—is both the key to understanding what made Maclaurin tick intellectually and to understanding the nature of his influence. In this paper, I want to demonstrate these conclusions.

In particular, I want to describe how Maclaurin applied "the Newtonian style" to areas ranging from the actuarial evaluation of annuities to the shape of the earth. Maclaurin seems to have thought that using this Newtonian style could guarantee success in any scientific endeavor. And I have another point to prove as well. Maclaurin's career illustrates and embodies the way mathematics and mathematicians, building on the historical prestige of geometry and

* Reprinted from Amer. Math. Monthly III, 10(Dec. 2004) 841–852.

the success of Newtonianism, were understood to exemplify certainty and objectivity during the eighteenth century. Using the Newtonian style invokes for your endeavor, whatever your endeavor is, all the authority of Newton, of whom Laplace said, "There is but one law of the cosmos, and Newton has discovered it," of whom Alexander Pope wrote, "Nature and Nature's Laws lay hid in night; God said, 'Let Newton be!' and all was light," and of whom Edmond Halley stated, "No closer to the gods can any mortal rise." The key word here is "authority." Maclaurin helped establish that authority.

2. What is the "Newtonian Style"?

The *Principia* presents Newton's methods in action. Of course I do not claim that nobody had ever used an approach in mathematical physics resembling his before. But our concern here is how Newton himself did his successful celestial mechanics in the *Principia*, because that is the source of what Maclaurin and many others internalized and applied. And the way Newton did his theoretical physics has best been described by I. Bernard Cohen, who first called this approach "the Newtonian style" [**5**, p. 132] [**6**, chap. 3].

In the *Principia*, according to Cohen, Newton first separated problems into their mathematical and physical aspects. A simplified or idealized set of physical assumptions was then treated entirely as a mathematical system. Then, the consequences of these idealized assumptions were deduced by applying sophisticated mathematical techniques. But since the mathematical system was chosen to duplicate the idealized physical system, all the propositions deduced in the mathematical system could now be compared with the data of experiment and observation. Perhaps the mathematical system was too simple, or perhaps it was too general and a choice had to be made. Anyway, the system was tested against experience. And then—this is crucial—the test against experience often required modifying the original system. Further mathematical deductions and comparisons with nature would then ensue. Success finally comes, in Cohen's words, "when the system seems to conform to (or to duplicate) all the major conditions of the external world" [**5**, p. 139]. Let me emphasize this: *all* the major conditions. What makes this approach nontrivial is the sophistication of the mathematics and the repeated improvement of the process. It is sophisticated mathematics, not only a series of experiments or observations, that links a mathematically describable law to a set of causal conditions.

In order to fully appreciate the process just described and thereby understand the argument of this paper, we need some examples. The best example of how the Newtonian style worked is the way Newton treated planetary motion in Book I of the *Principia*. He began with one body, a point mass, moving in a central-force field. But even the full solution to that problem is not adequate to the phenomena, so he continued. First he introduced Kepler's laws, then a two-body system with the bodies acting mutually on each other, then many bodies, then bodies that are no longer mass points but extended objects. The force that explained this entire system was universal gravitation, a force that Newton argued "really exists." So now we have found what causes both the fall of objects on the earth and the motions of the solar system.

Here is another, simpler example of Newton's use of the Newtonian style. He considered bodies moving in circles. Once he had derived the law of centripetal force, Newton proved mathematically that the times taken to go around the circle vary according to the nth powers of the radii (that is, as R^n) if and only if the centripetal force varies inversely as R^{2n-1}. One consequence of this general mathematical relation is that the periods vary with the $3/2$ power of the radii (Kepler's Third Law) if and only if the force varies inversely with the square of the radii [**27**, Book I, Theorem 4 and Scholium, pp. 449–452]. Thus the test of the general mathematical theory for central forces against Kepler's specific observation establishes that the inverse-square force—which had been suggested by others before Newton—must be the right one. Even more important: the *causal* relationship between these two pre-Newtonian conclusions (Kepler's Third Law, the inverse-square force) is revealed by the mathematical system that includes Newton's laws of motion.

Maclaurin knew the Newtonian style intimately, and explained examples like those we have just described in his own exposition of Newton's work in physics. Maclaurin beautifully expressed his deep understanding of the Newtonian style when he wrote: "Experiments and observations . . . could not alone have carried [Newton] far in tracing the causes from their effects, and explaining the effects from their causes: a sublime geometry . . . is the instrument, by which alone the machinery of a work [the universe], made with so much art, could be unfolded" [**20**, p. 8].

Throughout his career, I shall show, this is is the methodology Maclaurin followed.

3. Maclaurin's First Use of the Newtonian Style

A sort of trial run of the Newtonian style was Maclaurin's youthful attempt—he was sixteen—to build a calculus-based mathematical model for ethics. In a Latin essay still (perhaps mercifully) unpublished today, "De Viribus Mentium Bonipetis" ("On the Good-Seeking Forces of Mind"), Maclaurin mathematically analyzed the forces by which our minds are attracted to different morally good things. Although he didn't publish this essay, he liked it enough to send it to the Reverend Colin Campbell, in whose papers it survives at the University of Edinburgh [**21**].

In "On the Good-Seeking Forces of Mind," Maclaurin postulated that the "forces with which our minds are carried towards different good things are, other things being equal, proportional to the quantity of good in these good things." Also, the attractive force of a good one hour in the future would exceed that of the same good several hours in the future. And so on. Maclaurin represented the total quantity of good as the area under a curve whose x-coordinate gives the duration and y-coordinate the intensity of the good at a particular instant. He said that one could find the maximum and minimum intensities of any good or evil using Newtonian calculus. Maclaurin graphed the total attraction of a good under various assumptions about how the intensity varies over time, and, by integration, derived equations for the total good. One conclusion supported by his mathematical models was that good men need not complain "about the miseries of this life" since "their whole future happiness taken together" will be greater. Maclaurin thus tested his mathematical model against the doctrine of the Church of Scotland, and found that the results fit. He had shown mathematically that the Christian doctrine of salvation maximized the future happiness of good men.

Of course, "maximizing" and "minimizing" are important techniques in the calculus. Applications of this technique abound in eighteenth-century physics, from curves of quickest descent to the principle of least action. In fact, it was Maclaurin who, in his *Treatise of Fluxions*, gave the first sophisticated account of the theory of maxima and minima, using Maclaurin series to characterize maxima, minima, and points of inflexion of curves in terms of the signs, or equality to zero, of first, second, third, and nth derivatives [**22**, pp. 694–703]. This work was highly praised by Lagrange, who gave a similar theory enriched by the Lagrange Remainder for the Taylor series [**19**, pp. 233–236] (see also [**12**, pp. 136–137, p. 217 n. 65]). Maclaurin also applied the techniques of finding maxima and minima in many novel situations and then compared his results with the best data: for instance, the best design for waterwheels and windmills, how bees build the three-dimensional cells in honeycombs, and the most economical way to build a barn [**20**, pp. 149, 172–178], [**22**, pp. 733–742], [**23**, pp. 386–391, 397–400].

Still, one wonders what could possibly have inspired Maclaurin to write that theological essay. I think it likely that the idea was suggested by the Scots mathematician John Craige's 1699 *Theologiae Christianae Principia Mathematica*, whose title translates as "Mathematical Principles of Christian Theology." In this work, Craige graphed the intensity of pleasures as various functions of time, calculated the total pleasure by integration, and concluded, after an argument so Newtonian that nineteenth-century readers called it "an insane parody of Newton's *Principia*," that one should forego the finite pleasures of this world in favor of the infinite pleasures of the world to come. But before we moderns laugh too readily, remember that this was the first generation after the *Principia*. Who knew for sure what the limitations of the Newtonian style were?

In fact the Newtonian style did not automatically produce valid results, even in the physical sciences, even for Newton himself. G. E. Smith has shown how Book II of the *Principia* was Newton's attempt to use the Newtonian style to find, from the phenomena of motion in fluid media, the resistance forces acting on bodies [**29**, p. 251]. But unlike the situation in the discovery of universal gravitation that I sketched earlier, in which each successive mathematical idealization dropped some assumption that had simplified the mathematics, Newton's attempt to model the inertial resistance to motion and its relation to viscosity did not yield to his approach. Indeed, some of the key quantities (e.g., the drag coefficient and the Reynolds number [the ratio of inertial to viscous effects in a flow]) still cannot be functionally related from theoretical principles. In establishing universal gravitation, the sequences from point mass to extended body, or from one-body to two-body systems, work because each approximation suggests unresolved questions that the next stage can address. But this did not work for resisted motion. In Smith's words, "this time . . . the empirical world did not cooperate" [**29**, p. 288].

Still, for Maclaurin, the Newtonian style seemed to have a good track record, and Maclaurin completely committed himself to it. But there was more to Maclaurin's Newtonianism than what we have examined so far. Let us now turn to another Newtonian theme: the relationship between the authority of mathematical physics and religion.

4. Religion, Authority, and Mathematics for Newton and Maclaurin

A major theme in Maclaurin's work is the way the order of the universe demonstrates the existence and nature of God. Newton had written in the concluding section of his *Principia*, "This most elegant system of the sun, planets, and comets could not have arisen without the design and dominion of an intelligent and powerful being" [**27**, General Scholium, p. 940]. Maclaurin agreed, saying, "Such an exquisite structure of things could only arise from the contrivance and powerful influences of an intelligent, free, and most potent agent" [**20**, p. 388; compare p. 381]. Maclaurin took Newton to mean that it is the mathematically-based natural philosophy of the *Principia* that proves the existence of God, and that Newton's mathematically-based methods guarantee his conclusions about the nature of God's world.

Why was the authority of mathematics so attractive in the eighteenth century? Of course religion had authority, but there were theological disputes aplenty—Catholic versus Protestant, for instance. So for many eighteenth-century thinkers part of the authority of mathematics and science came, as Maclaurin said in the preface to his *Treatise of Fluxions*, from the belief that mathematical demonstration—unlike, say, theology or politics—produced universal agreement, leaving "no place for doubt or cavil" [**22**, p. 1].

George Berkeley, however, did have doubts, and attacked mathematics. Maclaurin felt obliged to defend it. Berkeley saw Newtonian science as entailing only a natural-order sort of God, and thus undermining the authority of Scripture. But besides criticizing the Newtonian philosophy in general, Berkeley, in *The Analyst, or a Discourse Addressed to an Infidel Mathematician* (1734), attacked the logical validity of the calculus. In particular, using well-chosen examples, Berkeley argued that, on the evidence of the way the calculus was actually being explained, mathematicians reasoned worse than theologians. And he ridiculed vanishing increments, which Newton had used to explain his calculus, as "ghosts of departed quantities" [**2**, sec. 35].

This is not the place to describe in detail how Maclaurin refuted Berkeley (see [**28**] and [**14**]). But I will say here that Maclaurin showed how rigorous Newtonian calculus could be made, and Maclaurin's improved algebraic and inequality-based understanding of Newton's limit concept played a role in the eventual complete rigorization of the calculus in the nineteenth century [**12**], [**14**]. For the present, the key point is that Maclaurin's refutation of Berkeley strongly reinforced the authority of mathematics in Britain.

5. Maclaurin's Mature Use of the Newtonian Style

The Shape of the Earth

We turn now to the heart of the current paper: how Maclaurin used the Newtonian style to become a successful scientist. Maclaurin's monumental *Treatise of Fluxions* is much more than an answer to Berkeley; it contains a wealth of important results both in mathematics and in physics. In particular, let us look at Maclaurin's treatment of the shape of the earth, for it exemplifies his use of the Newtonian style, especially the sophisticated use of mathematical models, to solve problems of great importance for physics and astronomy.

The earth is not a sphere, Newton argued in the *Principia*, since its rotation brings about real forces that cause it to be flattened at the poles and to bulge at the equator. He stated, though he did not prove, that the shape of the earth was an ellipsoid of revolution, and used the method of balancing columns to predict its dimensions assuming very small ellipticity [**27**, Book I, Prop. 91]. And he applied his theoretical discussion to the existing data on the period of pendulums at different latitudes [**27**, Book III, Prop. 19]. Maclaurin carried the theory of this subject substantially farther. He produced the first rigorously exact theory of homogeneous figures shaped like ellipsoids of revolution whose parts attract according to the inverse-square law. Among other things, he proved geometrically that a homogeneous ellipsoid, revolving around its axis of symmetry, is indeed a possible figure of equilibrium, and he established that the perpendicular columns balance. Maclaurin's geometric method was based on his deep knowledge of the conic sections and worked for figures whose ellipticities were finite as well as infinitesimal. He demonstrated how to calculate the gravitational attraction at any point on the surface of such an ellipsoid. Maclaurin also developed a theory of the attraction of such ellipsoids that were variable in density, in the process proving results about the gravitational attraction of confocal ellipsoids. He developed further, and exploited, many now-standard techniques (besides balancing columns, the idea of level surfaces) in his theory of these rotating bodies. All this was important and new and, especially through his correspondence with Alexis-Claude Clairaut, was fully integrated into what was taking place on the Continent [**15**, pp. 412–425, 585–601], [**22**, pp. 522–566], [**23**, pp. 342–344, 347–354, 359–370, 372–380].

Furthermore, besides the mathematical theory of rotating bodies, their shapes, and their attractions, Maclaurin took a keen interest in getting real data—the latest news from the expeditions to various parts of the world to measure the earth's shape, whether by his fellow Scotsman George Graham or by the various French academicians. Maclaurin, then, developed mathematically precise predictions about the earth's shape, and tested these predictions against the latest observations to refine his theory, from homogeneous ellipsoids of uniform density to stratified ones of variable density—an impressive and influential application of the Newtonian style [**22**, pp. 551–566]. Of course he did not solve all the problems in this subject, but he pioneered the serious mathematical study of the rotation of astronomical bodies, a subject carried further by Clairaut, d'Alembert, Lagrange, Jacobi, and on up to the twentieth century by Chandrasekhar (whose modern history of the topic, which devotes a full chapter to Maclaurin spheroids, can be consulted in [**4**]).

"Gauging": Finding the Volumes of Barrels

From the heavens, we move to earthly applications. Maclaurin, employing the fashionable rhetoric of his time that was based on the philosophy of Sir Francis Bacon, rejoiced that advances in science could produce useful knowledge and enthusiastically participated in that production. One of his contemporaries tells us that Maclaurin, before his untimely death in 1746, "had resolved . . . to compose a course of practical mathematics" [**24**, p. xix], the existing ones being of low quality. Perhaps as a first step toward this task, Maclaurin in 1745 published, with his own comments, a manuscript by David Gregory on finding the volumes of barrels, or as it was called at the time, "gauging."

Gauging was an important subject for eighteenth-century society, since the wooden barrel was the universal shipping container throughout the Atlantic economies of both Europe and America. And gauging was the subject of one of Maclaurin's most detailed contributions to applied science. In 1735, Maclaurin wrote a ninety-four-page memoir for the Scottish Excise Commission explaining the most accurate way to find the volume of molasses in the barrels in the port of Glasgow [**10**]. In 1998 I presented this story of Maclaurin's success in mathematically gauging molasses barrels as a case study in the use of mathematical authority to achieve consensus about a problem—taxation—clearly rife with disagreement between parties with vastly different interests [**13**]. But in the present context, Maclaurin's solution of the barrel-gauging problem is another example of his use of the Newtonian style.

Maclaurin's memoir provided a set of clear rules for finding the volumes of real barrels. To derive these rules, Maclaurin began with the case where the barrels had precise mathematical shapes. Many eighteenth-century manuals on gauging treated barrels as solids generated by rotating conic sections about their axes. Further, the manuals often approximated the volumes of these solids of revolution by imagining them made up of slices perpendicular to the axis of rotation, with each slice approximated by a cylinder of the same height (typically about ten inches) such that the diameter of each cylinder was the diameter at the midpoint of the altitude of the corresponding slice. But this method was far from accurate. First, the precision of the approximation using cylindrical slices depends not only on the height of the slices but also on the type of conic section whose rotation generates the solid. Maclaurin was the first to calculate—and to prove—exactly what the differences are between these approximating cylinders and the corresponding slices of solids of revolution. (For Maclaurin's beautiful geometric result, see [**10**, p. 193]; for his derivation, using Newtonian calculus, see [**10**, pp. 229–235]; for his geometric proof, see [**22**, pp. 24–27].) Second, real barrels are not solids of revolution. In the eighteenth century, wooden barrels were assembled out of individual staves and hoops, with no two barrels being identical. So besides giving the mathematical theory of solids of revolution and working out how each type of solid differed from its approximation by cylindrical slices, Maclaurin showed how the actual barrels deviated from the mathematical solids of revolution, then showed how new calculations could deal with those deviations, and then even addressed deviations from those deviations (for an example, see [**10**, pp. 209–215]). Once again, Maclaurin used the Newtonian style. He began with a mathematical model and corrected it repeatedly according to observation to produce an authoritative, precise, and realistic solution. And the Excise officers in Scotland used his method for many years [**24**, p. xix]. That he was serving his society must have further reinforced his commitment to the Newtonian style.

Social Agreement and Scientific Authority

Maclaurin recognized that scientists, whether pursuing pure research or solving society's problems, do not work alone. In fact he himself played a leading role in organizing the Philosophical Society of Edinburgh, Scotland's first real scientific society, in 1737. And social agreement and consensus—for Maclaurin, as we will see presently, these too come from Newtonianism—are part of the Society's rhetoric. The eighteenth-century ideal of consensus and universal agreement appears clearly in the Edinburgh Philosophical Society's stated rules: "Religious or Political Disputes" were forbidden, and members were warned that "in their Conversations, any Warmth that might be offensive or improper for Philosophical Enquiries is to be avoided" [**11**, p. 154].

But the avoidance of political disputes need not, and in the eighteenth century did not, mean a lofty lack of involvement in society. The Philosophical Society's members served as consultants to Scottish development agencies, from the British Linen Company to the Royal Bank of Scotland. They helped map Scotland and her coasts, made astronomical observations that could assist navigation and determine longitude, pumped water out of mines, designed and built canals, investigated the distribution of minerals in Scotland, and measured the forces of winds at sea. All these activities embody the eighteenth-century idea that science and technical expertise should be applied to solve problems of economic and social importance for the nation. Scientists were much sought after by progressive investors, whom the Scots called "improvers," both in Britain and on the Continent [**30**, p. 361]. And the scientists actually could help.

But why, in the eighteenth-century view, does science help? The crucial mark of the validity of the science to be applied, many eighteenth-century thinkers asserted, came from the fact that everybody could agree about it. Universal agreement gave eighteenth-century science its authority. It made science a collective, not just an individual, endeavor. And this universal agreement was, for Maclaurin, the achievement of Newton and of the Newtonian style. Maclaurin called the successful Newtonian style Newton's "right path." Newton, Maclaurin claimed, "had a particular aversion to disputes" (a hagiographical comment indeed) and "weighed the reasons of things impartially and coolly" [**20**, p. 13]. Moreover, just as Maclaurin had used the universal agreement historically possessed by mathematics to motivate his refutation of Berkeley, so he used his idealized picture of Newton to formulate his own version of the social achievement of eighteenth-century science. Maclaurin wrote [**20**, p. 62]: "We are now arrived at the happy aera of experimental philosophy; when men, having got into the right path, prosecuted useful knowledge... the arts received daily improvements; when not private men only, but societies of men, with united zeal, ingenuity and industry, prosecuted their enquiries into the secrets of nature, *devoted to no sect or system*" [italics added]. So it is the Newtonian style that lets scientists and scientific societies produce national prosperity.

Actuarial Science

Now let us turn to an episode that combines our main themes, the mathematician's authority and the Newtonian style, applied together to serve society: Maclaurin's actuarial work for the Scottish Ministers' Widows' Fund in 1743. First, consider Maclaurin's authority. Robert Wallace, Moderator of the General Assembly of the Church of Scotland, wrote Maclaurin that, when objections to the soundness of the pension scheme were raised in Parliament, "I answered them that the Calculations had been revised by you... this entirely satisfied them" [**1**]. Or, as Maclaurin's contemporary biographer Patrick Murdoch put it, "The authority of [Maclaurin's] name was of great use... removing any doubt" [**24**, p. xix]. The calculations as revised by Maclaurin were indeed satisfactory; the fund continued even into the twenty-first century to provide for some of the "widows and fatherless children of ministers of the Church of Scotland, the Free Church of Scotland and some of the professors of the four old Scottish Universities" [**9**, p. xii].

The Church's goal in developing the Fund was to keep its ministers' widows and orphans out of poverty by providing them an annuity. Schemes like this had been tried before, but the funds tended to run out of money. Everyone involved in the 1743 scheme recognized that making it work required finding data and creating a mathematical model based on that data. As it turned out, the task also involved checking and refining data, and then revising the model to fit that better data accurately. The men who did the first two steps, finding data and making a model, were Alexander Webster and Robert Wallace. Maclaurin, by carrying out the two last steps—checking theory against data and revising the model to deal with the discrepancies—improved their scheme and made it work. The Newtonian style is apparent even from the title of the printed account of the Fund: "Calculations, with the Principles and Data on which they are instituted: relative to a late Act of Parliament, intituled, an Act for raising and establishing a Fund for a Provision for the Widows and Children of the Ministers of the Church, and the Heads, Principals, and Masters of the Universities of Scotland, shewing The Rise and Progress of the Fund." Maclaurin thus used the Newtonian style once again, this time to become a pioneer of actuarial science.

To start planning for the Fund, Webster collected the relevant data by sending questionnaires to every parish of the Church, asking how many ministers there were, how many widows there were, how many orphans there were, how old these people all were, and so on. Wallace then made a mathematical model predicting the changes in the widow and orphan populations. He sent a copy of the proposed scheme to Maclaurin for possible criticism. And he got it.

Using Wallace's model, Maclaurin prepared tables to predict the future of the scheme according to probability theory. Maclaurin showed that the fund would run out of money unless it reduced the children's benefits. Why? Because Wallace, after consulting mortality tables, had assumed that 1/18 of the widows would die each year, a mathematically tractable assumption since his model began with eighteen widows, but an assumption that ignored the varying age distribution of the group.

Maclaurin sought empirical data to check Wallace's assumption and found it by a study, more careful than Wallace's, of the mortality tables from Breslau published by Edmond Halley in 1673. According to Maclaurin's refined model based on his analysis of the age distribution in this data, the average annual mortality rate of the widows would at first be lower, and thus more payouts would be needed. Even after the first year, the two models yield different predictions. Wallace's model has 35 widows; Maclaurin's has 35.42. After only thirty years, the predicted number of widows differs by almost 20 percent: 257 according to Wallace's model, 307 according to Maclaurin's [**7**, p. 53], [**17**, p. 63], [**23**, pp. 108–109].

Actuarial historians David Hare and William Scott assert that Maclaurin's recalculations were "the earliest actuarially-correct fund calculations ever carried out" because Maclaurin had used "a realistic and accurate life table" [**17**, p. 57, 68]. And, not incidentally, it worked; what actually happened corresponded remarkably well to the calculated figures.

Maclaurin's success with the Widows' Fund had great influence. In 1761, the first real life insurance plan in the United States, devised by the Presbyterian Church in Philadelphia, was based on it [**8**, pp. 19–20]. Likewise, the Fund's example influenced the first Scottish life insurance company to be established, as that company's name, Scottish Widows, indicates [**7**, p. 24]. Furthermore, the importance of this kind of successful mathematical modeling of society was enormous for the social sciences. Both Webster and Wallace later applied quantitative methods to study populations in broader contexts. For instance, Webster again used his parish survey methods to produce his *Account of the Number of People in Scotland*, published in 1755, a work whose focus on the "political arithmetic" of the nation makes it an important step toward the modern census. Wallace constructed mathematical models to study human populations in the books *A Dissertation on the Numbers of Mankind* (1753) and *Various Prospects of Mankind, Nature and Providence* (1761), which are often recognized as forerunners of the work of Thomas Malthus. And the mathematical models of agricultural growth (linear) and population growth (exponential) made Malthus conclude in his *Essay on Population* (1798), also in the Newtonian style, that the finite empirical world means that human populations cannot continue to grow exponentially forever. These conclusions became quite important, not least in their role in Darwin's argument for his population-based theory of evolution by natural selection.

"Method" and Authority

From these examples, let us turn back again to the relationship between Maclaurin and Newton. Maclaurin's most influential public explanation of Newtonianism was his book, published posthumously but based on his lectures at Edinburgh, *An Account of Sir Isaac Newton's Philosophical Discoveries* [**20**]. In this book, Maclaurin sought to consolidate Newtonianism and establish its authority beyond all doubt, "in order to proceed with perfect security, and to put an end forever to disputes" [**20**, p. 8]. Book I of the *Account* begins with a chapter on the scientific method done Newton's way, and then shoots down any differing approaches. Book II introduces elementary classical mechanics. Then the major books, Books III and IV, expound the Newtonian system of the world in the Newtonian style. Look at the titles of these books: Book III is called "Gravity demonstrated by analysis" and Book IV is entitled "The effects of the general power of gravity deduced synthetically." But to see the Newtonian style in action one more time, we need to explain the terms "analysis" and "synthesis" to which Maclaurin gave such prominence.

Newton defined his own version of "analysis" in natural science by referring to the Greek usage, where problem solving in mathematics is done by "analysis," literally "solution backwards." We discover how to construct a line, for instance, by assuming that the line has been constructed and working backwards from that until we find something we already know how to construct. Only afterwards can we *prove* that the original construction can be made, and the proof is obtained by reversing the order of the steps in the "analysis." If "analysis" was the method of discovery, the method of proof was called "synthesis," or, in Latin, "composition."

What, then, might "analysis" be in *science*, or as people in the eighteenth century called it, "natural philosophy"? Newton had addressed this question in his *Opticks*. After saying "As in Mathematicks, so in Natural Philosophy, the Investigation of difficult Things by the Method of Analysis, ought ever to precede the method of Composition" he explained that in natural philosophy "This Analysis consists in making Experiments and Observations, and in drawing general Conclusions from them by Induction, and admitting of no Objections against the Conclusions, but such as are taken from Experiments, or other certain Truths. For Hypotheses are not to be regarded in Experimental Philosophy." As for synthesis in natural philosophy, it "consists in assuming the Causes discover'd, and establish'd as Principles, [like the three Laws of Motion, which Newton actually called "axioms"] and by them explaining the Phaenomena proceeding from them, and proving the Explanations" [**26**, 31st Query, pp. 404–405].

This passage from Newton's *Opticks* is one of the most often quoted statements of Newton's methodology. Scholars often say that Newton was contrasting his more Baconian method, which requires extensive empirical observation and experiment, to what he regarded as the premature jumping to conclusions of Descartes's physics. But whatever Newton's reasons were for wording the passage as he did, Maclaurin gave it a different slant.

In his own gloss on this passage, Maclaurin emphasized the certainty that was achieved by obeying Newton's precepts. And to demonstrate this certainty, he identified the success of Newtonian physics with its following the methods of mathematics. It was "in order to proceed with perfect security, and to put an end for ever to disputes," wrote Maclaurin, that Newton says to us, "as in mathematics, so in natural philosophy, the investigation of difficult things by the method of analysis ought ever to precede the method of composition, or synthesis. For in any other way, we can never be *sure* that we assume the principles *which really obtain in nature*" [**20**, pp. 8–9; italics added]. According to Maclaurin, Newton's scientific achievement came from impartiality, not disputation; mathematical proof of causes, not hypotheses; and above all, success in finding the true laws of nature by the right use of mathematics.

6. Religious Authority Revisited

But for Maclaurin, there were even higher goals. Maclaurin said that natural philosophy's chief value is to lay "a sure foundation for natural religion and moral philosophy" [**20**, p. 3]. Natural religion proves God's existence arguing from the order of Nature; moral philosophers inquire into human and divine nature to support conclusions about ethics. The mature Maclaurin was much more sophisticated about how to do this than he was when he wrote that essay at age sixteen.

Maclaurin wanted this "sure foundation" for religion and morality, but this required a natural philosophy that was true. How to achieve this? A naive empiricism, for Maclaurin, cannot produce the true laws of nature. All we can obtain from experiments and observations alone, he said, are natural history and description; this does not take us from observed phenomena "to the powers or causes that produce them" [**20**, p. 221]. For that, we need something more.

Let us, then, return to an earlier quotation, which gives Maclaurin's answer to the question raised by philosophers like Thomas Hobbes and David Hume as to how a set of correlated observations could produce any understanding of causal connections: that is, true natural laws capable of supporting science, technology, moral philosophy, and theology. Recall how Newton had used mathematics to derive the causal connection between the centrally-directed, inverse-square force law and the three laws of Kepler. And now again read Maclaurin [**20**, p. 8]: "Experiments and observations," he wrote, "could not alone have carried [Newton] far in tracing the causes from their effects and explaining the effects from their causes: a sublime geometry was his guide in this nice and difficult enquiry." The "sublime geometry," the mathematics, was for Maclaurin, "the instrument, by which alone the machinery of a work [the universe], made with so much art, could be unfolded." Only the Newtonian style could reveal the true nature of the universe that God had made.

7. Conclusion

The "Newtonian style" seemed to Maclaurin to guarantee success, not only in physics, but also in areas ranging from gauging barrels to insurance to theology. Maclaurin's successes and reputation helped others expect comparable achievements from approaches that were or claimed to be Newtonian, from the quantified ethics of Maclaurin's Glasgow classmate Francis Hutcheson, to the geological world-machine of Maclaurin's student James Hutton, to the optimal economic outcomes of Hutcheson's student Adam Smith.

Maclaurin's work embodied the following key aspects of his Newtonianism: Newtonianism's mathematical prowess, its sophisticated use of mathematical models to solve problems, its links with particular social classes to advance and stabilize a particular system of government, Newtonianism's harmony with religion, and, above all, its authority. The mathematical core of the Newtonian style helped inspire Maclaurin's quest to gain, for Newtonian natural philosophy, all the authority of mathematics. The story as I have told it is set in Scotland. I could tell a different story for the Continent, with the key players including Euler, Lagrange, and Laplace, but the overall outcome there was in many ways the same [**16**] [**18**]. At the core of eighteenth-century science are Newtonian physics put in modern mathematical dress and the mathematization of many new areas. The perceived success of eighteenth-century science vastly and permanently increased the authority of mathematical methods in *all* endeavors. The success and prestige of modern science reflects and embodies the triumph of the Newtonian style.

Acknowledgments

I dedicate this paper to the memory of my friend and colleague Barbara Beechler. An early version was delivered as an address at the 2001 MathFest in Madison. For the present version,

I thank the Mathematics Institute of the University of Copenhagen for its hospitality and excellent library, and Professor Jesper Lützen for his suggestions.

References

1. Anonymous, The second paper of memoirs of Mr MacLaurin beginning 1725, Glasgow University Library MS Gen 1378/2.
2. G. Berkeley, *The Analyst, or a Discourse Addressed to an Infidel Mathematician*, in *The Works of George Berkeley*, vol. 4, A. A. Luce and T. R. Jessop, eds., T. Nelson, London, 1951, pp. 65–102.
3. A. Broadie, ed., *The Scottish Enlightenment: An Anthology, Edited and Introduced by Alexander Broadie*, Canongate, Edinburgh, 1997.
4. S. Chandrasekhar, *Ellipsoidal Figures of Equilibrium*, Yale University Press, New Haven, 1969.
5. I. B. Cohen, Newton's method and Newton's style, in *Newton: Texts, Background, Commentaries*, I. B. Cohen and R. S. Westfall, eds., Norton, New York, 1995, pp. 126–143.
6. ———, *The Newtonian Revolution: With Illustrations of the Transformation of Scientific Ideas*, Cambridge University Press, Cambridge, 1980.
7. J. B. Dow, Early actuarial work in eighteenth-century Scotland, in [**9**], pp. 23–55.
8. A. I. Dunlop, Provision for ministers' widows in Scotland—eighteenth century, in [**9**], pp. 3–22.
9. ———, ed., *The Scottish Ministers' Widows' Fund, 1743–1993*, St. Andrew Press, Edinburgh, 1992.
10. J. V. Grabiner, A mathematician among the molasses barrels: MacLaurin's unpublished memoir on volumes, *Proc. Edinburgh Math. Soc.* **39** (1996) 193–240.
11. ———, Maclaurin and Newton: The Newtonian style and the authority of mathematics, in [**31**], pp. 143–171.
12. ———, *The Origins of Cauchy's Rigorous Calculus*, M. I. T. Press, Cambridge, 1981; reprinted by Dover, New York (in press).
13. ———, "Some disputes of consequence": Maclaurin among the molasses barrels, *Social Studies of Science* **28**, 1998, 139–168.
14. ———, Was Newton's calculus a dead end? The continental influence of Maclaurin's *Treatise of Fluxions*, this MONTHLY **104** (1997) 393–410.
15. J. L. Greenberg, *The Problem of the Earth's Shape from Newton to Clairaut*, Cambridge, Cambridge University Press, 1995.
16. T. Hankins, *Science in the Enlightenment*, Cambridge University Press, Cambridge, 1985.
17. D. J. P. Hare and W. F. Scott, The Scottish Ministers' Widows' Fund of 1744, in [**9**], pp. 56–76.
18. V. Katz, *A History of Mathematics: An Introduction*, 2nd ed., Addison-Wesley, Reading, MA, 1998.
19. J.-L. Lagrange, *Théorie des functions analytiques*, Imprimérie de la République, Paris, 1797; second edition, 1813, in *Oeuvres de Lagrange*, Gauthier-Villars, Paris, 1867–1892, reprinted as vol. 9 by Georg Olms Verlag, New York, 1973.
20. C. Maclaurin, *An Account of Sir Isaac Newton's Philosophical Discoveries*, For the Author's Children, London, 1748; reprinted by Johnson Reprint Corporation, New York, 1968.
21. ———, De viribus mentium bonipetis, MS 3099.15.6, The Colin Campbell Collection, Edinburgh University Library.
22. ———, *A Treatise of Fluxions in Two Books*, T. Ruddimans, Edinburgh, 1742.
23. S. Mills, ed., *The Collected Letters of Colin Maclaurin*, Shiva Press, Nantwich, 1982.
24. P. Murdoch, An account of the life and writings of the author [Colin Maclaurin], in [**20**], pp. i–xx.

25. R. Nash, *John Craige's Mathematical Principles of Christian Theology*, Southern Illinois University Press, Carbondale, IL, 1991.

26. I. Newton, *Opticks, or a treatise of the reflections, refractions, inflections & colours of light*; based on the 4th edition, William Innys, London, 1730; reprinted by Dover, New York, 1952.

27. ———, *The Principia: Mathematical Principles of Natural Philosophy, A New Translation by I. B. Cohen and A. Whitman* (preceded by "A Guide to Newton's Principia" by I. B. Cohen), University of California Press, Berkeley, 1995.

28. E. L. Sageng, *Colin MacLaurin and the Foundations of the Method of Fluxions*, unpublished Ph. D. thesis, Princeton University, 1989.

29. G. E. Smith, The Newtonian Style in Book II of the *Principia*, in *Isaac Newton's Natural Philosophy*, J. Z. Buchwald and I. B. Cohen, eds., M. I. T. Press, Cambridge, 2000, pp. 249–313.

30. L. Stewart, *The Rise of Public Science: Rhetoric, Technology, and Natural Philosophy in Newtonian Britain, 1660–1750*, Cambridge University Press, Cambridge, 1992.

31. C. W. J. Withers and P. B. Wood, eds., *Science and Medicine in the Scottish Enlightenment*, Tuckwell Press, East Linton, 2002.

9

Why Should Historical Truth Matter to Mathematicians? Dispelling Myths while Promoting Maths*

Who is the audience for scholarship in the history of mathematics? Historians of mathematics, perhaps, or maybe some historians of science. But a much larger group is mathematicians and students of mathematics. I talk to mathematicians a lot, and I believe that we are in fact doing something of interest to mathematicians. We have an advantage over historians of other subjects. A distinguished mathematician asked me recently what Euclid had said about a particular result. I doubt that many historians of physics are asked by physicists, "Tell me, how did Aristotle prove that the earth can't move?" Mathematicians use history to enliven the subject, or to motivate topics, or to humanize mathematicians to a general audience, or to justify mathematics to government and public. But is the history right? As historians, we think getting it right matters. Do mathematicians think that?

Now there are a lot of myths out there in the mathematics community. I will walk the reader through seven of the myths I have encountered in my career as a historian of mathematics. I will suggest how to debunk them, and, more important, what is at stake, for the teaching and understanding of mathematics, in getting that particular thing right. At the end, I will bring this together by telling you again how important the historian's work can be.

A. *Myth: The social history of mathematics is easy; just determine what nation or group your mathematician comes from and generalize.*

Consider how Eric Temple Bell, in his book *Men of Mathematics*, describes some controversies between the 19th-century mathematicians Leopold Kronecker and Georg Cantor. Writing in 1937–you may want to think about European history around that date—Bell then remarks, "There

* Reprinted with permission from *Bulletin of the British Society for the History of Mathematics* 22:2 (2007), 78–91. This article is based on a talk Judith Grabiner gave at Queen's College, Oxford on February 24, 2007.

is no more vicious academic hatred than that of one Jew for another when they disagree on purely scientific matters . . . Gentiles either laugh these hatreds off or go at them in an efficient, underhand way which often enables them to accomplish their spiteful ends under the guise of sincere friendship." (Bell, p. 562) I'm afraid I defaced the library copy I was reading by writing in the margin "You mean like Newton and Leibniz?"

Georg Cantor, though his last name and his choice of the Hebrew letter "aleph" to denote infinite cardinals may suggest otherwise, was a Lutheran, as was his father; his mother was a Catholic. But Cantor's religion is not the myth I want to address here.

Here is another instance, less emotionally charged, which underlines the point I do want to make. At a conference I went to in Germany in the 1980s, one scholar explained something by distinguishing between "French mathematics" and "German mathematics," and one of the German historians objected, "That sounds like Bieberbach, with his 'Ayran mathematics' vs. 'Jewish mathematics.' We need to be careful."

Why does it matter?

The fact that it was in relatively modern societies that we saw the enslavement of people of African descent and Hitler's attempt to murder all the Jews in Europe should give us pause about using ethnic designations as explanatory categories. It is of course important to place mathematics in cultural context. I do some social history of mathematics myself. But sociological explanations need to demonstrate causality, not just note correlations.

B. *Second Myth: All Modern Mathematics Comes from Men, Mostly White Christian Men in the Graeco-European Tradition*

Presumably this list will take care of the "men" part:

WOMEN IN WORLD MATHEMATICS: PARTIAL LIST*

Hypatia of Alexandria (ca. 355–415)
Gabrielle Émilie le Tonnelier de Breteuil, Marquise du Châtelet (1706–1749)
Maria Gaetana Agnesi (1718–1799)
Caroline Herschel (1750–1848)
Sophie Germain (1776–1831)
Ada Lovelace (1815–1852)
Florence Nightingale (1820–1910)
Christine Ladd-Franklin (1847–1930)
Sofia Kovalevskaia (1850–1891)
Charlotte Angas Scott (1858–1931)
Grace Chisholm Young (1868–1944)
Emmy Noether (1882–1935)
Anna Johnson Pell Wheeler (1883–1966)
Dame Mary Cartwright (1900–1998)

* This list is taken from the poster "Women of Mathematics," Mathematical Association of America, 2008.

Mina Rees (1902–1997)
Ruth Moufang (1905–1977)
Olga Taussky-Todd (1906–1995)
Grace Hopper (1906–1992))
Emma Lehmer (1906–2007)
Cora Ratto de Sadosky (1912–1981)
Hanna Neumann (1914–1971)
Julia Bowman Robinson (1919–1985)
Olga Alexandrovna Ladyzhenskaya (1922–2004)
Olga Arsen'evna Oleinik (1925–2001)
Etta Zuber Falconer (1933–2002)

Over 30% of new U. S. Ph. Ds. in Mathematics (up from 0% in 1876. Biological change? Oh, sure.)

Now for the Graeco-Roman part. We are all sophisticated enough to know that every culture ever has had mathematics. Many of us have learned that the Incas calculated with base-10 place-value numbers faster than the priests from Spain could calculate, that the medieval Chinese studied congruences and formulated the Chinese Remainder Theorem, that medieval Jewish mathematicians used what we now call "expected value" in solving the problem of fair division, and that mathematicians in India around 1500 had the infinite-series expansions for sine and cosine. So all these people are smart. Still, the myth goes, our own mathematics comes from the Greeks; it was preserved, perhaps, in the Islamic world, but nothing important—important in the sense that it is taught for A-levels—really originated until the revival of Greek mathematical learning in the European Renaissance.

But there are clear-cut counterexamples: here are three.

First, there is base-10 place-value arithmetic, including decimal fractions, all the algorithms for working with them, from multiplication to square roots, and the realization that any real number can be approximated by decimal-fraction expansions to any given degree of accuracy. The base-10 integers were from India. But the whole developed system, with fractions and approximations to irrationals [al-Samawa'al, 12th century], comes from the Islamic world. This is partly through the Latin translation of a book on the base-10 place-value system by Muhammad ibn Musa al-Khwarizmi (c. 780–850).

Europeans Latinized his name and called the Hindu-Arabic number system and the rules for computing with it "the method of Algorism" and then, confusing it with the word "arithmos" for number, "the method of Algorithm." This is the origin of our term "algorithm" for any systematic method of calculating something by a set of mechanically applicable rules. That in itself is an important new idea.

Second, there is algebra—another Arabic word, as the prefix "al" for "the" indicates. Algebra is not general alphabetic symbolic notation, which comes from 16th-century France. In the Islamic world, algebra is the systematic study of the classification and, when possible, solution of equations, of second, third, and higher degrees; methods of solving cubics of all classes by intersecting conics or by approximations; and the proving, in a Euclidean way, that the methods of solution are valid. All this comes into Europe from the Islamic world. Europeans learned Islamic algebra partly from the 12th-century Latin translation of al-Khwarizmi's "Book of al-jabr and al-muqabala" (al-jabr, literally "restoration," is the term used for adding the same thing to both sides of an equation; al-muqabala for subtracting the same thing). We still call the subject "algebra"; if it had not been new, there would have been a Greek or Latin word to translate the term "algebra" into. Europeans also learned algebra from the 13th-century writings of Leonardo Fibonacci of Pisa. Leonardo traveled all over the Arabic-speaking world, and wrote about work from the Islamic world on decimal numbers, quadratic, cubic, and higher degree equations, Diophantine problems, and magic squares. His books became a major source of Italian algebra; this was the tradition that in the 16th century produced the Italian solutions of the general cubic and quartic equations—solutions also devised without general symbolic notation.

Third, there is trigonometry. Greek trigonometry used the chord, while Indian trigonometry used the half-chord, our sine. It is fun to tell people that our word "sine" is a Latin translation of the mis-read Arabic transliteration of the original Sanskrit word for the half-chord. (jyā-ardha = chord-half, abbreviated as jyā or jīvā; transliterated into Arabic as jiba; read without vowels (erroneously) as jaib; but the Arabic word "jaib" translates into Latin as fold, cavity, etc., or "sinus"—hence our word "sine." The history of the word demonstrates trigonometry's multicultural origins.

In the Islamic world by the 10th century, the sine and cosine are joined by the other four trigonometric functions, all the basic trigonometric identities, and precise methods for determining trigonometric functions of all angles by using these identities and by interpolation. In the 15th century al-Kashi has tables of sines from 1 degree via 15-minute steps, to many decimal places. How does he get the sine of one degree? From Euclid's treatment of the pentagon, we can find the sine and cosine for 108 degrees; then, from half-angle formulas, for 54 and then 27 degrees; 30 degrees is easy, so the formula for the sine of the difference between two angles applied to 30–27 degrees gives the sine of 3 degrees; then the cubic identity for sin 3A in terms of sin A can be solved, by excellent approximation techniques for cubics, for sin A, yielding the sine of 1 degree.

This is a whole intellectual technology, the trigonometry developed in the Islamic world, and it comes into Europe in Arabic-language texts—a fact illustrated by the etymology of our term "sine." Because of the importance of trigonometry to astronomy, and astronomy for reform of the Christian calendar, not to mention map-making, medieval and Renaissance Europeans enthusiastically adopted the plane trigonometry they had learned from the Islamic world. (A similar story could be told about spherical trigonometry, which was advanced greatly in the Islamic world because of its application to finding the exact direction of Mecca for prayer.)

What difference does it make if people believe that modern mathematics all comes from Greece and Europe and men? And many people do believe it: I recently asked my calculus class, "What language is the word 'algebra' from?" They said Latin.

The myth makes people whose ancestors come from other parts of the world, and women, feel that mathematics does not belong to them, that it is somebody else's creation. Not so. Don't speak of "western mathematics"; it's *modern* mathematics. It belongs to all of us.[1]

---

[1] See Victor J. Katz, *A History of Mathematics: An Introduction*, 2d edition, Addison-Wesley, 1998; J. L. Berggren, *Episodes in the Mathematics of Medieval Islam*, Springer, 1986.

C. Third Myth: There Was No Real Mathematics in the European Middle Ages. After the Decline of Greek Mathematics, Nothing Much Happened Mathematically in Europe Until the Renaissance

It is true that in the early Middle Ages in Europe few intellectuals worked seriously on mathematics. But beginning in the 12th century with the rediscovery of Greek learning from Arabic-language sources, and the discovery of Islamic learning, European scientific thought "took off," including mathematically.

From Aristotle and his Arabic commentators came the idea of "form"—form and matter together helping make up things in the world. From Christianity came an interest in the infinite, since God is infinite. Also, Aristotle's physics is all about change. Medieval thinkers combine the ideas of form, change, and infinity, into what is unquestionably mathematics.

Consider the medieval discussion of the intension and remission of forms. Suppose you are becoming more charitable. How is the form of charity changing over time, and what is your total amount of charitableness over a fixed time period? Such a form can be uniform, but if it changes it is said to be "difform." If it is difform, its change can be "uniformly difform" or "nonuniformly difform." And so on, in the typical Aristotelian two-valued classification scheme. Mathematicians get interested when you tell them: we retain this idea still when we, like Galileo, speak of "uniformly accelerated motion." But medieval thinkers were interested in the properties of all kinds of variation, with little consideration about whether they appear precisely in nature.

In the 14th century at Merton College, Oxford, Richard Swineshead, Thomas Bradwardine, William of Heytesbury, and others studied the different ways a form could change and the overall effect of that change. Suppose a form is changing so that it is uniformly difform. Can we describe the total amount of this form that we have over some given time period? In what is now called the Merton mean-speed theorem, William of Heytesbury, in 1335, said, "Every latitude, provided that it is terminated at some finite degree, and is acquired or lost uniformly, will correspond to its mean degree." For instance, the total distance covered by a body with a uniformly difform speed equals the distance that would be covered during the same time at the mean speed. A diagram helps:

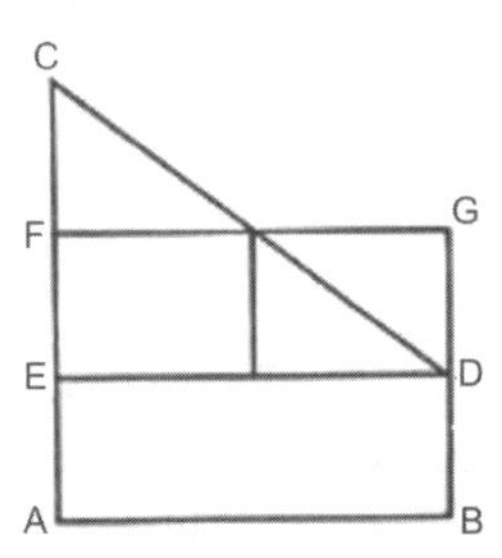

This diagram is also medieval. It is due to the 14th-century French cleric, Nicole Oresme, who got the idea of graphing the changes over time. This same diagram appears in the *Two New Sciences* of Galileo in his discussion of uniformly accelerated motion.

Here is something to say (perhaps a slight overstatement) that will excite mathematicians about this topic: Galileo took the medieval analysis of uniformly accelerated change in general and applied it to the real-world motion of falling bodies—and thereby founded modern mathematical physics.

The scholars at Merton College explored other mathematically interesting ways that a form could vary. For instance, suppose a form has the intensity of 1 during the first half of a period of time, 2 the first half of the time remaining, 3 the first half of the time yet remaining, and so on ad infinitum. Then its total intensity would be given by, in our notation, the sum of terms of the form $k/2^k$ from $k = 1$ to infinity. That is:

$$1/2 + 2/4 + 3/8 + 4/16 + 5/32 + \cdots .$$

What is this total? I gave this problem to a mathematical audience, and a distinguished topologist quickly reinvented Swineshead's method. Swinsehead did it in words; here is the solution in the fraction and sum notation we use now:

$$\begin{aligned}
&1/2 + 2/4 + 3/8 + 4/16 + 5/32 + \cdots\cdots\cdots \\
&= \\
&1/2 + 1/4 + 1/8 + 1/16 + 1/32 + \cdots\cdot = 1 \\
&\qquad +1/4 + 1/8 + 1/16 + 1/32 + \cdots\cdot = 1/2 \\
&\qquad\qquad +1/8 + 1/16 + 1/32 + \cdots\cdot = 1/4 \\
&\qquad\qquad\qquad +1/16 + 1/32 + \cdots\cdot = 1/8 \\
&\qquad\qquad\qquad\qquad +1/32 + \cdots\cdot = 1/16 \\
&\qquad\qquad\qquad\qquad\qquad \cdots
\end{aligned}$$

Adding these separate sums vertically gives the total $= \mathbf{2}$.

Oresme gave the same problem, drew a diagram for it, and did the sum by an ingenious geometrical argument. Then he solved a similar geometric problem using the same time intervals, only this time the motion was continuous, uniform in the odd-numbered intervals, and uniformly accelerated in the even-numbered ones.

Finally, Oresme investigated another pattern of variation that gives rise to an infinite series, now called the harmonic series:

$$1/2 + 1/3 + 1/4 + 1/5 + 1/6 + 1/7 + 1/8 + \cdots$$

And he noted that if one does this term-by-term comparison:

$$\begin{aligned}
1/2 &\geq 1/2 \\
1/3 + 1/4 &\geq 1/4 + 1/4 = 1/2 \\
1/5 + 1/6 + 1/7 + 1/8 &\geq 1/8 + 1/8 + 1/8 + 1/8 = 1/2,
\end{aligned}$$

and so on, one sees that the harmonic series, since it is bounded below by an infinite sum of terms each equal to $1/2$, increases without any bound. (This result got lost, so it had to be rediscovered by Jakob Bernoulli in the 1690s.)

Why does this matter? What difference does it make if we wrongly say there was no medieval European mathematics? And many mathematical books do say it, or imply it by omitting the Middle Ages from their run through mathematical history.

First, it perpetuates the 19th-century prejudice that science and religion are always at war. This is far from the case, the examples of Darwin and Galileo notwithstanding. It is important that the Jesuits taught science to many generations, that Aristotelian Catholic physicists focused the attention of mathematicians on change when Plato had said that mathematics is about what does not change—leading ultimately to the mathematics of change, namely the calculus. It is important to remain aware, also, that mathematics receives ideas and suggestions from sources external to mathematics as well as within it—even including theology. (Besides the example of spherical trigonometry and the direction of Mecca, Cantor's infinites provide an interesting modern example.) And conversely: looking at the specifics of medieval mathematics—especially the mean-speed theorem—reminds us that things that we think of as applied mathematics, like Galileo's analysis of uniformly accelerated motion, were often undertaken for reasons that seem purely theoretical.

The Renaissance conceit that between the Graeco-Roman world and its "rebirth" there was a barren desert called "the Middle Ages" implies that history is marked by great explosions of ideas owing nothing to one's immediate predecessors. It also neglects Europe's debt, via medieval Europe, to the Islamic world. Mathematics is built on past mathematics. One should not disrespect one's teachers.

D. *Fourth Myth: Newton Invented the Calculus Just to Do His Mathematical Physics*

This one has a certain plausibility. After all, didn't Newton give his three laws of motion in the *Principia*, and then work out their consequences mathematically?

Yes, he did. But this myth is still wrong, on two counts. First, although Newton claimed late in his life that he worked out the *Principia* using calculus and only later presented it in the geometric form it now has, this seems to be false. The foremost scholar of Newton's mathematics, D. T. Whiteside of Cambridge, who edited Newton's mathematical manuscripts in eight volumes, says that there is no evidence that the *Principia* was ever done in any form than the geometric one in which we have it. Newton's claim to have done it by calculus seems to be part of his campaign to assert priority over Leibniz in the invention of the calculus. Still, it is also false to say that the *Principia* contains no calculus-like ideas. There certainly are such ideas: instantaneous velocity, limits, infinite series, the areas enclosed by curvilinear figures, and so on.

Nevertheless, the answer to the question, "Did Newton invent calculus to do physics?" is "no," because we can read Newton's manuscripts and actually follow his invention of the calculus in the early 1660s.

Newton had been reading Descartes' *Geometry* with the extensive commentary of the Dutch mathematician van Schooten. Newton also was studying the algebraic determination of curvilinear areas by John Wallis, and other 17th-century mathematical work on tangents, areas, arc lengths, and infinite series. It was Newton's desire to push these techniques further and solve a wider class of problems that led him to the general method of what he called fluxions and fluents, what we call derivatives and integrals. Reading his manuscripts not only demonstrates this, it lets us appreciate the excitement of his discoveries.[2]

As Whiteside summarizes the matter: "Descartes' [analytic] Geometry gave Newton his first true vision of the universalizing power of the algebraic free variable, and its capacity to generalize the particular and lay bare its inner structure to outward inspection." With a fully

---

[2] Recommended reading: D. T. Whiteside, ed., *Mathematical Papers of Newton*, Cambridge University Press, 1967. Volume I covers the period 1664–1666.

developed algebra, writes Whiteside, "the way was open [for Newton] to a universal calculus of *all* that is spatially definable."[3]

Why does it matter if people believe this myth about Newton?

Mathematics sometimes arises from scientific problems, but sometimes it does not. Mathematical ideas, for whatever reason they may be first considered, take on a life of their own, and mathematicians often pursue them with no concern about their applications. Amazingly enough, the mathematics often turns out to have applications anyway. It is important to understand this historical fact about mathematics and its applications—it is a strong argument for government funding of pure mathematics. Pure mathematics ultimately can pay off. Don't let them cite Newton—arguably the greatest scientist in history—as a counterexample.

E. Fifth Myth: Colin Maclaurin, Because of His Old-Fashioned Geometrical Approach to the Calculus, Halted Mathematical Progress in 18th-Century Britain

In 1742 Maclaurin published a 754-page book called *Treatise of Fluxions*. He started the book to refute Bishop Berkeley's attack on the rigor of the calculus (more on this later) by proving the truths of the calculus using methods modeled on the rigor of Greek geometry. Many historians have seen Maclaurin's book as a giant step backwards. For instance, we are told that Maclaurin's *Treatise* was "of little use for the researcher" (Guicciardini), that it had "no influence on the development of mathematics" (Mahoney) and that after his treatise, English mathematics became undistinguished for a century (H. W. Turnbull).

Is that so? Did Maclaurin really kill British mathematics because of his geometrical approach to analysis? Not at all. Maclaurin used his deep geometrical insight to study, among other things, convexity, concavity, inflection points—to study them analytically. It is because of his geometric motivation that Maclaurin asked the right questions. He was the first to characterize maxima and minima of functions in terms of the values—zero, negative, or positive—of first, second, up to the nth derivatives. How? He used the Maclaurin series to answer the question, "What happens analytically when a geometric curve is concave up, concave down, has a maximum, has a minimum?" And these pieces of geometry, translated into analysis, turn into statements about inequalities.

Maclaurin treats all the derivatives in the Taylor series as bounded. Then, in modern notation:

Let $f(x)$ be a maximum, greater than either $f(x+h)$ or $f(x-h)$, for which he draws a picture. From the series expansion, we have (in more modern notation):

$$f(x) > f(x+h) = f(x) + hf'(x) + h^2/2!\, f''(x) + h^3/3!\, f'''(x) + \cdots$$

and

$$f(x) > f(x-h) = f(x) - hf'(x) + h^2/2!\, f''(x) - h^3/3!\, f'''(x) + \cdots$$

Now, he says, for small enough h, the linear term in h can be made to exceed the sum of all the rest of the terms. So the only way both these inequalities can be true, with one adding

[3] See D. T. Whiteside, "Newton the Mathematician" [1982], reprinted in I B Cohen and R S Westfall, *Newton*, Norton, 1995, pp. 406–413.

$hf'(x)$ to f and one subtracting $hf'(x)$ from $f(x)$, is for $f'(x)$ to be zero. (This is his necessary condition for the maximum.) If $f'(x) = 0$, then h can be taken sufficiently small so that the h^2 term exceeds the sum of al the rest, and the only way $f(x)$ can still exceed both of those series is for $f''(x)$ to be negative (if f'' is not zero, this is a sufficient condition for the maximum). Unless of course f'' too is zero, and then the higher-order terms come into play. This was a good job, praised and then improved upon by Lagrange, who gave the same argument but made it more precise by using the Lagrange remainder for the Taylor series.

And there is much more. Precisely because Maclaurin knew his Greek geometry, especially the conic sections, inside out, he was able to make a major contribution to the theory of the shape of the earth. He proved that the ellipsoid of revolution is a figure of equilibrium for a rotating body made up of particles obeying Newtonian gravitation—important in determining the shape of the earth then, and the behavior of rotating bodies now.[4]

What difference does it make if people believe the myth about Maclaurin?

The myth asserts that British mathematics died after Newton. Why should somebody say this? This view was promulgated by early 19th-century English scientists who wanted to increase government and industrial support for science and mathematics—the so-called "decline of science in England" of Charles Babbage. That is not a fight we need to be engaged in now.

In addition, the myth overstates British mathematical isolation, in the age of Maclaurin and Stirling in Scotland, and also Simpson and Waring in England, and thus overstates Continental independence. After all, Continental mathematicians, including Clairaut, Euler, Lagrange, Cauchy, and Jacobi, praised and cited Maclaurin's work. Dispelling the myth helps us see that mathematicians seek each other out, even across oceans and political differences; they want to form, and often do form, an international research community.

Also, dispelling this myth teaches us something about progress: things can be lost as well as gained. The myth says, geometry is old, analysis is new, so geometry holds you back. But Maclaurin's geometric insight led him to new and important results. The triumph of analysis removed much of that geometry from mathematical education. This is not always an advantage.

F. *Sixth Myth: Lagrange was a Formalist. He Tried to Rigorize the Calculus, But Failed Because of His Unreflective Reliance on Formal Power Series*

In 1797, Lagrange tried to solve the problem of the foundations of the calculus, without any of the older ideas he thought were contradictory or vague: that is, without infinites, infinitesimals, or limits. But Lagrange did it by assuming that all functions had power series, and were uniquely defined by those power series, and that is obviously wrong and so he is wrong. Historians—Bell, Bourbaki—say that he was a formalist and thus in error.

Here is a little more of the story. In 1734, Bishop Berkeley had criticized the calculus for, among other things, lacking rigor. Part of his argument is that the computation of fluxions is logically inconsistent. To make this clear, using an even simpler example than he does, let us compute the fluxion of x^2.

---

[4] Judith V. Grabiner, "Was Newton's Calculus a Dead End? The Continental Influence of Maclaurin's *Treatise of Fluxions*," *Amer. Math. Monthly*, 1997.

Erik Sageng, "1742. Colin MacLaurin's Treatise of Fluxions," in I. Grattan-Guinness, ed., *Landmark Writings in Western Mathematics, 1640–1940* (Elsevier, 2005), pp. 143–158.

The ratio of differences is $[(x+h)^2 - x^2]/h = [2xh + h^2]/h = 2x + h$, which somehow, as h vanishes, becomes $2x$. Now either h is zero or it is not, says Berkeley. If it is zero, how can we set up this ratio, with top and bottom both zero? But if it is not zero, what right do we have to throw it away at the end?

Eighty years later Cauchy would answer this objection by, in effect, reinterpreting *equality* between a sequence and its limit as expressing instead an infinite set of *inequalities*: for any epsilon, we can find a delta such that if $|h| <$ delta, f' and the ratio differ in absolute value by less than epsilon. But Lagrange did not have Cauchy's limit concept. Instead, Lagrange believed that there was an algebra of infinite series. This idea was based on 18th-century work by people like Euler. It included conversions between infinite series and infinite products, by analogy with that for polynomials and finite products, a technique which led to correct results. Also, there were power-series representations derived—Euler would think, algebraically—for transcendental functions like sine, cosine, exponentials, and logarithms. Lagrange believed this was all algebraic. And, assuming—thinking he had proved—that every function had a power series expansion except at isolated points, he defined the coefficient of the linear term in the power-series expansion of $f(x+h)$ to be what he called the "derived function" of $f(x)$—the origin of our term "derivative"—and writing it as $f'(x)$.

That is, if

$$f(x+h) = f(x) + hp(x) + h^2q(x) + h^3r(x) + \cdots,$$

then Lagrange defined $f'(x) = p(x)$.

The higher-order derivatives were defined recursively by Lagrange. Thus $f''(x)$ is the coefficient of the linear term in the power-series expansion for $f'(x)$, and so on.

Lagrange's f' notation, he says, has a twofold purpose. It avoids using df/dx, which makes one erroneously think of infinitesimals. And it emphasizes that the derivative is not a ratio, but a function, and one derived from the original $f(x)$.

Thereafter, by term-by-term equating of coefficients in power series, Lagrange derives Taylor's series:

$$f(x+h) = f(x) + hf'(x) + h^2/2!\, f''(x) + h^3/3!\, f'''(x) + \cdots$$

It is almost always now given as I have given it, in Lagrange's notation, save for our using h for the increment instead of Lagrange's i, since the latter is now used for $\sqrt{-1}$.

One reason Lagrange finds his definition of $f'(x)$ as the coefficient of h in the Taylor-series expansion for $f(x+h)$ is how it answers objections like Berkeley's. Look at our simple example again.

$$(x+h)^2 = x^2 + 2xh + h^2.$$

The coefficient of h is exactly $2x$, so $2x$ is the derivative of x^2. There is no need to worry about what happened to the vanishing quantity. And of course his definition works for many other functions.

But so far, the myth seems to be right. In fact, it looks as though Lagrange has assumed the existence of infinitely many derivatives just so he can define the first derivative.

Lagrange, however, knows more about the derivative than that. It is true that he used this algebraic machinery, and had perhaps an unwarranted confidence in it because it seemed to him pure algebraic reasoning, owing nothing to intuition, to words, or to pictures. But he uses this machinery also because he is after something. He wanted also to justify an *inequality*, an inequality that would support proofs of all the major results of the calculus. And Lagrange had precisely the inequality that supports proofs of the major results of the calculus. Here it is:

Given any D (for "donnée), we can find h such that

$$h[f'(x) - D] \leq f(x+h) - f(x) \leq h[f'(x) + D].$$

Perhaps we are more used to seeing it divided by h:

$$f'(x) - D \leq [f(x+h) - f(x)]/h \leq f'(x) + D.$$

Lagrange got this inequality from his own power-series definition of the derivative. Borrowing the inequality technique we saw Maclaurin use, which was also used by Euler, and implicitly assuming the derivatives all to be bounded, and given the series

$$f(x+h) = f(x) + hf'(x) + h^2/2!\, f''(x) + h^3/3!\, f'''(x) + \cdots,$$

Lagrange said we can choose h sufficiently small so that any given term will exceed the sum of the remainder of the series. This leads immediately, if we take the second term to exceed the rest, to what I have called the Lagrange Property of the derivative:

$$f(x+h) = f(x) + hf'(x) + hV,$$

where V, each term of which has h as a factor, goes to zero with h.

The Lagrange property is key to his proofs of theorems of the calculus, since it is from the Lagrange property that Lagrange gets the inequalities given above. The inequality "Given any D, we can find h such that $h[f'(x) - D] \leq f(x+h) - f(x) \leq h[f'(x) + D]$" is how Lagrange translated the phrase "V goes to zero with h" in his proofs.

Lagrange used this inequality, and its higher-order analogues which are derived from Taylor series in the same way, to prove many important results, including that a function with positive derivative on an interval is increasing there—he was the first person to conceive that this even needed proof—the mean-value theorem for derivatives, Taylor's theorem with Lagrange remainder, and the fundamental theorem of calculus.

Cauchy saw something else: that Lagrange's inequality can be seen as a translation of the old, imprecise verbal statement that the derivative is the limit of the quotient of differences. So Cauchy gave a new definition of the derivative precisely to embody Lagrange's property. Cauchy defined the derivative as the limit, when it exists, of the quotient of differences, with limit understood in inequality terms. And Cauchy was then able to take over the structure of many of Lagrange's proofs, to within an alphabetical isomorphism. Cauchy used epsilon for Lagrange's D, and the corresponding value of h is chosen less than delta. Lagrange's whole structure of proved theorems then became, not the result of formalism, but the heart and soul of Cauchy's differential calculus. Cauchy lifted up Lagrange's entire structure of proofs, and put it on his new foundation.[5]

What difference does it make for us to get it right about Lagrange?

This story illustrates that the progress of mathematics is far from being linear; it proceeds by fits and starts, sometimes by means of blind alleys that nonetheless have treasure at the

[5] Judith V. Grabiner, *The Origins of Cauchy's Rigorous Calculus*, MIT Press, 1981.

end of them. One can be right for the wrong reasons. Mathematical invention cannot yet be programmed. Maybe this should give mathematicians—in fact, all of us—tolerance for our students and colleagues when they come at things from nontraditional perspectives.

One more thing. Getting the Lagrange story right reminds us also that foundations and definitions are not something just for chapter one, but are for use. They are chosen to support real mathematics, not just to answer critics like Berkeley (though they do that) and not just to placate students at the beginning of a course (though they do that too). Lagrange wants to prove the major results of calculus from unassailable first principles. So does Cauchy. And so does Weierstrass, who actually did it. Thus the final definitive formulation comes at the end of research on a subject, not at the start. The Lagrange story helps make this point.

G. *Last Myth, Held by a Number of Past and Present Mathematicians: The Mathematical Approach Can be Applied to Solve Almost Any Major Question*

Descartes, in his *Discourse on Method* of 1637, says his new mathematically based method can be used to discover "all things knowable by man." Many people agreed. For instance, the 18th-century Newtonian philosopher Francis Hutcheson wrote an essay called "On computing the morality of actions," in which he says, "that action is best, that produces the greatest happiness for the greatest number." There are many similar examples, including Jeremy Bentham, a Queen's College man. But first, let us look again at Cauchy.

If anybody since Descartes had the right to say he had brought about a methodological revolution in mathematics, it was Cauchy. In a series of beautifully logical textbooks, Cauchy developed the theory of convergence in 1821, and in 1823 gave rigorous definitions of limit, derivative, and integral and proved the key theorems about them (although at this time without distinguishing between pointwise and uniform convergence). Cauchy attacked virtually all his 18th-century predecessors on the subject of methods. They reason, he said, from the "generalness of algebra." This is unjustified, said Cauchy; formulas are only true under particular conditions and for specified values of the variables. So much for Lagrange. Similarly, one cannot simply jump from the real to the complex, or from the finite to the infinite. And, whatever Euler might have thought, Cauchy declared categorically, "A divergent series has no sum."

Was Cauchy, then, a dogmatist? Since many mathematicians who knew what rigor is have felt that they could legislate for all other fields as well, Cauchy's consistent religious dogmatism—he was a very conservative Catholic; his contemporary Neils Henrik Abel said Cauchy was "infinitely Catholic and bigoted"—might well suggest that he was one of those people. Note too that some of Cauchy's most eminent contemporaries were among those who used their mathematical authority to try to dictate to society. For instance, Laplace and Poisson favored using probabilistic models to evaluate human judges to the supposed betterment of justice. Also, just as Lagrange had tried to reduce the calculus to algebra, Lagrange's disciple Auguste Comte favored a reductionist model of all of the sciences in which physics is reduced to mathematics, as in Lagrange's *Analytical Mechanics*; then chemistry would be reduced to physics, biology to chemistry, psychology to biology, and finally Comte's new subject, sociology, could be reduced to psychology. So if we know mathematics we can know everything. And after all, Cauchy himself had reduced the calculus to algebra.

But I will allow Cauchy himself to dispel the myth of mathematical thought as the only way to truth. Religious dogmatist or not, Cauchy, while advocating methodological rigor within mathematics, expressed humility about other fields. His words in the *Cours d'analyse* are quite clear: "I have sought to perfect mathematical analysis; but I am far from claiming that this analysis can serve for all sciences that employ reasoning. . . . We cannot try to prove the existence of Augustus or Louis XIV by analysis, but every sensible man is as certain of their existence as of the square on the hypotenuse or the theorem of Maclaurin. . . . Let us assiduously cultivate the mathematical sciences, without wanting to extend them beyond their domain, and let us not imagine that we can attack history with formulas, nor justify morality with theorems of algebra or integral calculus."

Why does knowing that the master of rigor was not a methodological imperialist help us here?

Just because we have a hammer, everything else is not a nail. Success in mathematics does not imply that the mathematical method works everywhere. We—all of us, not just mathematicians—should learn from Cauchy's words in the *Cours d'analyse* to be aware of the limitations of our tools as well as their power. In many ways, the fate of the world depends on this lesson.

H. *A conclusion in four parts*

1) Mathematics is incredibly rich and mathematicians are unpredictably ingenious. Therefore, the history of mathematics is not rationally reconstructible. It must be the subject of empirical investigation.
2) Mathematics sometimes develops because it is working on problems posed by other fields. And sometimes it proceeds without any attention to those fields. Nonetheless, those fields may ultimately benefit.
3) Honest history reflects real mathematical practice. Students need to know that their crooked paths to problem-solution are the way professionals work. They need to know that all kinds of people do mathematics, and that mathematicians do not work alone, but form communities, even though the members of those communities need not all think in the same way. And mathematicians need to know that their success in doing mathematics does not justify equal confidence in the conclusions of other types of *a priori* reasoning.
4) Myths tend to serve somebody's interests. But truth should always be the more compelling interest.

Cauchy is correct; history and sociology are not mathematics. Still, to get them right matters. Mathematics is an important part of civilization and of modern society. People need to understand it and its many roles. But if members of our society are to understand mathematics, we need historians of it—because people need to know how mathematics actually developed, not how mathematicians or philosophers wish it had developed, or how other interested parties want it to develop. Thus, dispelling myths and promoting math are two complementary—and overwhelmingly important—tasks for anybody interested in the history of mathematics.

10

Why Did Lagrange "Prove" the Parallel Postulate?*

1. Introduction

We begin with an often-told story from the *Budget of Paradoxes* by Augustus de Morgan: "Lagrange, in one of the later years of his life, imagined" that he had solved the problem of proving Euclid's parallel postulate. "He went so far as to write a paper, which he took with him to the [Institut de France], and began to read it."

But, De Morgan continues, "something struck him which he had not observed: he muttered 'Il faut que j'y songe encore' [I've got to think about this some more] and put the paper in his pocket" [**8**, p. 288].

Is De Morgan's story true? Not quite in that form. But, as Bernard Cohen used to say, "Truth is more interesting than fiction." First, according to the published minutes of the Institut for 3 February 1806, "M. Delagrange *read* an analysis of the theory of parallels" [**25**, p. 314; italics added]. Those present are listed in the minutes: Lacroix, Cuvier, Bossut, Delambre, Legendre, Jussieu, Lamarck, Charles, Monge, Laplace, Haüy, Berthollet, Fourcroy—a most distinguished audience!

Furthermore: Lagrange did not throw his manuscript away. It survives in the library of the Institut de France [**32**]. There is a title page that says, in what looks to me like Lagrange's handwriting, "On the theory of parallels: memoir read in 1806," together with the signatures of yet more distinguished people: Prony and Poisson, along with Legendre and Lacroix. The first page of text says, again in Lagrange's handwriting, that it was "read at the Institut in the meeting of 3 February 1806."

It is true that Lagrange never did publish it, so he must have realized there was something wrong. In another version of the story, told by Jean-Baptiste Biot, who claims to have been there (though the minutes do not list his name), everybody there could see that something was wrong, so Lagrange's talk was followed by a moment of complete silence [**2**, p. 84]. Still, Lagrange kept the manuscript with his papers for posterity to read.

This episode raises the three questions I will address in this article. First, what did Lagrange actually say in this paper? Second, once we have seen how he "proved" the parallel postulate, why

* Reprinted from Amer. Math. Monthly 116, 1(Jan. 2009) 3–18.

did he do it the way he did? And last, above all, why did Joseph-Louis Lagrange, the consummate analyst, creator of the *Analytical Mechanics*, of Lagrange's theorem in group theory and the Lagrange remainder of the Taylor series, pioneer of the calculus of variations, champion of pure analysis and foe of geometric intuition, why did Lagrange risk trying to prove Euclid's parallel postulate from the others, a problem that people had been unsuccessfully trying to solve for 2000 years? Why was this particular problem in geometry so important to him?

I think that the manuscript is interesting in its own right, but I intend also to use it to show how Lagrange and his contemporaries thought about mathematics, physics, and the universe. As we will see, this was not the way we view these topics today.

2. The Contents of Lagrange's 1806 Paper

First, we look at the contents of the paper Lagrange read in 1806. The manuscript begins by asserting that the theory of parallels is fundamental to all of geometry. Notably, that includes the facts that the sum of the angles of a triangle is two right angles, and that the sides of similar triangles are proportional. But Lagrange agreed with both the ancients and moderns who thought that the parallel postulate should not be assumed, but needed to be proved.

To see why people wanted to prove the parallel postulate, let us recall Euclid's five geometric postulates [**9**, pp. 154–155]. The first is that a straight line can be drawn from any point to any other point; the second, that a finite straight line can be produced to any length; the third, that a circle can be drawn with any point as center and any given radius; the fourth, that all right angles are equal; and the fifth, the so-called parallel postulate, which is the one in question. Euclid's parallel postulate is not, as a number of writers wrongly say (e.g., [**5**, p. 126]), the statement that only one line can be drawn parallel to a given line through an outside point. Euclid's postulate states that, if a straight line falls on two straight lines making the sum of the interior angles on the same side of that line less than two right angles, then the two straight lines eventually meet on that side. Euclid used Postulate 5 explicitly only once: to prove that if two lines are parallel, the alternate interior angles are equal. Of course, many later propositions rest on this theorem, and thus presuppose the parallel postulate.

Already in antiquity, people were trying to prove Postulate 5 from the others. Why? Of course one wants to assume as little as possible in a demonstrative science, but few questions were raised about Postulates 1–4. The historical focus on the fifth postulate came because it felt more like the kind of thing that gets proved. It is not self-evident, it requires a diagram even to explain, so it might have seemed more as though it should be a theorem. In any case, there is a tradition of attempted proofs throughout the Greek and then Islamic and then eighteenth-century mathematical worlds. Lagrange followed many eighteenth-century mathematicians in seeing the lack of a proof of the fifth postulate as a serious defect in Euclid's *Elements*. But Lagrange's criticism of the postulate in his manuscript is unusual. He said that the assumptions of geometry should be demonstrable "just by the principle of contradiction"—the same way, he said, that we know the axiom that the whole is greater than the part [**32**, p. 30R]. The theory of parallels rests on something that is not self-evident, he believed, and he wanted to do something about this.

Now it had long been known—at least since Proclus in the fifth century—that the "only one parallel" property is an easy consequence of Postulate 5. In the eighteenth century, "only one parallel" was adopted as a postulate by John Playfair in his 1795 textbook *Elements of Geometry* and by A.-M. Legendre in his highly influential *Elements of Geometry* [**34**]. So this equivalent to

Postulate 5 had long been around; in the 1790s people focused on it, and so did Lagrange. But Lagrange, unlike Playfair and Legendre, didn't assume the uniqueness of parallels; he "proved" it. Perhaps now the reader may be eager to know, how did Lagrange prove it?

Recall that Lagrange said in this manuscript that axioms should follow from the principle of contradiction. But, he added, besides the principle of contradiction, "There is another principle equally self-evident," and that is Leibniz's principle of sufficient reason. That is: nothing is true "unless there is a sufficient reason why it should be so *and not otherwise*" [**42**, p. 31; italics added]. This, said Lagrange, gives as solid a basis for mathematical proof as does the principle of contradiction [**32**, p. 30V].

But is it legitimate to use the principle of sufficient reason in mathematics? Lagrange said that we are justified in doing this, because it has already been done. For example, Archimedes used it to establish that equal weights at equal distances from the fulcrum of a lever balance. Lagrange added that we also use it to show that three equal forces acting on the same point along lines separated by a third of the circumference of a circle are in equilibrium [**32**, pp. 31R–31V].

Now we are ready to see how Lagrange deduced the uniqueness of parallels from the principle of sufficient reason.

Suppose DE is drawn parallel to the given line AB through the given point C. Now suppose the parallel DE isn't unique. Then we can also draw FG parallel to AB. (See Figure 1a.) But everything ought to be equal on each side, Lagrange said, so there is no reason that FG should make, with DE, the angle ECG on the right side; why not also on the left side? So the line HCI, making the angle DCH equal to angle ECG, ought also to be parallel to AB. One can see why the argument so far seemed consistent with Lagrange's views on sufficient reason.

By the same procedure, he continued, we can now make another line KL that makes angle HCK equal to angle ICG, but placed on the other side of the new parallel line HI (see Figure 1b); and we can keep on in this way to make arbitrarily many in this fashion (see Figure 1c), which, as he said, "is evidently absurd" [**32**, pp. 32V–33R].

The modern reader may object that Lagrange's symmetry arguments are, like the uniqueness of parallels, equivalent to Euclid's postulate. But the logical correctness, or lack thereof, of Lagrange's proof is not the point. (In this manuscript, by the way, Lagrange went on to give an analogous proof—also by the principle of sufficient reason—that between two points there is just one straight line, because if there were a second straight line on one side of the first, we could

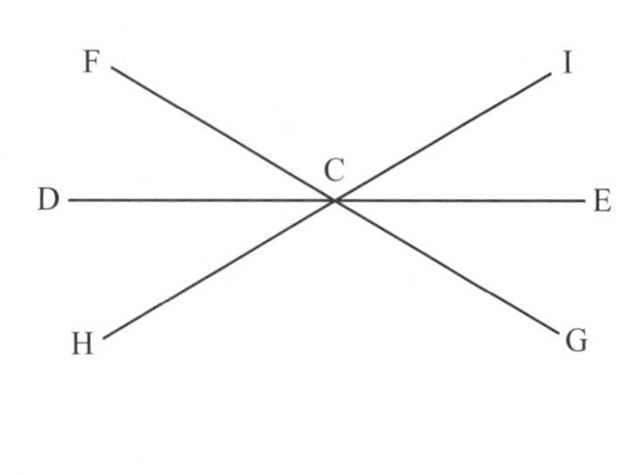

Figure 1a Lagrange's proof, step 1.

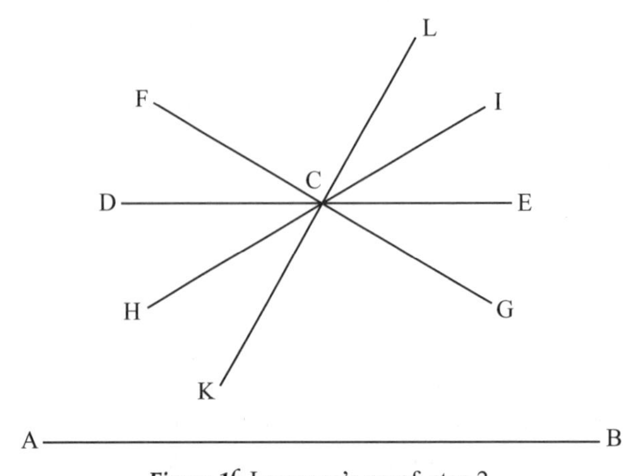

Figure 1b Lagrange's proof, step 2.

then draw a third straight line on the other side, and so on [**32**, pp. 34R–34V]. Lagrange, then, clearly liked this sort of argument.)

3. Why Did He Attack the Problem This Way?

It is now time to address the second, and more important question: Why did he do it in the way he did?

I want to argue this: Lagrange's arguments from sufficient reason were shaped by properties of space, space as it was believed to be in the seventeenth and eighteenth centuries. These properties are profoundly Euclidean. To eighteenth-century thinkers, space was infinite, it was exactly the same in all directions, no direction was privileged, it was like the plane in having no curvature, and symmetrical situations were equivalent. Lagrange himself explicitly linked his symmetry arguments to Leibniz's principle of sufficient reason, but—as we will see—these ideas are also historically linked to Giordano Bruno's arguments for the infinite universe, Descartes' view of space as indefinite material extension, the projective geometry used to describe perspective in

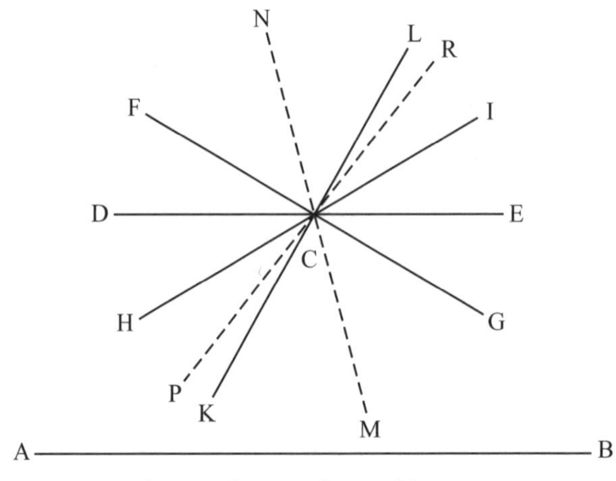

Figure 1c Lagrange's proof, last step.

Renaissance art, various optimization arguments like "light travels in straight lines because that is the shortest path," and, above all, the Newtonian doctrine of absolute space. As we will soon see, these properties were essential to physical science in the seventeenth and eighteenth centuries: both physics and philosophy promoted the identification of space with its Euclidean structure.

This goes along with a shift in emphasis concerning what Euclidean geometry is about. Geometry, in ancient times, was the study of geometric figures: triangles, circles, parallelograms, and the like, but by the eighteenth century it had become the study of space [**41**, chapter 5]. The space eighteenth-century geometry was about was, in Henri Poincaré's words, "continuous, infinite, three-dimensional, homogeneous and isotropic" [**44**, p. 25]. Bodies moved through it preserving their sizes and shapes. The possible curvature of three-dimensional space did not even occur to eighteenth-century geometers. Their space was Euclidean through and through.

Why did philosophers conclude that space had to be infinite, homogeneous, and the same in all directions? Effectively, because of the principle of sufficient reason. For instance, Giordano Bruno in 1600 argued that the universe must be infinite because there is no reason to stop at any point; the existence of an infinity of worlds is no less reasonable than the existence of a finite number of them. Descartes used similar reasoning in his *Principles of Philosophy*: "We recognize that this world... has no limits in its extension.... Wherever we imagine such limits, we... imagine beyond them some indefinitely extended space" [**28**, p. 104]. Similar arguments were used by other seventeenth-century authors, including Newton. Descartes identified space and the extension of matter, so geometry was, for him, about real physical space. But geometric space, for Descartes, had to be Euclidean. This is because the theory of parallel lines is crucial for Descartes' analytic geometry—not for Cartesian coordinates, which Descartes did not have, but because he needed the theory of similar figures in order to give meaning to expressions of arbitrary powers of x [**23**, p. 197]. Descartes was the first person to justify using such powers. But an expression like x^4 for Descartes is not the volume of a 4-dimensional figure, but a line, which can be defined as the fourth proportional to the unit line, x, and x^3. That is, $1/x = x^3/x^4$. They are all lines, and since all powers of x are lines, they can all be constructed geometrically—but only if we have the theory of similar triangles, for which we need the theory of parallels.

Now let us turn from seventeenth-century philosophy to seventeenth-century physics. Descartes, some 50 years before Newton published his first law of motion, was a co-discoverer of what we call linear inertia: that in the absence of external influences a moving body goes in a straight line at a constant speed. Descartes called this the first law of nature, and for him, this law follows from what we now recognize as the principle of sufficient reason. Descartes said, "Nor is there any reason to think that, if [a part of matter] moves... and is not impeded by anything, it should ever by itself cease to move with the same force" [**30**, p. 75]. And the straight-line motion of physical moving bodies obviously requires the indefinite extendability of straight lines and thus indefinitely large, if not infinite, space [**23**, p. 97].

Descartes' contemporary Pierre Gassendi, another co-discoverer of linear inertia, used "sufficient reason" to argue for both inertia and the isotropy of space. Gassendi said, "In principle, all directions are of equal worth," so that in empty spaces, "motion, in whatever direction it occurs... will neither accelerate nor retard; and hence will never cease" [**29**, p. 127].

Artists, too, helped people learn to see space as Euclidean. We see the space created in the paintings and buildings of the Renaissance and later as Euclidean. Renaissance artists liked to portray floors with rectangular tiles and similar symmetric architectural objects—to show how good they were at perspective. These works of art highlight the observations that parallel lines

Figure 2 Piero della Francesca (1410/1420–1492), "The Ideal City."

are everywhere equidistant, that two lines perpendicular to a third line are parallel to each other. And our experience of perspective in art and architecture helps us shape the space we believe we live in. (See Figures 2, 3, and 4.) We have seen pictures like these many times, but consider them now as conditioning people to think in a particular way about the space we live in: as Euclidean, symmetric, and indefinitely extendible—going on to infinity [**11**].

Artist-mathematicians like Piero della Francesca began the development of the subject of projective geometry, but the first definitive mathematical treatise on it is that of Girard Desargues in the 1630s. Seventeenth-century projective geometry used the cone (like the artist's rays of sight or light) to prove properties of all the conic sections as projections of the circle. For instance, geometers treated the ellipse as the circle projected to a plane not perpendicular to the cone. And the parabola, as Kepler pointed out, behaves projectively like an ellipse with one focus at infinity. So projective geometry explicitly brought infinity into Euclidean geometry: planes and lines go to infinity; parallel lines meet at the point at infinity.

And the geometry of perspective and projective geometry reinforced Euclideanness in a wide variety of other ways, from the role of Euclid's *Optics* in the humanistic classical tradition to

Figure 3 Leonardo da Vinci (1452–1519), "The Last Supper."

Figure 4 Raphael (1483–1520), "The School of Athens."

the use of the theory of parallels to draw military fortifications from 2-dimensional battlefield sketches [**12**, p. 24].

Although these Euclidean views prevailed, perhaps they didn't have to. There were alternatives suggested even in the eighteenth century [**22**]. Is visual space Euclidean? Not necessarily. Bishop Berkeley, for instance, said that we don't "see" distance at all; we merely infer it from the angles we do see. And Thomas Reid pointed out that a straight line right in front of you looks exactly like a circle curved with you at the center—or even a circle curved away from you in the other direction. Reid gave a set of rules for visual space—he called this the "geometry of visibles"—which clearly are not Euclid's rules; a modern philosopher has called Reid's geometry of visibles "the geometry of the single point of view" [**46**, p. 396].

And there are other alternatives to Euclideanness. Cultures other than the western often speak about space differently and order their perceptions differently: particular directions have special connotations, and "closeness" can be cultural as well as metrical. Many cultures do not use the idea of an outside abstract space at all; instead—as Leibniz did—they recognize only the relations between bodies [**3**], [**35**], [**36**]. So, as a matter of empirical fact, abstract Euclidean space is not something that all human thinkers do use, let alone that all humans must use.

In the twentieth century, experimental psychologists showed that when people in a dark room are asked to put luminous points into two equidistant lines, or two parallel lines, the people are satisfied when the lines in fact curve away from the observer. As a result, Rudolf Luneburg in the 1940s claimed that visual space is a hyperbolic space of constant curvature;

Figure 5 Parmigianino (1503–1540), "Self Portrait in a Convex Mirror."

later psychological experiments suggest that visual space is not represented by any consistent geometry [**48**, pp. 30–31].

Even in the Renaissance, some painters portrayed what we now recognize as 3-dimensional non-Euclidean spaces, using reflections in a convex mirror, notably Parmigianino's (1524) "Self Portrait in a Convex Mirror," and, most famously, the "Arnolfini Wedding" by Jan van Eyck (1434). (See Figures 5, 6a, and 6b.) In the spaces in these mirrors, parallel lines are not everywhere equidistant.

A modern physicist, John Barrow, has said that if people had paid more attention to these mirrors, non-Euclidean geometry might have been discovered much sooner [**1**, p. 176]. But I am not so sure. I think that these artists viewed convex mirrors as presenting an especially difficult problem in portraying 3-dimensional Euclidean space on a 2-dimensional Euclidean canvas; for instance, J. M. W. Turner included such drawings in his strongly Euclidean lectures [**45**] on perspective. (See Figures 7a, 7b.)

The winning view, I think, is that expressed by the Oxford art historian Martin Kemp, who says that from the Renaissance to the nineteenth century, the goal of constructing "a model of the world as it appears to a rational, objective observer" [**27**, p. 314] was shared by scientists and artists alike. Virtually unanimously, artists, armed with Euclid's *Optics*, have long helped teach us to "see" a Euclidean world.

4. The Crucial Argument: Newtonian Physics

Let us now return to physics and to the most important seventeenth-century argument of all for the reality of infinite Euclidean space: Newtonian mechanics. Newton needed absolute space as a reference frame, so he could argue that there is a difference between real and apparent accelerations. He wanted this so he could establish that the forces involved with absolute (as opposed to relative) accelerations are real, and thus that gravity is real. Newton's absolute space is infinite and uniform, "always similar and immovable" [**38**, p. 6], and he described its properties in Euclidean terms. And it is real; it has a Platonic kind of reality.

Figure 6a Jan van Eyck (c. 1390–1441), "The Arnolfini Wedding."

Leibniz, by contrast, did not believe in absolute space. He not only said that spatial relations were just the relations between bodies, he used the principle of sufficient reason to show this. If there were absolute space, there would have to be a reason to explain why two objects would be related in one way if East is in one direction and West in the opposite direction, and related in another way if East and West were reversed [**24**, p. 147]. Surely, said Leibniz, the relation between two objects is just one thing! But Leibniz did use arguments about symmetry and sufficient reason—sufficient reason was his principle, after all. Thus, although Descartes and

Figure 6b Detail from "The Arnolfini Wedding."

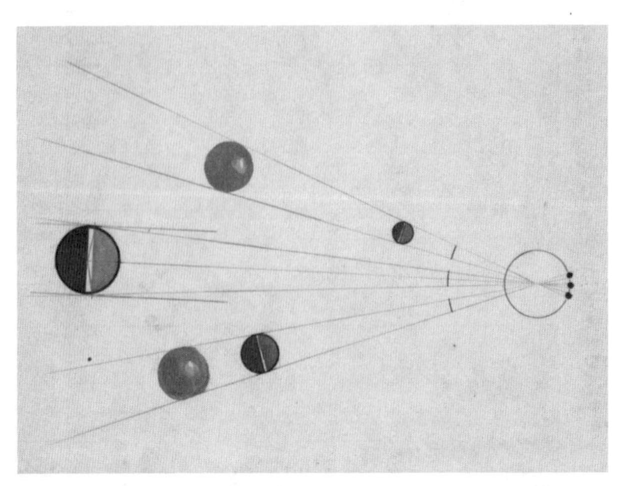

Figure 7a J. M. W. Turner (1775–1851), “Spheres at Different Distances from the Eye.” © Tate, London 2008.

Leibniz did not believe in empty absolute space and Newton did, they all agreed that what I am calling the Euclidean properties of space are essential to physics.

In the eighteenth century, the Leibniz-Newton debate on space was adjudicated by one of Lagrange’s major intellectual influences, Leonhard Euler. In his 1748 essay “Reflections on Space and Time,” Euler argued that space must be real; it cannot be just the relations between bodies as the Leibnizians claim [**10**]. This is because of the principles of mechanics—that is, Newton’s first and second laws. These laws are beyond doubt, because of the “marvelous” agreement they have with the observed motions of bodies. The inertia of a single body, Euler said, cannot possibly depend on the behavior of other bodies. The conservation of uniform motion in the same direction makes sense, he said, only if measured with respect to immovable space, not to various other bodies. And space is not in our minds, said Euler; how can physics—real physics—depend on something in our minds? So space for Euler is real.

The philosopher Immanuel Kant was influenced by Euler’s analysis [**14**, pp. 29, 207]. Kant agreed that we need space to do Newtonian physics. But in his *Critique of Pure Reason* of 1781, Kant placed space in the mind nonetheless. We order our perceptions in space, but space itself is in the mind, an intuition of the intellect. Nevertheless, Kant’s space turned out to be Euclidean too. Kant argued that we need the intuition of space to prove theorems in geometry. This is

Figure 7b J. M. W. Turner, “Reflections in a Single Metal Globe and in a Pair of Polished Metal Globes.” © Tate, London 2008.

because it is in space that we make the constructions necessary to prove theorems. And what theorem did Kant use as an example? The sum of the angles of a triangle is equal to two right angles, a result whose proof requires the truth of the parallel postulate [**26**, "Of space," p. 423].

Even outside of mathematics and physics, explicit appeals to sufficient reason, symmetry, parallels, and infinity pervade eighteenth-century thought, from balancing chemical equations to symmetry in architecture to the balance of powers in the U.S. Constitution.

Let me call one last witness from philosophy: Voltaire. Like many thinkers in the eighteenth century, Voltaire said that universal agreement was a marker for truth. Religious sects disagree about many things, he said, so on these topics they are all wrong. But by contrast, they all agree that one should worship God and be just; therefore that must be true. Voltaire pointed out also that "There are no sects in geometry" [**47**, p. 195]. One does not say, "I'm a Euclidean, I'm an Archimedean." What everyone agrees on: that is what is true. "There is but one morality," said Voltaire, "as there is but one geometry" [**47**, p. 225].

5. The Argument from Eighteenth-Century Mathematics and Science

Now let us turn to eighteenth-century mathematics and science. Eighteenth-century geometers tended to go beyond Euclid himself in assuming Euclideanness. As a first example, look at the 1745 *Elémens de Géométrie* by Alexis-Claude Clairaut. Clairaut grounded geometry not on Euclid's postulates but on the capacity of the mind to understand clear and distinct ideas. For instance, Euclid had defined parallel lines as lines in the same plane that never meet. Clairaut, less interested in proof than in Euclidean plausibility, defined parallel lines as lines that are everywhere equally distant from one another [**6**, p. 10]. The great French *Encyclopedia* [**7**, vol. 11, pp. 905–906] defined parallel lines as "lines that prolonged to infinity never get closer or further from one another, or *that meet at an infinite distance*" [italics added] assuming, then, a uniform, flat Euclidean space infinitely extended. The "equidistant" definition of parallels is reinforced by ordinary language, as we speak of parallel developments, or, more geometrically, ships on parallel courses, and even of parallels of latitude.

And speaking of latitude raises the question of why the fact that the geometry on the surface of a sphere, with great circles serving as "lines," is not Euclidean—there are no parallels, for example—did not shake mathematicians' conviction that all of Euclid's postulates are true and mutually consistent. Lagrange himself is supposed to have said that spherical trigonometry does not need Euclid's parallel postulate [**4**, pp. 52–53]. But the surface of a sphere, in the eighteenth-century view, is not non-Euclidean; it exists in 3-dimensional Euclidean space [**20**, p. 71]. The example of the sphere helps us see that the eighteenth-century discussion of the parallel postulate's relationship to the other postulates is not really about what is logically possible, but about what is true of real space.

Now, let us turn to eighteenth-century physics. As we will see, Euclideanness, especially the theory of parallels and the principle of sufficient reason, was essential to the science of mechanics in the eighteenth century, not only to its exposition, but to its progress.

Johann Heinrich Lambert was one of the mathematicians who worked on the problem of Postulate 5. Lambert explicitly recognized that he had not been able to prove it, and considered that it might always have to remain a postulate. He even briefly suggested a possible geometry on a sphere with an imaginary radius. But Lambert also observed that the parallel postulate is related to the law of the lever [**20**, p. 75]. He said that a lever with weightless arms and with equal weights at equal distances is balanced by a force in the opposite direction at the center equal to the sum of the weights, and that all these forces are parallel. So either we are using the parallel postulate, or perhaps, Lambert thought, some day we could use this physical result to prove the parallel postulate.

Lagrange himself in his *Analytical Mechanics* [**31**, pp. 4–5] gave an argument about balancing an isosceles triangle similar to, but much more complex than, Lambert's discussion. Lagrange himself did not explicitly link the law of the lever to the parallel postulate, but the geometry of the equilibrium situation that Lagrange was describing nonetheless requires it [**4**, pp. 182–183]. In a similar move, d'Alembert had tried to deduce the general law of conservation of momentum purely from symmetry principles [**13**, pp. 821–823]. And in the 1820s, J.-B. Fourier, from a very different philosophical point of view, also said that the parallel postulate could be derived from the law of the lever. From this Fourier concluded that geometry follows from statics and so geometry is a physical science [**20**, pp. 78–79]. But note that it is still Euclidean geometry.

Let us now concentrate further on Lagrange's mechanics. His deepest conviction was that a subject must be seen in its full generality. Like many Enlightenment thinkers only more so, Lagrange wanted to reduce the vast number of laws and principles to a single fundamental general principle, preferably one that is independent of experience. "Sufficient reason" was such a principle.

Although he did not explicitly cite Leibniz's principle in his *Analytical Mechanics* [**43**, p. 146], Lagrange used it frequently. For instance, he wrote, "The equilibrium of a straight and horizontal lever with equal weights and with the fulcrum at its midpoint is a self-evident truth because there is no reason that either of the weights should move."

Another key example of Euclideanness as physical argument is the use of parallelograms to find resultant forces. Lagrange, in his *Analytical Mechanics*, used the principle of sufficient reason, Euclid's theory of parallels, and the infinity of space and its Euclidean nature to discuss the composition of forces [**31**, p. 17]. He said that a body which is moved uniformly in two different directions simultaneously must necessarily traverse the diagonal of the parallelogram whose sides it would have followed separately. So parallels are needed. Lagrange continued, "with regard to the direction in the case of two equal forces, it is obvious that there is no reason that the resultant force should be nearer to one than the other of these two equal forces; therefore it must bisect the angle formed by these two forces" [**31**, p. 21]. Lagrange also used the principle of composition of forces to get the conditions of equilibrium when two parallel forces are applied to the extremities of a straight lever. He suggested that we imagine "that the directions of the forces extend to infinity" and then, on this basis, we can prove that "the resultant force must pass through the point of support." In effect this is the parallelogram argument with the corner of the parallelogram at infinity. Lagrange tried to reduce even his own fundamental physical principle—the principle of virtual velocities—to levers and parallelograms of forces, and after him Ampère, Carnot, Laplace, and Poisson tried to do the same [**39**, p. 218].

Pierre-Simon Laplace, too, related *a priori* arguments, including sufficient reason and Euclid's theory of parallels, to argue that physical laws had to be the way they were. For instance he said that a particle on a sphere moves in a great circle because "there is no reason why it should deviate to the right rather than the left of that great circle"—notice, "not a word about the forces acting on this particle on this sphere" [**15**, p. 104]. Laplace also asked why gravitation had to be inverse-square, and gave a geometric answer [**4**, p. 53], [**16**, p. 310]. He said that inverse-square gravitation implies that if the size of all bodies and all distances in the whole universe were to decrease proportionally, the bodies would describe the same curves that they do now, so that universe would still look exactly the same. The observer, then, needs to recognize only the ratios. So, Laplace said, even though we haven't proved Euclid's fifth postulate, we know it must be true, and so the theorems deduced from it must also be true. For Laplace, then, the idea of space includes the following self-evident property: *similar figures have proportional sides* [**4**, p. 54]. The Newtonian physical universe requires similar figures to have proportional sides, and this of course requires the theory of parallels, and thus geometry must be Euclidean [**33**, p. 472].

These men did not want to do mechanics, as, say, Newton had done. They wanted to show not only that the world was this way, but that it necessarily had to be. A modern philosophical critic, Helmut Pulte, has said that Lagrange's attempt to "reduce" mechanics to analysis strikes us today as "a misplaced endeavour to mathematize . . . an empirical science, and thus to endow it with infallibility" [**39**, p. 220]. Lagrange would have responded, "Right! That's just exactly what we are all doing." Lagrange thought these two things: Geometry is necessarily true; mechanics is mathematics. He needed them both.

6. Why Did It Matter so Much?

And now, we are ready for the last question. Why did actually proving Postulate 5 matter so much to Lagrange and to his contemporaries? I trust I have convinced the reader of the central role of Euclideanness in the eighteenth century. But still, if it were just a matter of simple logic, surely after 2000 years people should have concluded: we have been trying as hard as possible, we cannot imagine how to prove this, so let us just concede defeat. It can't be done. Euclid was right in deciding that it had to be assumed as a postulate. Why did eighteenth-century geometers not settle for this, and, in particular, why didn't Lagrange?

Because there was so much at stake. Because space, for Newtonian physics, has to be uniform, infinite, and Euclidean, and because metaphysical principles like that of sufficient reason and optimality were seen both as Euclidean and as essential to eighteenth-century thought. How could all of this rest on a mere assumption? So, many eighteenth-century thinkers believed that it was crucial to shore up the foundations of Euclid's geometry, and we can place Lagrange's manuscript in the historical context of the many attempts in the eighteenth century to cure this "blemish" in Euclid by proving Postulate 5.

Also, Lagrange was not just any eighteenth-century mathematician. Lagrange was, mathematically speaking, a Cartesian and a Leibnizian. His overall philosophy of mathematics was to reduce each subject to the most general possible principle. In calculus, as I have argued at length in two books [**17**], [**18**], Lagrange wanted to reduce all the ideas of limits and infinites and infinitesimals and rates of change or fluxions to "the algebraic analysis of finite quantities." In algebra, Lagrange said that even Newton's idea of algebra as "universal arithmetic" wasn't general enough; algebra was the study of systems of operations. In mechanics, his goal was to reduce everything to the principle of virtual velocities—and then to use "only algebraic operations subject to a regular and uniform procedure" [**31**, preface]. Lagrange even composed his *Analytical Mechanics* without a single diagram, precisely so he could show he had reduced physics to pure analysis. Geometry, then, ought also to be reducible to self-evident principles, to clear, distinct, and general ideas.

Finally, there are social causes to be considered. First, the social background will help answer this question: Why was Lagrange doing this in 1806, as opposed, say, to the 1760s when he taught mathematics at the military school in Turin or in the 1770s and 1780s when he was the leading light of the Berlin Academy of Sciences? One reason is that in about 1800 there was a revival of interest in synthetic geometry in France. There was a Parisian school in synthetic geometry including Monge, Servois, Biot, Lacroix, Argand, Lazare Carnot and his students, and Legendre. Important reasons for this were partly practical, partly ideological [**40**, p. 450]. The practical needs are related to Monge's championing of descriptive geometry, so clearly useful in architecture and in military planning. Monge also helped directly to pique Lagrange's interest, writing two letters to him in the early 1790s soliciting his assistance on problems involving the geometry of perspective [**37**].

As for the ideology promoting geometry in France after the Revolution, as Joan Richards has written, "the quintessentially reasonable study of universally known space had a central role to play in educating a rational populace" [**40**, p. 454]. Lagrange himself articulated such views throughout his lifetime, writing as early as 1775 that synthetic geometry was sometimes better than analytic because of "the luminous clarity that accompanies it" [**19**, p. 135], and, near the end of his life, telling his friend Frédéric Maurice that "geometric considerations give force and clarity to judgement" [**19**, p. 1295].

So the French geometers would not have been favorably disposed to inventing a non-Euclidean geometry. It is no wonder that only comparative outsiders like the Hungarian Janos Bolyai and the Russian Nikolai Ivanovich Lobachevsky were the first to publish on this topic. Even Gauss, who out of fear of criticism did not publish his own invention of the subject that he christened "non-Euclidean geometry," was somewhat outside the French mathematical mainstream.

The British would not have found inventing non-Euclidean geometry enticing either. Even William Rowan Hamilton, who in the 1840s was to devise the first noncommutative algebra, wrote in 1837, "No candid and intelligent person can doubt the truth of the chief properties of Parallel Lines, as set forth by Euclid in his Elements, two thousand years ago. . . . The doctrine involves no obscurity nor confusion of thought and leaves in the mind no reasonable ground for doubt" [**21**, p. 354]. In fact, even after Hermann von Helmholtz and W. K. Clifford had introduced non-Euclidean geometry into Victorian Britain, some British thinkers continued to maintain that real space had to be Euclidean. There was a great deal at stake in Britain: the doctrine of the unity of truth, the established educational program based on the Euclidean model of reason, and the attitudes toward authority that this entailed.

The authority and rigor of Euclid, both in Britain and in France, were part and parcel of the established intellectual order. And—one last social point—non-Euclidean geometry even in the twentieth century was culturally seen as anti-establishment, partly through its association with relativity theory. For instance, surrealist artists used it that way: misunderstood, perhaps, but still explicitly part of their assault on traditional artistic canons. Two examples are Yves Tanguy's "Le Rendez-vous des parallèles" (1935) and Max Ernst's "Young Man Intrigued by the Flight of a Non-Euclidean Fly" (1942–1947).

And Figure 8 shows an example from an artist who really did understand what a 3-dimensional non-Euclidean space might look like.

Figure 8 M. C. Escher (1898–1972), "Hand with Reflecting Sphere."

7. Conclusion

I cannot explain why Lagrange initially thought that his proof was a good one, but I hope it is clear why he thought he needed to prove the parallel postulate, and why he tried to prove it using the techniques that he used.

The story I have told reminds us that, although the great eighteenth-century mathematicians are our illustrious forbears, our world is not theirs. We no longer live in a world of certainty, symmetry, and universal agreement. But it was only in such a world that the work of Lagrange and Laplace, Fourier and Kant, Euler and d'Alembert could flourish. That space must be Euclidean was part of the Cartesian, Leibnizian, Newtonian, symmetric, economical, and totally rationalistic world view that underlies all of Lagrange's mathematics and classical mechanics—ideas that, from Newton and Leibniz through Kant and Laplace, buttressed the whole eighteenth-century view of the universe and the laws that govern it. And the certainty of Euclidean geometry was the model for the whole Enlightenment program of finding universally-agreed-upon truth through reason.

Thus, though Lagrange's illustrious audience in Paris may have realized that his proof was wrong, their world-view made them unable to imagine that the parallel postulate couldn't be proved, much less to imagine that the world itself might be otherwise.

Acknowledgments

I thank the Department of the History and Philosophy of Science, University of Leeds, England, for its hospitality and for vigorous discussions of this research. I also thank the Bibliothèque de l'Institut de France for permission to study Lagrange's manuscripts, the donors of the Flora Sanborn Pitzer Professorship at Pitzer College for their generous support, the Mathematical Association of America for inviting me to talk about this topic at MathFest 2007, and Miss Kranz, my trigonometry teacher at Fairfax High School in Los Angeles, who once on a slow day in class revealed to us all that there was such a thing as non-Euclidean geometry.

For the right to reproduce the works of art in this paper, I thank the following:

Figures 2, 3, 4, 5, 6a, 6b: Erich Lesser/Art Resource;

Figures 7a and 7b: The Tate Gallery, London;

Figure 8: M.C. Escher's "Hand with Reflecting Sphere" © 2008 The M.C. Escher Company-Holland: All rights reserved. www.mcescher.com

References

1. J. D. Barrow, Outer space, in *Space: In Science, Art and Society*, F. Penz, G. Radick, and R. Howell, eds., Cambridge University Press, Cambridge, 2004, 172–200.
2. J.-B. Biot, Note historique sur M. Lagrange, in *Mélanges Scientifiques et Littéraires*, vol. III, Michel Lévy Frères, Paris, 1858, 117–124.
3. P. Bloom et al., *Language and Space*, MIT Press, Cambridge, MA, 1996.
4. R. Bonola, *Non-Euclidean Geometry*, Dover, New York, 1955.
5. R. Carnap, *An Introduction to the Philosophy of Science*, Dover, New York, 1995; reprint of Basic Books, New York, 1966.
6. A.-C. Clairaut, *Éléméns de Géométrie*, Par la Compagnie des Libraires, Paris, 1765.
7. J. D'Alembert and D. Diderot, eds., *Encyclopédie, ou Dictionnaire Raisonné des Sciences, des Arts et des Métiers*, Briasson, Paris, 1762–1772.
8. A. De Morgan, *A Budget of Paradoxes*, Longmans Green, London, 1872.
9. Euclid, *The Thirteen Books of Euclid's Elements*, T. L. Heath, ed., Cambridge University Press, Cambridge, 1956.
10. L. Euler, Reflexions sur l'espace et le tems, *Mémoires de l'académie de Berlin* **4** (1750) 324–333.
11. J. V. Field, *The Invention of Infinity: Mathematics and Art in the Renaissance*, Oxford University Press, Oxford, 1997.
12. J. V. Field and J. J. Gray, *The Geometrical Work of Girard Desargues*, Springer, New York, 1987.
13. J. Franklin, Artifice and the natural world: Mathematics, logic, technology, in *Cambridge History of Eighteenth Century Philosophy*, K. Haakonssen, ed., Cambridge University Press, Cambridge, 2006, 815–853.

14. M. Friedman, *Kant and the Exact Sciences*, Harvard University Press, Cambridge, MA 1992.

15. E. Garber, *The Language of Physics: The Calculus and the Development of Theoretical Physics in Europe, 1750–1914*, Birkhäuser, Boston, 1999.

16. C. C. Gillispie, *Pierre-Simon Laplace, 1749–1827: A Life in Exact Science*, in collaboration with R. Fox and I. Grattan-Guinness, Princeton University Press, Princeton, 1997.

17. J. V. Grabiner, *The Calculus as Algebra: J.-L. Lagrange, 1736–1813*, Garland, New York, 1990.

18. ———, *The Origins of Cauchy's Rigorous Calculus*, MIT Press, Cambridge, MA, 1981.

19. I. Grattan-Guinness, *Convolutions in French Mathematics, 1800–1840*, 3 vols., Birkhäuser, Basel, 1990.

20. J. Gray, *Ideas of Space: Euclidean, Non-Euclidean and Relativistic*, 2nd ed., Clarendon Press, Oxford, 1989.

21. T. L. Hankins, Algebra as pure Time: William Rowan Hamilton and the foundations of algebra, in *Motion and Time, Space and Matter*, P. Machamer and R. Turnbull, eds., Ohio State University Press, Columbus, OH, 1976, 327–359.

22. R. J. Herrnstein and E. G. Boring, *A Source Book in the History of Psychology*, Harvard University Press, Cambridge, MA, 1966.

23. L. Hodgkin, *A History of Mathematics*, Oxford University Press, Oxford, 2005.

24. N. Huggett, ed., *Space from Zeno to Einstein: Classic Readings with a Contemporary Commentary*, MIT Press, Cambridge, MA, 1999.

25. Institut de France, Académie des Sciences, *Procès-Verbaux des Séances de l'Académie, 1804–1807*, vol. III, Académie des Sciences, Hendaye, 1913.

26. I. Kant, *Critique of Pure Reason* (trans. F. M. Müller), Macmillan, New York, 1961.

27. M. Kemp, *The Science of Art: Optical Themes in Western Art from Brunelleschi to Seurat*, Yale University Press, New Haven, CT, 1990.

28. A. Koyré, *From the Closed World to the Infinite Universe*, Johns Hopkins University Press, Baltimore, MD, 1957.

29. ———, *Metaphysics and Measurement*, Harvard University Press, Cambridge, MA, 1968.

30. ———, *Newtonian Studies*, University of Chicago Press, Chicago, IL, 1965.

31. J.-L. Lagrange, *Analytical Mechanics*, A. Boissonnade and V. N. Vagliente, trans. and eds., Kluwer, Dordrecht, 1997; from J.-L. Lagrange, *Mécanique analytique*, 2nd ed., Courcier, Paris, 1811–1815. In *Oeuvres de Lagrange*, Gauthier-Villars, Paris, 1867–1892, vol. XI.

32. ———, Sur la Théorie des Parallèles, Mémoire lu en 1806. Manuscript in the Bibliothèque de l'Institut de France, Inst MS 909, ff 18–35. Transcribed and published in the original French, together with a useful historical introduction in Italian, in Borgato, M. T., and Pepe, L., Una memoria inedita di Lagrange sulla teoria delle parallele, *Bollettino di Storia delle Scienze Matematiche* **8** (1988) 307–335.

33. P.-S. Laplace, *Exposition du système du monde*, Cercle-Social l'An IV, Paris, 1796, in *Ouevres complètes de Laplace*, Gauthier-Villars, Paris, 1878–1912, vol. VI.

34. A.-M. Legendre, *Eléments de Géométrie*, Didot, Paris, 1794.

35. S. C. Levinson, Language and mind: Let's get the issues straight, in *Language in Mind: Advances in the Study of Language and Thought*, D. Gertner and S. Goldin-Meadow, eds., MIT Press, Cambridge, MA, 2003, 25–46.

36. ———, Frames of reference and Molyneux's question: Crosslinguistic evidence, in *Language and Space*, MIT Press, Cambridge, MA, 1996, 109–170.

37. G. Monge, Two letters to Lagrange, n. d., *Oeuvres de Lagrange*, Gauthier-Villars, Paris, 1867–1892, vol. XIV, 308–310, 311–314.

38. I. Newton, *Sir Isaac Newton's Mathematical Principles of Natural Philosophy and His System of the World*, trans. A. Motte, rev. F. Cajori, University of California Press, Berkeley, 1960.

39. H. Pulte, 1788: Joseph Louis Lagrange, *Mechanique analitique*, in *Landmark Writings in Western Mathematics, 1640–1940*, I. Grattan-Guinness, ed., Elsevier, Amsterdam, 2005, 208–224.

40. J. L. Richards, The Geometrical Tradition: Mathematics, Space, and Reason in the Nineteenth Century, in *The Modern Physical and Mathematical Sciences*, M. J. Nye, ed., Vol. 5 of the *Cambridge History of Science*, Cambridge University Press, Cambridge, 2003, 449–467.

41. B. A. Rosenfeld, *A History of Non-Euclidean Geometry: Evolution of the Concept of a Geometric Space*, Springer, Berlin and Heidelberg, 1988.

42. B. Russell, *A Critical Exposition of the Philosophy of Leibniz*, Allen and Unwin, London, 1937.

43. R. Taton, Lagrange et Leibniz: De la théorie des functions au principe de raison suffisante, in *Beiträge zur Wirkungs- und Rezeptionsgeschichte von Gottfried Wilhelm Leibniz*, A. Heinekamp, ed., Franz Steiner Verlag, Stuttgart, 1986, 139–147.

44. R. Torretti, *Philosophy of Geometry from Riemann to Poincaré*, D. Reidel, Boston, 1978.

45. J. M. W. Turner, drawings, in A. Fredericksen, *Vanishing Point: The Perspective Drawings of J. M. W. Turner*, Tate, London, 2004.

46. J. Van Cleve, Thomas Reid's geometry of visibles, *Philosophical Review* **111** (2002) 373–416.

47. F. M. Arouet de Voltaire, *Philosophical Dictionary, articles* "Sect" and "Morality," excerpted in *The Portable Voltaire*, B. R. Redman, ed., Viking, New York, 1949.

48. M. Wagner, *The Geometries of Visual Space*, Erlbaum, Mahwah, NJ, 2006.

Index

About the Author

Judith Victor Grabiner received her B. S. in Mathematics from the University of Chicago, M. A. in the History of Science from Radcliffe College, and Ph. D. in the History of Science from Harvard in 1966. She is a member of both Phi Beta Kappa and Sigma Xi. She was a Sigma Xi Society National Lecturer in 1988–90.

Among the many awards she has received are three Carl Allendoerfer Awards (for the best article in the *Mathematics Magazine*, 1984, 1988, 1996), and four Lester Ford Awards (for the best article in the *American Mathematical Monthly*, 1984, 1998, 2005, 2009).

In 2003, she won the national Deborah and Franklin Tepper Haimo Award for Distinguished College or University Teaching of the Mathematical Association of America. Professor Grabiner has held a Woodrow Wilson Fellowship, an American Council of Learned Societies Fellowship, and two National Science Foundation fellowships.

Professor Grabiner has had Visiting Scholar positions in England, Scotland, Australia, and Denmark. She currently teaches at Pitzer College, one of the Claremont Colleges in Claremont, California, where she is the Flora Sanborn Pitzer Professor of Mathematics.

She is the author of numerous articles in the history of mathematics, and her books include: *The Origins of Cauchy's Rigorous Calculus* (MIT Press, 1981), which was reprinted by Dover in 2005. She is also the author of *The Calculus as Algebra: J.-L. Lagrange, 1736–1813* (Garland, 1990), which is reprinted as part of the present volume.